RENEWALS 458-4574

| DATE DUE | | | |
|---|---|---|---|
| APR 8 | | | |
| APR 2 3 | | | |
| OCT 27 | | | |
| NOV 1 3 | | | |
| NOV 2 8 2007 | | | |
| | | | |
| | | | |
| | | | |
| | | | |
| | | | |
| | | | |
| | | | |
| | | | |
| | | | |
| GAYLORD | | | PRINTED IN U.S.A. |

# HEAT PUMP TECHNOLOGY

# Systems Design, Installation, and Troubleshooting

## Second Edition

BILLY C. LANGLEY

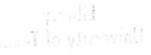

PRENTICE HALL, Englewood Cliffs, New Jersey 07632

Langley, Billy C., 1931-
    Heat pump technology : systems design, installation, and
troubleshooting / Billy C. Langley. -- 2nd ed.
        p.    cm.
    "A Reston book."
    Includes index.
    ISBN 0-13-385766-2
    1. Heat pumps.    I. Title.
TJ262.L36 1989
621.402'5--dc19

88-22430
CIP

Editorial/production supervision and
  interior design: Eileen M. O'Sullivan
Cover design: George Cornell
Manufacturing buyer: Bob Anderson
Page layout: Diane Koromhas

Printed in the United States of America

10  9  8  7  6  5  4  3  2  1

ISBN  0-13-385766-2

Prentice-Hall International (UK) Limited, *London*
Prentice-Hall of Australia Pty. Limited, *Sydney*
Prentice-Hall Canada Inc., *Toronto*
Prentice-Hall Hispanoamericana, S.A., *Mexico*
Prentice-Hall of India Private Limited, *New Delhi*
Prentice-Hall of Japan, Inc., *Tokyo*
Simon & Schuster Asia Pte. Ltd., *Singapore*
Editora Prentice-Hall do Brasil, Ltda., *Rio de Janeiro*

# Contents

## 3  HEAT PUMP COMPONENTS                                                25

## 4  HEAT PUMP CONTROLS                                                   57

# Preface

*Heat Pump Technology,* Second Edition, is intended for use as a curriculum guide or a textbook, or for an independent study course. It covers practical fundamentals as well as recommended service and startup procedures, and serves as a comprehensive textbook for beginning students and a valuable reference for experienced service technicians. The material is organized so that simple heat pump system concepts are presented before those that are more complex.

There are ten chapters in the book, each of which covers a specific area of this exciting industry. Each chapter begins with an introduction to a particular phase of the industry and advances through the more technical aspects of that phase. A summary and questions that cover the minimum material with which readers should be familiar conclude each chapter, where applicable. English–metric equivalents are presented throughout to aid the reader in converting to the metric system. Troubleshooting charts and standard service procedures are included, along with representative heat pump wiring diagrams to aid in servicing these systems. The new technology on earth-coupled water-source heat pump systems is included.

Upon completion of this text, the reader will have the knowledge and confidence necessary to properly install and service heat pump systems.

*Billy C. Langley*

# 1

# Heat Pump Systems

## INTRODUCTION

During the early days of heat pump design, manufacturers believed that a heat pump was a simple system—essentially a conventional air conditioning unit with a few valves installed to reverse the flow of refrigerant. Consequently, they manufactured units that were constant sources of trouble, failing to produce the required amount of heat while operating at very high cost. Because of these faults, heat pump systems acquired a bad name.

However, through these mistakes many manufacturers learned that heat pump systems have specific operating conditions that must be included in the design of the equipment. Among these are the following:

1. The indoor coil must have an extra amount of surface area to prevent the condensing temperaturers from becoming too high during the heating cycle.

2. The air-handling capacity of the indoor unit and duct system must allow sufficient air flow to assure adequate condensing of the refrigerant.

3. The compressor must have a special design for heat pump application because it operates all year long at completely different operating pressures and conditions from those of a standard air-conditioning compressor.

4. Heat pump systems must be equipped with a suction line accumulator to prevent liquid refrigerant from entering the compressor.

5. A crankcase heater is required to prevent refrigerant migration to the compressor lubricating oil during the offcycle and low starting temperatures.

6. A defrost cycle is required to keep the unit operating at peak efficiency.

7. Auxiliary heating elements must be incorporated to aid the unit during periods of extreme cold or when the system might be malfunctioning.

## BENEFITS OF HEAT PUMP SYSTEMS

There are many advantages in using heat pump systems to heat and cool buildings. Two of them are:

1. *Heat pump systems provide a more even temperature throughout a building without blowing a blast of hot air on the occupants on system startup.* Because of this lower discharge air temperature, a higher relative humidity is present, which provides better health conditions for the occupants and is better for the construction materials and furnishings used in homes, offices, and factories.

2. *Heat pump systems cost less to operate than electric resistance heating units.* Because the equipment is in use all year long, a lower cost per hour of use on the original equipment purchase is attained. Hence the sales, installation, and maintenance of heat pump systems tends to level out the air-conditioning business by providing work for persons in these areas on a year-long basis, rather than just seasonally.

## WHAT IS A HEAT PUMP?

A heat pump, in the most basic sense, is little more than a conventional air-conditioning system that is equipped with the necessary components to cause it to reverse its running cycle. While the heat pump is operating in reverse, it absorbs heat from the air outside and releases it inside the building.

The question might arise: How can a heat pump get heat from cold, outside air? The answer is that even though the outside air is cold, it still contains a great amount of heat. Heat is present in all substances down to a temperature of $-459\,°F$ ($-273.15\,°C$), or absolute zero on the Kelvin scale. Therefore, air at $0\,°F$ ($-17.8\,°C$) contains more heat than air at $-20\,°F$ ($-28.9\,°C$) and less heat than air at $40\,°F$ ($4.4\,°C$).

It can be seen, then, that since absolute zero is $-459\,°F$ ($-273.15\,°C$), there is 4.59 times the amount of heat in air at $0\,°F$ ($-17.8\,°C$) than there is between a $0\,°F$ and a $100\,°F$ ($37.8\,°C$) day. Thus there is still a great amount of heat in the air at $0\,°F$. This heat is removed from the air by the heat pump by evaporating refrigerant in the outdoor coil. The coil operates at a temperature below the temperature of the outside air, causing the heat to flow from the air to the refrigerant.

The efficiency of the heat pump is determined by the amount of heat it can transfer from the outside air into the conditioned space. A great amount of research is presently being done to find ways to increase the amount of heat transferred without putting more energy into the system.

## HOW A HEAT PUMP WORKS

A heat pump is a total electric system with the capabilities of providing both heating and cooling. This process is accomplished by moving the heat with a refrigerant rather than by generating the desired amount by using a strip heater or a gas-fired furnace.

To understand how a machine can move heat in two directions—from inside to outside, as in the cooling mode, and from outside to inside, as in the heating mode—the refrigeration cycle must be understood. The name *heat pump* is certainly descriptive of its operation. *Heat* is a physical property with two factors, level and amount. Both factors are important in producing the desired comfort. The level or intensity of heat is indicated by its temperature as measured with a thermometer. The amount or quality is measured in British thermal units (Btu; Figure 1-1).

A *pump* is a device used to make something move in a direction that it ordinarily would not go. For example, a pump can be used to cause water to run uphill (Figure 1-2). The pump or compressor used in a heat pump serves a similar function because it forces heat to move from a cool place to a warmer place.

An outside force is required to raise the heat or energy from a lower to a higher temperature. A *compressor* provides this force. The amount of force required is determined by how far the heat must be raised. For example, it is easier to absorb heat from air at 40 °F (4.4 °C) and release it at 75 °F (23.9 °C) than it is to absorb heat at 40 °F and release it at 95 °F (35 °C).

When the unit is operating in the cooling mode and the thermostat demands cooling, heat from the air inside the building is absorbed by the vaporizing refrigerant in the indoor coil. The indoor air is thus cooled, dehumidified, and redistributed through the duct system. The heat-laden refrigerant is pumped outdoors by the compressor, where the heat is rejected by the outdoor coil (Figure 1-3). The refrigerant vapor is cooled and changed to a liquid, which is pumped back to the indoor coil, and the cycle is repeated.

When the unit is operating in the heating mode and the thermostat demands heating, the reversing valve changes position to reverse the refrigerant flow. The heat is now absorbed into the system by the outdoor coil. The refrigerant is changed from a low-temperature liquid to a low-temperature, low-pressure vapor, and then flows to the compressor. In the compressor, the refrigerant is compressed into a high-pressure, high-temperature vapor and is discharged to the indoor coil (Figure 1-4). The heat picked up in the outdoor coil, plus the heat of compression, is removed from the refrigerant in the indoor coil. The air passing through the coil absorbs this heat, and its temperature is raised. The air is then forced through the duct system and distributed to the conditioned space.

While the unit is in the heating mode, the outside air passes through the outdoor coil and gives up its heat to the refrigerant. The air temperature is reduced below the freezing temperature of water (32 °F) (0 °C), even though the air temperature may be 35 or 40 °F (1.7 or 4.4 °C); the reduction in temperature, approximately 10 °F (5.56 °C), will cause the moisture contained in the air to freeze and form frost on the surface of the outdoor coil. This frost

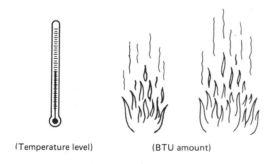

(Temperature level)          (BTU amount)

**Figure 1-1**  Temperature and Heat

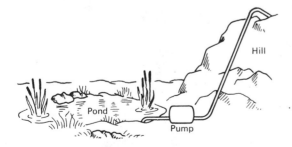

**Figure 1-2**  Pumping Water up a Hill

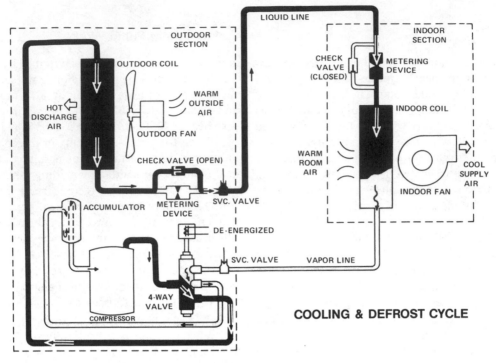

**Figure 1-3** Heat Pump in Cooling Mode (Courtesy of Carrier Air Conditioning.)

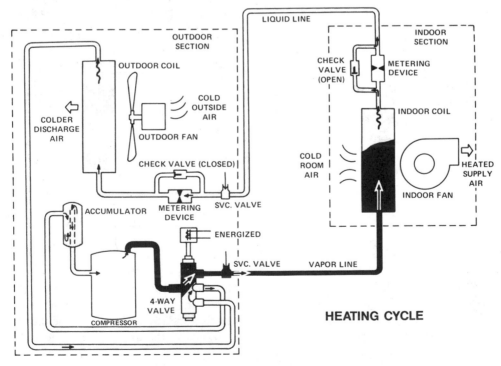

**Figure 1-4** Heat Pump in Heating Mode (Courtesy of Carrier Air Conditioning.)

will continue to form, and when the coil is sufficiently iced over, it will render the system useless unless some means of removing it is provided. Several methods are used to defrost the unit, and each manufacturer will use the method that best suits his needs. Some manufacturers use a resistance heater on the outdoor coil to melt the ice. However, the most popular method is to reverse the refrigerant flow, thus putting the unit back in the cooling mode. The outdoor coil then becomes the condenser, and the hot gas melts the frost buildup. To increase the defrosting action, the outdoor fan is stopped during the defrost period. Since the unit is now operating in the cooling mode, supplemental electric heaters located in the indoor air stream are energized to temper the indoor air, thus offsetting the cooling effect and preventing cold drafts. When the frost has all melted, the defrost controls cause the revers-

ing valve to switch and put the unit back in the heating mode (Figure 1-5).

## TYPES OF HEAT PUMP SYSTEMS

There has been much experimentation with heat pump systems in an effort to find the most efficient type. This experimentation has led to four basic types: (1) water–water, (2) water–air, (3) ground–air, and (4) air–air. The first part of these combinations is referred to as the source and indicates the source of heat for the outdoor coil during the heating cycle. The second part refers to the medium treated by the refrigerant in the indoor coil. Also, heat pump systems can be divided into split and packaged units.

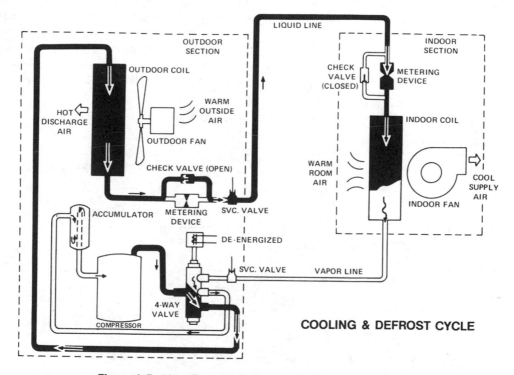

**Figure 1-5**  Heat Pump in Defrost Cycle (Courtesy of Carrier Air Conditioning.)

*Water-Water:* This type of system uses water to treat the outdoor coil and water is treated by the indoor coil (Figure 1-6). There are several sources of water for the outdoor coil. Some of these are wastewater, single well, double well, lake or pond, and cooling tower. Water-water heat pumps use less electricity than other heat pumps when they are properly maintained. However, without proper maintenance the operating costs increase dramatically.

*Wastewater.* Water from the city water main is used and this water is expelled to the sewage system (Figure 1-7). This system has several disadvantages. First, it is very expensive to use water in this manner. Second, the treatment of outdoor coil water is always difficult and in some cases impossible, thus allowing the outdoor coil to accumulate scale faster. The indoor coil water is recirculated and can be treated with the conventional methods used to treat water chillers and boilers.

*Single well.* The use of well water as a source of heat for heat pump systems has received much experimentation. However, local authorities should be consulted before attempting this method because of possible violation of laws governing wells used by the public. One method is to use a single well with one pipe extending farther into the water table than the other (Figure 1-8). One pipe is extended far enough to reduce the possibility of recirculating the heated water, especially during the cooling season.

*Two wells.* The two-well system is the preferred method of using wells to obtain heat for heat pump units. The wells are located far enough apart so that there is no danger of recirculating the water (Figure 1-9). Each well has its own funnel separate from the other. The water from one, therefore, does not mix with the other. A disadvantage of the two-well system is that many times a large plot of land is required in order to properly separate the wells. There cannot be any water treatment on the outdoor coils because of contamination of the public water system. The indoor coil water on systems using well water is treated with the

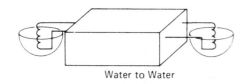

Water to Water

*Figure 1-6* Water-water Heat Pump

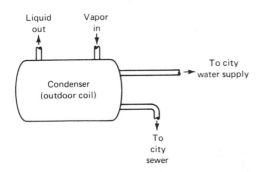

Liquid out    Vapor in

Condenser (outdoor coil)

To city water supply

To city sewer

*Figure 1-7* Waste-water Condenser

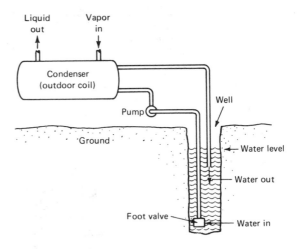

Liquid out    Vapor in

Condenser (outdoor coil)

Pump

Well

Ground

Water level

Water out

Foot valve

Water in

*Figure 1-8* Single-well Outdoor Piping

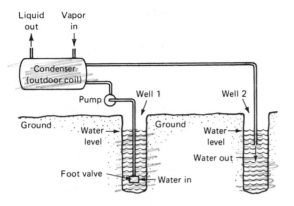

**Figure 1-9** Two-well Installation

conventional methods used to treat water chillers and boilers.

*Lake or pond.* Lakes or ponds are good sources of heat for the heat pump if they are deep enough to prevent water temperature equalization from top to bottom. When this method is used, the return water must be far enough away from the intake to prevent recirculating the water (Figure 1-10). This method can also be used where a swimming pool is available. The water will be heated during the cooling season. If the pool is covered with a clear plastic or glass canopy during the winter, the water will usually remain warm enough to satisfy heat pump requirements. The water in-

take to the unit should be located in the deep end of the pool, while the return is located in the shallow end (Figure 1-11).

*Water-Air:* This type of system uses water in the outdoor coil and air is treated by the indoor coil (Figure 1-12). The same sources of water for the outdoor coil as for water–water systems are used with these units. The only difference is the indoor section and its method of treating the air. The air passes through a direct expansion coil, where it is cooled and dehumidified. This system is less efficient than water–water units because air does not make as good contact with the indoor coil as does water. The outdoor coil requires the same type of maintenance as the water–water units for optimum efficiency.

*Ground-Air:* In these systems the outdoor coil is buried underground. The heat is extracted from the ground (Figure 1-13). In most

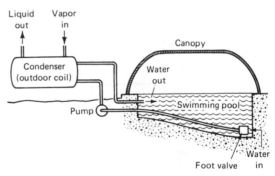

**Figure 1-11** Swimming Pool Installation

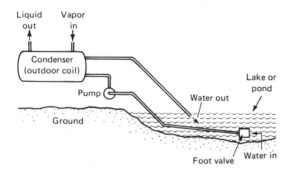

**Figure 1-10** Lake or Pond Installation

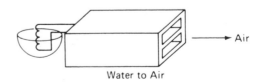

**Figure 1-12** Water-to-air System

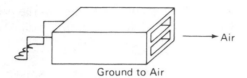

**Figure 1-13** Ground-to-air System

climates the temperature of the earth remains a fairly constant 65 °F (18.33 °C) at about 6 ft (1.824 m) below ground level. This system requires the burial of several feet of pipe per ton of refrigeration, thus requiring a large amount of land. Some of the disadvantages of this system are as follows:

1. Finding refrigerant leaks is a problem, and when a leak is found to be in the outdoor coil, the earth must be removed from around the tube before repairs can be made.
2. The ground tends to become loose around the tube, due to expansion and contraction of the pipe, thus reducing the heat transfer surface.
3. These units do not operate as efficiently as the other types of systems.

*Air-Air:* These are the most popular systems for several reasons. They are easier and more economical to install, and maintenance costs are less than for water–water or water–air units. However, air–air systems use more electricity than do water-source heat pumps.

Air–air systems use air on both coils (Figure 1–14). These systems must include defrost controls and periods to maintain the

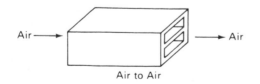

**Figure 1-14** Air-to-air System

highest efficiency possible. This is true especially at low outdoor temperatures. Overall unit efficiency is reduced because of the defrost period, because the equipment is operating along with supplemental heat strips with little or no heat being added to the conditioned space. This factor, along with the reduced efficiency at lower outdoor temperatures, increases the overall costs of operating air–air units.

*Split and packaged units.* The type of building generally dictates whether to use a split or packaged unit. *Split-systems units* are usually preferred on residences and other such buildings (Figure 1–15). In these units the indoor and outdoor units are separate and connected with copper refrigerant lines. The indoor unit has a coil and blower mounted in a cabinet. The unit is then connected to the required ductwork, which distributes the treated air throughout the conditioned space (Figure 1-16). The outdoor unit is made up of a coil, compressor, and fan mounted in a cabinet along with the necessary controls. The air is drawn through the coil and discharged back into the atmosphere. There is no air-duct connection from the outside unit to the conditioned space.

*Packaged, or self-contained, units have* all the components housed in one cabinet (Figure 1–17). When these units are used, a hole must be cut in the building to allow for the treated air to enter the building or for the outside air

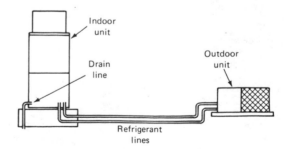

**Figure 1-15** Split System

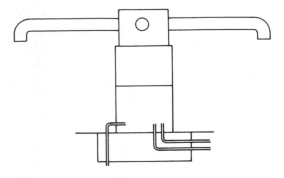

*Figure 1-16* Indoor Unit and Ductwork

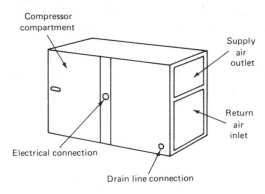

*Figure 1-17* Packaged Unit

to be brought into and out of the building (Figure 1–18). The hole around the required ductwork must be properly sealed to prevent air and rain from entering the building. Horizontal-type self-contained units may be installed on the ground outside the building or on the roof (Figure 1–19). These units are most popular when installation is required in existing buildings or when floor space is needed for other reasons.

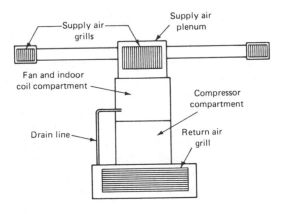

*Figure 1-18* Packaged Installation

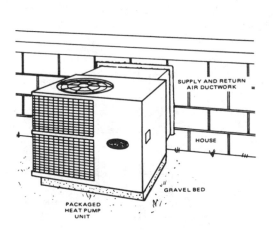

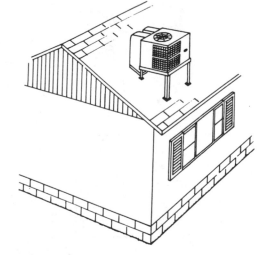

*Figure 1-19* Horizontal Packaged Unit (Courtesy of Carrier Air Conditioning.)

# SOLAR-ASSISTED HEAT PUMPS

Solar energy can be used in conjunction with a heat pump system in several ways. The most popular are as follows:

1. By adding an additional coil to the outdoor unit on the air inlet side of the coil
2. By adding a hot-water coil to the indoor unit
3. A combination of the preceding two methods

The method chosen will depend on the installation requirements, the finances available, and the customer's preference.

*Coil on Air Inlet of Outdoor Coil:* A hot-water coil is mounted on the air inlet side of the outdoor coil (Figure 1-20). The solar-heated water is pumped through the coil, which gives the same effect as a higher outdoor air temperature, adding more heat to the refrigerant during the heating season. When this method is used, a higher-volume blower is required in the outdoor unit to overcome the additional resistance of the added coil. Also, extra attention must be given to the cleanliness of these coils because they will stop more dirt on the added fin and coil surfaces.

The heat pump is operated during the complete heating season when this method of assistance is used. Unit efficiency will be maximum during the complete season even during periods of little or no sun, because the lower-temperature water will still provide more heat for the unit than the cooler outdoor air.

*Coil on Indoor Unit:* This method of solar assist uses a hot-water coil in the indoor unit (Figure 1-21). The heat pump unit does not operate when the solar system will provide adequate heating for the building, thus providing energy savings during periods of plentiful sunshine. When the solar-heated water drops to a point that it will no longer heat the structure, the solar pump is stopped and the heat pump unit is started. This method is still more efficient than heating with resistance heat strips.

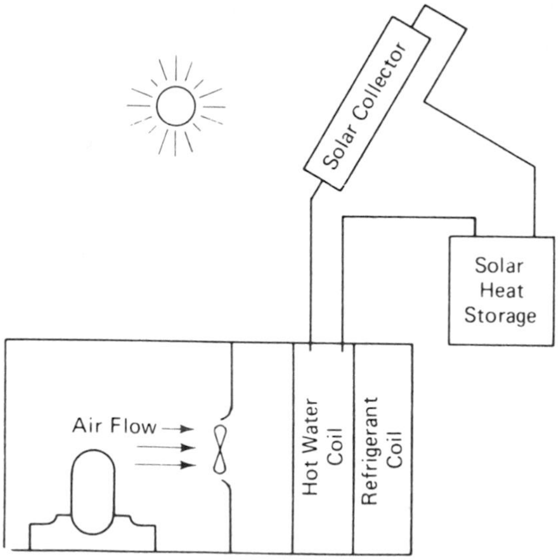

**Figure 1-20** Hot-water Coil on Air Inlet Side Outdoor Coil

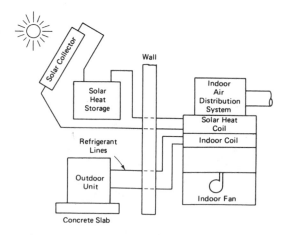

**Figure 1-21** Hot-water Coil in Indoor Unit

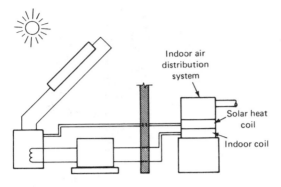

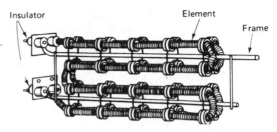

*Figure 1-23* Auxiliary Heat Strips (Courtesy of Tutco, Inc.)

*Figure 1-22* Hot-water Coil on Both Indoor and Outdoor Units

2. They are used to temper the indoor air during the defrost cycle and prevent cold drafts inside the building.

3. They can be used to heat the building should the refrigeration part of the system fail.

A blower capable of overcoming the added resistance of the additional coil must be used on the indoor unit.

*Coil on Both Indoor and Outdoor Units:* Using hot-water coils on both the indoor and outdoor units increases the flexibility of the equipment. However, the original cost of the equipment also increases because of the extra coils and heavier-duty fan motors required. The indoor coil can be used when the solar energy is sufficient to heat the building. The outdoor coil can be put to use by adding heat to the outdoor coil during periods of low solar energy (when the heat pump is used; Figure 1-22). The changeover is normally done automatically with the required controls.

Auxiliary heat strips are located in the indoor unit on the discharge air outlet (Figure 1-24). These components should never be located so that the air will pass through them before passing through the refrigeration coil. This will prevent overheating of the indoor coil, which could cause the compressor-compression ratio to increase to an unsafe level.

The heat strips are controlled from two points: (1) the second stage of the indoor thermostat, and (2) outdoor thermostats, when used. Both thermostats must demand before the strips are energized, but either one will de-energize them. The outdoor thermostats

## AUXILIARY HEAT STRIPS ____

There are three purposes for heat strips (Figure 1-23) on heat pump systems:

1. They are used to aid the heat pump system when the outdoor ambient temperatures drop so low that the heat pump does not have enough heating capacity for the building.

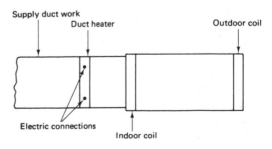

*Figure 1-24* Auxiliary Heat Strip Location.

are usually set at temperatures corresponding to the system balance points, thus preventing accidental operation of the strips and providing more economical operation of the system. (The *system balance point* is the temperature below which the operating part of the refrigeration system will no longer meet the load demand.)

## METHODS OF DEFROST CONTROL

When the heat pump is operating in the heating mode, refrigerant is evaporating in the outdoor coil. If the temperature of the coil falls below 32°F (0°C), frost will begin to form on the coil. Eventually, the frost can build up enough on the coil to restrict the air passing through the coil, causing reduced efficiency. The frost will also act as an insulator on the finned surface and reduce heat transfer, thus reducing coil efficiency even further. When the coil efficiency is reduced enough to appreciably affect system capacity, the frost must be removed.

Frost is removed by causing the reversing valve to switch positions, putting the system in the cooling cycle and stopping the outdoor fan. This directs the hot discharge refrigerant vapor to the outdoor coil to melt the frost.

There are several methods of automatic defrost initiation and termination. The most popular are (1) air-pressure differential across the outdoor coil, (2) outdoor coil temperature, (3) time, (4) time and temperature, and (5) solid-state defrost control.

*Air-Pressure Differential:* This method of automatically initiating the defrost cycle uses a diaphragm-type control to measure the air pressure across the outdoor coil (Figure 1–25). As frost forms on the heat pump outdoor coil, the air pressure drops across the coil (Figure 1–26). As the pressure differential increases to a predetermined difference, the diaphragm

**Figure 1-25** Air-pressure Differential Defrost Control (Courtesy of Ranco Controls Division.)

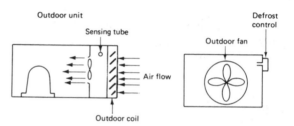

**Figure 1-26** Air Pressure Differential Defrost Control Location

operates a pilot duty electrical switch, which provides the initiation signal to the defrost control. The defrost control puts the unit into the cooling mode and stops the outdoor fan.

As the frost is removed, the difference in pressure across the outdoor coil is decreased. As the pressure differential is reduced to a predetermined difference, the diaphragm operates a pilot duty switch to open and terminate the defrost cycle, thus putting the unit back into the heating mode and starting the outdoor fan. This is one of the most popular methods of defrost control.

*Outdoor Coil Temperature:* This method automatically initiates the defrost cycle by use of a thermostat (Figure 1–27). The defrost thermostat senses the reduction in unit efficiency and operates independently of time, outside

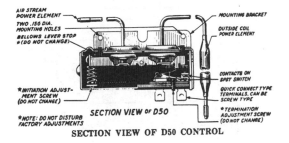

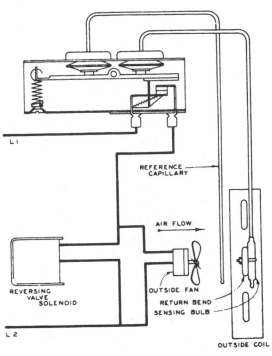

*Figure 1-27* Defrost Thermostat (Courtesy of Ranco Controls Division.)

temperature, wind, and atmospheric conditions.

The signal that operates the defrost control is caused by an increase in the temperature difference between the air temperature around the outside coil and the coil temperature itself with ice on it, as compared to the temperature difference of an ice-free coil at the same outside air temperature. When the defrost cycle is initiated, the outdoor fan stops and the reversing valve is switched to the cooling mode (Figure 1-28). The outdoor coil warms up, melting the frost from the coil. The heat pump continues to run in this mode until the outdoor coil reaches a sufficient temperature, usually 60°F (15.56°C) on the preset control, to terminate the defrost cycle. The coil temperature stays around 32°F (0°C) until all the frost is removed; then the temperature rises rapidly to the termination temperature. Once the termination temperature (60°F) is reached, the unit returns to the heating phase.

*Time:* This method automatically puts the unit into the defrost cycle after a predetermined amount of running time has elapsed. It uses a timer motor that incorporates contacts to provide the necessary switching action (Figure 1-29). When the predetermined amount of running time has elapsed, usually 90 minutes, the timer contacts close, which switches the unit into the defrost mode. When a predetermined

*Figure 1-28* Schematic Wiring Diagram of Thermostatic Defrost Control (Courtesy of Ranco Controls Division.)

*Figure 1-29* Defrost Timer (Courtesy of Lennox Industries, Inc.)

period of time has elapsed with the unit in defrost, usually 15 minutes, the timer contacts open and place the unit back in the heating mode.

The disadvantages of this method are that the unit may defrost when not necessary, may wait too long before initiating defrost, or may be kept in defrost too long or not long enough, depending on the outdoor conditions. Thus the unit will not operate at peak efficiency.

*Time and Temperature:* This method of automatically initiating the defrost cycle is very popular with heat pump manufacturers. It uses a clock timer in combination with a defrost thermostat to determine when a defrost cycle is needed and to initiate the cycle (Figure 1–30). The defrost thermostat contacts are normally open; they close on a fall in temperature. The thermostat sensing element is located on a refrigerant tube near the refrigerant outlet of the outdoor coil. When the coil temperature drops to 32°F (0°C), the contacts close. The thermostat contacts are in series with the normally open contacts of the clock timer (Figure 1–31). The clock motor is wired in parallel with the outdoor fan motor and runs when the compressor runs (Figure 1–32). It will help you to understand the electrical circuits if you make all the connections on the pictorial part of the diagram. Use the schematic part as an aid in making these connections.

In operation, after a predetermined period of time has passed, usually 90 minutes, the

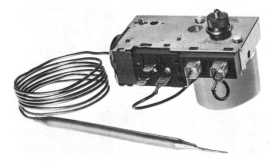

**Figure 1-30** Defrost Timer and Thermostat (Courtesy of Ranco Controls Division.)

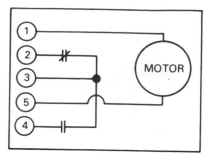

**Figure 1-31** Defrost Clock Timer Schematic (Courtesy of Lennox Industries, Inc.)

timer motor causes the normally open set of contacts to close. These contacts remain closed for only a few seconds during each cycle of the clock mechanism. If the defrost thermostat contacts are closed at the same time as the clock timer contacts, the defrost relay coil is energized and initiates a defrost cycle. If the defrost thermostat contacts are not closed, it indicates that the coil does not need to be defrosted, and the clock timer will start another cycle. The amount of time required for the clock timer to cycle can be reduced to 30 minutes if necessary to help keep the unit operating with a frost-free coil. Depending on the area weather conditions, the timing cycle may be varied to allow proper coil defrosting. The clock timer motor is deenergized along with the outdoor fan motor during the defrost cycle.

The unit will stay in the defrost cycle until the defrost thermostat warms up to approximately 65 °F (18.33°C). The thermostat contacts will then open, which deenergizes the defrost relay and terminates the defrost cycle.

*Solid-State Defrost Control:* This method of defrost uses a solid-state control and two thermistors (Figure 1–33). A *thermistor* is a device that changes electrical resistance as its temperature changes. One thermistor is located near the refrigerant outlet of the outdoor coil and senses the coil temperature. The other

MAKE ALL WIRING CONNECTIONS ON THE PICTORIAL DIAGRAM.

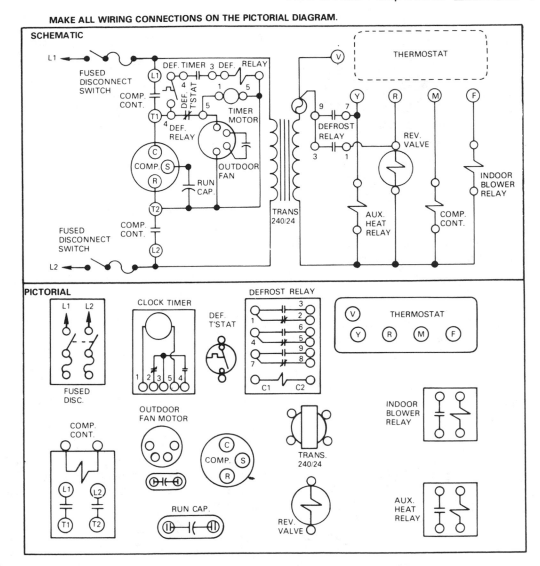

**Figure 1-32** Wiring Diagram of Heat Pump (Courtesy of Lennox Industries, Inc.)

thermistor senses the air temperature as it enters the outdoor coil.

When the outdoor air temperature falls below 45 °F (7.22 °C), frost will probably form on the outdoor coil. This causes an increase in the temperature difference between the coil

thermistor and the air thermistor. When this temperature difference reaches 15 to 25 °F (8.4 to 14 °C), the defrost cycle will be initiated.

In this control system the coil of the outdoor fan and defrost relay is a 24-V dc coil and is energized through terminals R and R of the

**Figure 1-33** Solid-state Defrost Control (Courtesy of Lennox Industries, Inc.)

**Figure 1-35** Defrost Termination Switch (Courtesy of Lennox Industries, Inc.)

the coil. This control is located in the liquid line in the outdoor unit.

## EFFICIENCY _____

Properly designed heat pump systems are more efficient than almost any other form of heating. Two rating methods are used to calculate heat pump efficiency: (1) coefficient of performance (COP), and (2) seasonal energy efficiency ratio (SEER). Regardless of the method used, you should be familiar with how they are determined.

*Coefficient of Performance (COP):* This method has been used for many years when discussing the efficiency of heat pump systems. To calculate the COP of a heat pump, use the following formula:

$$\text{COP} = \frac{\text{Btu out}}{\text{Btu paid for}}$$

$$= \frac{\text{Btu/h capacity}}{\text{unit wattage} \times 3.413 \text{ Btu/W}}$$

solid-state defrost control (Figure 1–34). A high-pressure switch terminates the defrost cycle by opening the electrical circuit to the control board when the refrigerant pressure in the outdoor coil reaches 275 psi (1896.059 kPa; Figure 1–35). This pressure indicates that the coil temperature is high enough (approximately 124 °F or 51.11 °C) to melt all the frost from

### EXAMPLE _____

Given a unit with 40,000 Btu/h heating capacity and consuming 4380 W/h at 40 °F (4.4 °C), the COP would be

$$\text{COP} = \frac{40,000}{4,380 \times 3.413 \text{ Btu/W}}$$

$$= \frac{40,000}{14,948.94} = 2.67$$

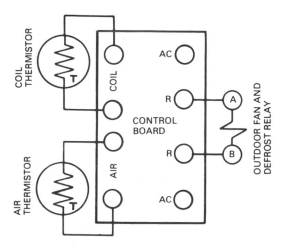

**Figure 1-34** Solid-state Defrost Control Location (Courtesy of Lennox Industries, Inc.)

This is an indication of how efficient the units are when compared to electric resistance heating. Electric resistance heating produces 3.413 Btu of heat for each watt consumed. Thus the COP of electric resistance heating is 1 at maximum output. When we compare the COP of a heat pump to an electric resistance heater, we see that the heat pump can provide more heat per watt of power consumed. The heat pump is, therefore, more efficient. With the operating conditions in our previous example, the heat pump will deliver 2.67 times the amount of heat delivered by a resistance heater using the same wattage.

*Seasonal Energy Efficiency Ratio (SEER):* The SEER is a new way of calculating the efficiency of heat pumps in the cooling mode and all other types of air-conditioning units. The SEER rating of a complete system is made available to all persons interested in unit efficiency. It is an addition to the older energy efficiency ratio (EER) that takes into consideration the average number of seasonal operating hours.

The EER calculates how many Btus are produced for every watt of electric power consumed. It is calculated by dividing the total unit capacity, in Btus, of the system by the total electric power in watts per hour consumed by the equipment. To calculate the SEER, it is easier to first calculate the EER of the unit.

**EXAMPLE** _____

Using the same unit as in the previous example, given a unit with 40,000 Btu/h cooling capacity and consuming 4380 W/h at 105°F (40.56°C), the EER would be

$$EER = \frac{Btu \ out}{power \ in} = \frac{40,000}{4380} = 9.13$$

The SEER would then be

$$
\begin{aligned}
SEER &= EER \times 0.885 + 1.237 \\
&= 9.13 \times 0.885 + 1.237 \\
&= 8.08 + 1.237 \\
&= 9.32
\end{aligned}
$$

where

$$
\begin{aligned}
EER &= \text{energy efficiency ratio} \\
0.885 &= \text{a constant} \\
1.237 &= \text{a constant}
\end{aligned}
$$

The heat pump, like any refrigeration system, will begin to lose capacity as the suction pressure drops. When the heat pump is operating in the heating mode, a capacity reduction happens as the outdoor temperature drops. This also causes a reduction in the SEER and COP of the unit. This capacity reduction occurs because the vapor density of the refrigerant decreases as the suction pressure decreases. Therefore, the refrigerant vapor weighs less per cubic foot and the compressor pumps fewer pounds of refrigerant through the system. A refrigerant is capable of absorbing a certain number of Btu (calories) per pound (0.4536 kg). When fewer pounds (kilograms) of refrigerant are circulated, the capacity is reduced accordingly.

Along with the decrease in vapor density at lower suction pressures, the reexpansion stroke of the compressor piston is longer. Thus a shorter effective stroke causes a decrease in the refrigerant flow rate.

## HEAT PUMP CAPACITY MEASUREMENT _____

Determining the Btu capacity of a heat pump is a fairly simple task. Use the following steps:

1. Set the room thermostat to the heating or automatic position.
2. Set the room thermostat temperature selector to 90°F (32.22°C).
3. Turn off all electricity to the resistance heat strips. Only the heat pump and indoor fan are to be operating during this test.
4. Measure the temperature rise through the indoor unit (Figure 1–36). Use the

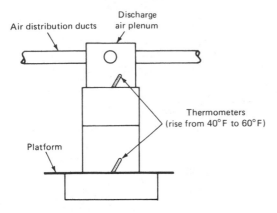

**Figure 1-36** Measuring Temperature Rise through a Unit

following steps when measuring the temperature rise:

a. Use the same thermometer for determining the return and supply air temperatures.

b. Do not measure in the lined areas. True air temperature cannot be measured in areas affected by radiant heat.

c. Make the temperature measurements within 6 ft (1.83 m) of the indoor unit. Measurements taken at the return and supply grills are inaccurate.

d. Use the average temperature when more than one duct is connected to the plenum.

e. Make sure that the air temperature is stable before taking measurements.

f. Take measurements downstream from any mixed air source.

g. Record the temperature difference of the return and supply air as $\Delta T$.

5. Determine the Btu output by using the following formula:

$$\text{Btu} = \text{cfm} \times \Delta T \times 1.08$$

where

cfm = total measured air flow (cubic feet per minute)

$\Delta T$ = supply air temperature minus return air temperature

1.08 = specific heat of air constant.

## EXAMPLE _____

The measured air flow of a unit is 1600 cfm and the measured temperature rise is 30 °F (16.8 °C). The Btu capacity is

$$\text{Btu} = 1600 \times 30 \times 1.08 = 51,840$$

## _____ SUMMARY _____

Heat pump systems provide a more even temperature throughout the building. Relative humidity is also higher in buildings equipped with these systems.

Heat pump systems operate more economically than electric resistance heating units.

Heat pump systems have a lower cost per hour of use on the original equipment purchase.

A heat pump, in the basic sense, is little more than a conventional air-conditioning system equipped with the necessary components to cause it to reverse its running cycle.

A heat pump is capable of providing both heating and cooling by moving the heat with a refrigerant.

The heat picked up in the outdoor coil plus the heat of compression during the heating mode is removed from the indoor coil by the air passing through it.

The most popular method of defrosting a heat pump outdoor coil is by reversing the refrigerant flow, thus putting the unit back in the cooling mode.

During the defrost cycle, supplemental electric heaters located in the indoor air stream are energized to temper the indoor air.

The four basic types of heat pumps are (1) water–water, (2) water–air, (3) ground–air, and (4) air–air. The first part of these combinations is referred to as the source and indicates the source of heat for the outdoor coil during the heating cycle. The second part refers to the medium treated by the refrigerant in the indoor coil.

Water–water heat pumps use less electricity than the other systems when they are properly maintained.

A lake or pond is a good source of heat for a heat pump if the water is deep enough to prevent water-temperature equalization from top to bottom.

The water–air units are less efficient than the water–water units because the air does not make as good a contact with the indoor coil as does the water.

Ground–air systems require the burial of several feet of pipe per ton of refrigeration, thus requiring a large amount of land.

Air–air systems are the most popular types of heat pump systems; however, they use more electricity than water source units. Their efficiency is reduced because of the required defrost cycle.

Split-system heat pumps are usually preferred on residences and other such buildings.

Packaged units are most popular when installation is required in existing buildings or when floor space is needed for other reasons.

Solar-assisted heat pumps add a hot-water coil to the outdoor unit, the indoor unit, or to both.

Auxiliary heat strips are used on heat pump units for three reasons: (1) to aid the heat pump system when the outdoor ambient temperature is so low that the heat pump does not have enough heating capacity for the building, (2) to temper the indoor air during the defrost cycle, and (3) to heat the building should the refrigeration system fail.

Heat strips are controlled from two points: (1) the second stage of the indoor thermostat, and (2) the outdoor thermostats, when used. Both thermostats must demand before the strips are energized.

The most popular methods of defrost initiation and termination are (1) air-pressure differential across the outdoor coil, (2) outdoor coil temperature, (3) time, (4) time and temperature, and (5) solid-state defrost control.

Heat pump systems must be equipped with a suction line accumulator to prevent liquid refrigerant entering the compressor.

A crankcase heater is required to prevent refrigerant migration to the compressor lubricating oil during the off cycle and under low starting temperatures.

Two rating methods are used to calculate heat pump efficiency: (1) coefficient of performance (COP), and (2) seasonal energy efficiency ratio (SEER). These are an indication of how efficient the units are when compared to electric resistance heating.

When we compare the COP of a heat pump to an electric resistance heater, we see that the heat pump can provide more heat per watt of power consumed.

When the heat pump is operating in the heating mode, a capacity reduction occurs as the outdoor temperature drops. This also causes a reduction in the SEER and the COP of the unit. This capacity reduction occurs because the vapor density of the refrigerant decreases as the suction pressure decreases. Also, the reexpansion stroke of the compressor piston is longer.

## ——— REVIEW QUESTIONS ———

1. Do electric resistance heaters or heat pump systems operate more efficiently?
2. What device is used to prevent refrigerant migration to the compressor during the system off cycle?
3. What is the purpose of a suction line accumulator?
4. Is the relative humidity inside a building higher or lower when a heat pump is used for heating?
5. From where does a heat pump system get heat?
6. What is used to bring heat from the source into the building?
7. What device reverses the direction of refrigerant flow?
8. What term is used to indicate when a heat pump goes into the defrost cycle?
9. What happens to the outdoor fan when the heat pump goes into a defrost cycle?
10. What type of heat pump system is the most efficient?
11. Where are auxiliary heat strips installed on heat pump units?
12. Name two control points for auxiliary heat strips.
13. What is the purpose of the defrost cycle of a heat pump unit?
14. Name the two most popular defrost control systems.
15. What are the two methods of rating the efficiency of heat pump systems?

# 2

# Heat Pump
# Refrigeration Cycles

## INTRODUCTION

To understand how a heat pump system works, it is necessary to know how the refrigeration system functions and the part it plays in the overall operation of the total system. The different refrigeration cycles involved are cooling, heating, and defrosting.

The basic concept of a heat pump system is to reverse the refrigerant flow from the cooling to the heating cycle. Each coil will act as an evaporator or a condenser, depending on the direction of the refrigerant flow. Therefore, the coils are referred to by their location rather than their function. The coil inside the building is termed the *indoor coil*, and the coil outside the building is termed the *outdoor coil* (Figure 2-1). This terminology is to prevent confusion when discussing heat pump systems.

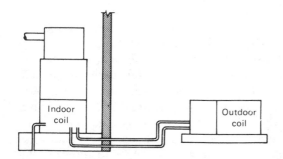

**Figure 2-1** Indoor and Outdoor Coil Locations

## COOLING CYCLE

In discussing the refrigeration cycle, we could start at any point in the system and return to that point on the completed cycle. However, we will start at the compressor discharge valve. The purpose of the cooling cycle is to remove heat from within the space to be cooled and to maintain that space at a temperature lower than the surroundings. This is accomplished through the use of a refrigeration system, in most cases.

In operation, the refrigerant leaves the compressor discharge valve in a hot, vaporous form (Figure 2-2). The hot vapor then flows through the reversing valve, which directs the refrigerant to the outdoor coil. In the outdoor coil the refrigerant is cooled by the outdoor air flowing over the coil tubes and is caused to condense; it becomes a warm, high-pressure liquid. In the outdoor coil the heat is picked up from inside the building, and also the heat of compression is removed. In some cases the liquid refrigerant is further cooled in the last few rows of tubes in the coil. This is known as *subcooling* the liquid.

As the warm, high-pressure liquid refrigerant leaves the outdoor coil, it encounters a flow-control device and a check valve. These two devices are arranged so that the warm, high-pressure liquid refrigerant is allowed to pass through the check valve and not through the flow-control device during the cooling cycle (Figure 2-3). The warm, high-pressure refrigerant then flows through the liquid line to the indoor unit. Before entering the indoor coil, the warm, high-pressure liquid refrigerant encounters another flow-control device and check valve. The check valve is closed, causing the refrigerant to pass through the flow-control device.

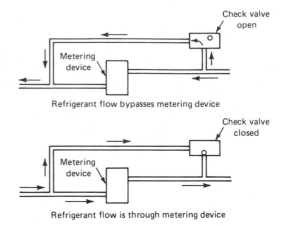

Refrigerant flow bypasses metering device

Refrigerant flow is through metering device

**Figure 2-3**  Flow through Outdoor Check Valve during Cooling Cycle

Upon passing through the flow-control device, the pressure is reduced, and the warm, high-pressure liquid becomes a cooled, low-pressure liquid–vapor mixture. At this point a large loss of refrigeration effects occurs, because some of the liquid refrigerant is evaporated to cool the remaining liquid down to the evaporator temperature. This is known as *flash gas* and equals about 20% of the total system capacity. The purpose of subcooling the liquid is to reduce this loss as much as is possible.

The cool, low-pressure liquid–vapor mixture then enters the indoor coil over which the indoor air is flowing. Heat is picked up by the refrigerant as it changes to a cool, low-pressure vapor. An additional amount of heat, known as *superheat*, is picked up by the refrigerant vapor in the indoor coil. This superheat is what causes the flow-control devices to admit the correct amount of refrigerant to the indoor coil.

The cool, low-pressure vapor then flows through the suction line to the reversing valve, where it is directed to the suction line accumulator. In the accumulator, the cool, low-pressure refrigerant is trapped to allow any liq-

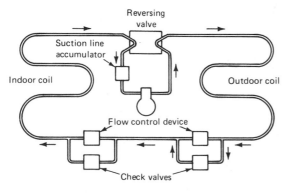

**Figure 2-2**  Heat Pump in Cooling Cycle

uid refrigerant to evaporate before entering the compressor. From the suction line accumulator, the cool, low-pressure refrigerant is drawn into the compressor, compressed and forced through the compressor discharge valve, and the refrigerant starts another cycle .

## HEATING CYCLE _____

The purpose of the heating cycle is to pick up heat outside the building and put it inside the building to heat the contents and occupants during the winter. To do this, the flow of refrigerant is reversed. It first flows to the indoor coil during this mode of operation.

Beginning our cycle at the discharge of the compressor, the hot, high-pressure refrigerant vapor flows through the reversing valve, which directs the refrigerant through the suction line to the indoor coil (Figure 2-4). In the indoor coil the refrigerant gives up its heat to the indoor air passing through the coil fins and over the tubes. In the indoor coil the refrigerant gives up the heat picked up outdoors, plus the heat of compression. The warmed air then flows into the building to provide the required heating function. In giving up its heat, the refrigerant is condensed.

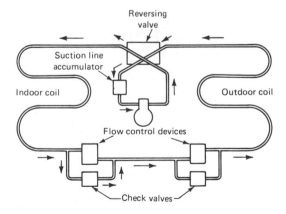

**Figure 2-4** Heat Pump in Heating Cycle

The warm, high-pressure liquid then leaves the indoor coil, where it encounters the flow-control device and check valve. The flow-control device opposes a flow in this direction, and the check valve opens to permit the warm, high-pressure liquid to pass into the liquid line.

After flowing through the liquid line, the warm, high-pressure liquid encounters the flow-control device and check valve located in the outdoor unit. The check valve closes, forcing the warm, high-pressure liquid to flow through the flow-control device; the pressure is reduced and the refrigerant becomes a low-temperature, low-pressure liquid–vapor mixture. At this point a large loss of refrigeration effect occurs, because some of the liquid refrigerant is evaporated to cool the remaining liquid down to the evaporator temperature. This is known as flash gas and equals about 20% of the total system capacity. The purpose of subcooling is to reduce this loss as much as possible.

This cool, low-pressure liquid–vapor mixture then enters the outdoor coil over which the outdoor air is flowing. Heat is picked up by the refrigerant as it changes to a cool, low-pressure vapor. An additional amount of heat, known as superheat, is picked up by the refrigerant vapor in the outdoor coil. This superheat is what causes the flow control devices to admit the correct amount of refrigerant to the outdoor coil.

The cool, low-pressure vapor then flows through the reversing valve into the suction line accumulator. In the accumulator, the cool, low-pressure refrigerant is trapped to allow any liquid refrigerant to evaporate before entering the compressor.

From the suction line accumulator, the cool, low-pressure refrigerant vapor enters the compressor crankcase and cylinder. The cool, low-pressure refrigerant is compressed and forced through the compressor discharge valve, and the refrigerant starts another cycle.

## DEFROST CYCLE _____

The purpose of the defrost cycle on a heat pump is to remove the accumulation of frost built up on the outdoor coil during the heating cycle. To accomplish this task, the reversing valve directs the flow of hot refrigerant vapor to the outdoor coil (Figure 2–5). This hot vapor causes the frost to melt.

When the system goes into defrost, the outdoor fan is stopped to permit the outdoor coil to heat up much faster. The indoor fan continues to run, and some of the indoor strip heaters are energized electrically to temper the air that is being distributed inside the building. This is necessary because the indoor coil is now absorbing heat from the indoor air to aid in the defrosting process.

The unit should not stay in the defrost cycle any longer than is required because energy is used without heat being discharged into the building. Thus, the overall cost of heating the building is increased.

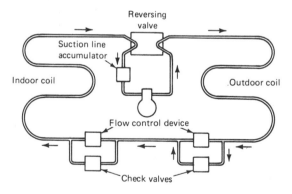

**Figure 2-5**  Heat Pump in Defrost Cycle

## _____ SUMMARY _____

The refrigerant coils on a heat pump system are referred to as indoor and outdoor coils.

The basic concept of a heat pump system is to reverse the refrigerant flow from the cooling to the heating cycle.

The purpose of the cooling cycle is to remove heat from within the space to be cooled and to maintain that space at a temperature lower than the surroundings.

The purpose of the heating cycle is to pick up heat outside the building and put it inside the building to heat the contents and occupants during the winter.

In both the indoor and outdoor coils the refrigerant gives up the heat absorbed when evaporating, plus the heat of compression gained during the compression process.

The purpose of the defrost cycle on a heat pump is to remove the accumulation of frost build up during the heating cycle.

During the defrost cycle, the outdoor fan is stopped and part of the auxiliary heat strips are energized.

A long defrost cycle will increase the operating costs of a heat pump system.

## _____ REVIEW QUESTIONS _____

1. What is the basic concept of a heat pump system?
2. What are the terms for the coils used on heat pump systems?
3. What component(s) is (are) responsible for reversing the refrigerant flow?
4. What constitutes a very large loss in refrigeration effect in a refrigeration system?
5. During the cooling cycle, which coil does the compressor pump into?
6. During the heating cycle, which coil does the compressor pump refrigerant from?
7. What is the last component in a heat pump system that the refrigerant enters before entering the compressor?
8. What components direct the refrigerant around the flow-control device(s) used on a heat pump system?
9. Why should a heat pump system be in the defrost cycle only as long as necessary?
10. What is the purpose of the defrost cycle on a heat pump system?

# 3

# Heat Pump Components

## INTRODUCTION

The different components used on heat pump systems are vital to the operation and efficiency of the total unit. Each component has a purpose and function, and these should be fully understood by persons who design, install, and service heat pump units. The more a person knows about these components, the easier his or her job will be.

## *COMPRESSOR*

The compressor circulates the refrigerant through the system. The reciprocating compressor is the most popular compressor used on heat pump systems (Figure 3–1). The operating principle of these compressors is the same as that for any other reciprocating compressor.

However, certain conditions are encountered in a heat pump system that are rarely, if ever, encountered in an air-conditioning or refrigeration application. The compression

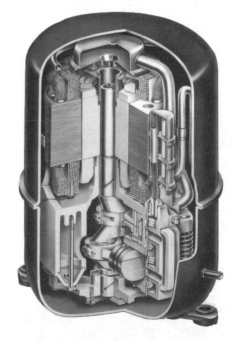

*Figure 3-1* Reciprocating Compressor (Courtesy of Lennox Industries, Inc.)

25

ratio (CR) is of great importance. During a normal cooling cycle, the unit could be operating with a 120°F (48.89°C) condensing temperature and a 40°F (4.44°C) suction temperature. On a system using refrigerant 22 (R-22), this would represent a 262-psig (1806.43-kPa) discharge pressure and a 69-psig (475.738-kPa) suction pressure (Table 3-1). To determine the compression ratio, use the following formula:

$$CR = \frac{\text{absolute discharge pressure}}{\text{absolute suction pressure}}$$

To convert to absolute pressure, add atmospheric pressure (14.7) to the gauge pressure readings.

In our example the answer would be

$$CR = \frac{262 + 14.7}{69 + 14.7} = 3.3$$

This compression ratio is in the satisfactory range of an air-conditioning type of compressor.

However, if the system is changed to the heating cycle, the heat pump compressor does not operate in the air conditioning range because of the lower outdoor air temperatures and the lower suction pressures encountered. It is now operating in the range in which refrigeration units normally operate. For example, if we were heating at 0°F (−17.78°C), the suction temperature could be as low as −20°F (−28.9°C) and the condensing temperature could be 100°F (37.8°C). This would represent a suction pressure of 10.3 psig (70.948 kPa) and a discharge pressure of 198 psig (1365.162 kPa). The compression ratio would now be

$$CR = \frac{198 + 14.7}{10.3 + 14.7} = 8.5$$

This is much too high a compression ratio for an air-conditioning compressor. Many hermetic compressors are not designed to operate with a compression ratio of 7.5 or greater. Compressors designed to operate at about the 8.5-CR range will not be able to withstand operating conditions above this level for a very long period of time. Any increase in the discharge pressure during the heating cycle will cause an increase in the compression ratio and cause the compressor to operate beyond the range it was designed for. Therefore, if the indoor air quantity is reduced during the heating cycle, compressor damage is very possible. This condition must be avoided for the protection of the compressor and the overall system efficiency.

Lubrication is another problem with compressors used in heat pump systems. Liquid refrigerant often enters the compressor during the heating cycle because of the lower operating temperatures and changes from the heating to the defrost cycle and back again. Usually, compressor lubricating oil can be mixed with up to 10% by volume of liquid refrigerant and operate satisfactorily. However, caution should be taken to allow as little liquid refrigerant as possible to enter the compressor, because when the lubricating oil is mixed with 14% liquid refrigerant, the oil starts breaking down and compressor failure is possible. When this situation cannot be avoided, excess oil can be added to the crankcase. Use caution not to add enough to damage the compressor valves. This excess oil allows for some oil trapping in the system while leaving enough for proper compressor lubrication. Use a low-foaming lubricant designed for heat pump use.

Heat pump compressor motors run hotter because of the low refrigerant density at low suction pressures. Some designs take this into account and re-route the refrigerant vapor through the motor windings.

The unit capacity is reduced to almost half at low suction and discharge pressures. Thus, the lower the outdoor air temperature the lower the system capacity.

**TABLE 3–1**
Temperature–Pressure Chart

| °F | R-12 | R-13 | R-22 | R-500 | R-502 | R-717 Ammonia |
|---|---|---|---|---|---|---|
| -100 | 27.0 | 7.5 | 25.0 | — | 23.3 | 27.4 |
| -95 | 26.4 | 10.9 | 24.1 | — | 22.1 | 26.8 |
| -90 | 25.7 | 14.2 | 23.0 | — | 20.7 | 26.1 |
| -85 | 25.0 | 18.2 | 21.7 | — | 19.0 | 25.3 |
| -80 | 24.1 | 22.2 | 20.2 | — | 17.1 | 24.3 |
| -75 | 23.0 | 27.1 | 18.5 | — | 15.0 | 23.2 |
| -70 | 21.8 | 32.0 | 16.6 | — | 12.6 | 21.9 |
| -65 | 20.5 | 37.7 | 14.4 | — | 10.0 | 20.4 |
| -60 | 19.0 | 43.5 | 12.0 | — | 7.0 | 18.6 |
| -55 | 17.3 | 50.0 | 9.2 | — | 3.6 | 16.6 |
| -50 | 15.4 | 57.0 | 6.2 | — | 0.0 | 14.3 |
| -45 | 13.3 | 64.6 | 2.7 | — | 2.1 | 11.7 |
| -40 | 11.0 | 72.7 | 0.5 | 7.9 | 4.3 | 8.7 |
| -35 | 8.4 | 81.5 | 2.6 | 4.8 | 6.7 | 5.4 |
| -30 | 5.5 | 91.0 | 4.9 | 1.4 | 9.4 | 1.6 |
| -28 | 4.3 | 94.9 | 5.9 | 0.0 | 10.6 | 0.0 |
| -26 | 3.0 | 98.9 | 6.9 | 0.7 | 11.7 | 0.8 |
| -24 | 1.6 | 103.0 | 7.9 | 1.5 | 13.0 | 1.7 |
| -22 | 0.3 | 107.3 | 9.0 | 2.3 | 14.2 | 2.6 |
| -20 | 0.6 | 111.7 | 10.1 | 3.1 | 15.5 | 3.6 |
| -18 | 1.3 | 116.2 | 11.3 | 4.0 | 16.9 | 4.6 |
| -16 | 2.1 | 120.8 | 12.5 | 4.9 | 18.3 | 5.6 |
| -14 | 2.8 | 125.7 | 13.8 | 5.8 | 19.7 | 6.7 |
| -12 | 3.7 | 130.5 | 15.1 | 6.8 | 21.3 | 7.9 |
| -10 | 4.5 | 135.4 | 16.5 | 7.8 | 22.8 | 9.0 |
| -8 | 5.4 | 140.5 | 17.9 | 8.8 | 24.4 | 10.3 |
| -6 | 6.3 | 145.7 | 19.3 | 9.9 | 26.0 | 11.6 |
| -4 | 7.2 | 151.1 | 20.8 | 11.0 | 27.7 | 12.9 |
| -2 | 8.2 | 156.5 | 22.4 | 12.1 | 29.5 | 14.3 |
| 0 | 9.1 | 162.1 | 24.0 | 13.3 | 31.2 | 15.7 |
| 2 | 10.2 | 167.9 | 25.6 | 14.5 | 33.1 | 17.2 |
| 4 | 11.2 | 173.7 | 27.3 | 15.7 | 35.0 | 18.8 |
| 6 | 12.3 | 179.8 | 29.1 | 17.0 | 37.0 | 20.4 |
| 8 | 13.5 | 185.9 | 30.9 | 18.4 | 39.1 | 22.1 |
| 10 | 14.6 | 192.1 | 32.8 | 19.8 | 41.1 | 23.8 |
| 12 | 15.8 | 198.6 | 34.7 | 21.2 | 43.3 | 25.0 |
| 14 | 17.1 | 205.2 | 36.7 | 22.7 | 45.5 | 27.5 |
| 16 | 18.4 | 211.9 | 38.7 | 24.2 | 47.8 | 29.4 |
| 18 | 19.7 | 218.8 | 40.9 | 25.7 | 50.1 | 31.4 |
| 20 | 21.0 | 225.7 | 43.0 | 27.3 | 52.5 | 33.5 |
| 22 | 22.4 | 233.0 | 45.3 | 29.0 | 55.0 | 35.7 |
| 24 | 23.9 | 240.3 | 47.6 | 30.7 | 57.5 | 37.9 |
| 26 | 25.4 | 247.7 | 49.9 | 32.5 | 60.1 | 40.2 |
| 28 | 26.9 | 255.5 | 52.4 | 34.3 | 62.8 | 42.6 |
| 30 | 28.5 | 263.2 | 54.9 | 36.1 | 65.4 | 45.0 |
| 32 | 30.1 | 271.3 | 57.5 | 38.0 | 68.3 | 47.6 |
| 34 | 31.7 | 279.5 | 60.1 | 40.0 | 71.2 | 50.2 |
| 36 | 33.4 | 287.8 | 62.8 | 42.0 | 74.1 | 52.9 |
| 38 | 35.2 | 296.3 | 65.6 | 44.1 | 77.2 | 55.7 |
| 40 | 37.0 | 304.9 | 68.5 | 46.2 | 80.2 | 58.6 |
| 45 | 41.7 | 327.5 | 76.0 | 51.9 | 88.3 | 66.3 |
| 50 | 46.7 | 351.2 | 84.0 | 57.8 | 96.9 | 74.5 |
| 55 | 52.0 | 376.1 | 92.6 | 64.2 | 106.0 | 83.4 |
| 60 | 57.7 | 402.3 | 101.6 | 71.0 | 115.6 | 92.9 |
| 65 | 63.8 | 429.8 | 111.2 | 78.2 | 125.8 | 103.1 |
| 70 | 70.2 | 458.7 | 121.4 | 85.8 | 136.6 | 114.1 |
| 75 | 77.0 | 489.0 | 132.2 | 93.9 | 148.0 | 125.8 |
| 80 | 84.2 | 520.8 | 143.6 | 102.5 | 159.9 | 138.3 |
| 85 | 91.8 | — | 155.7 | 111.5 | 172.5 | 151.7 |
| 90 | 99.8 | — | 168.4 | 121.2 | 185.8 | 165.9 |
| 95 | 108.3 | — | 181.8 | 131.2 | 199.7 | 181.1 |
| 100 | 117.2 | — | 195.9 | 141.9 | 214.4 | 197.2 |
| 105 | 126.6 | — | 210.8 | 153.1 | 229.7 | 214.2 |
| 110 | 136.4 | — | 226.4 | 164.9 | 245.8 | 232.3 |
| 115 | 146.8 | — | 242.7 | 177.3 | 262.6 | 251.5 |
| 120 | 157.7 | — | 259.9 | 190.3 | 280.3 | 271.7 |
| 125 | 169.1 | — | 277.9 | 203.9 | 298.7 | 293.1 |
| 130 | 181.0 | — | 296.8 | 218.2 | 318.0 | 315.0 |
| 135 | 193.5 | — | 316.6 | 233.2 | 338.1 | 335.0 |
| 140 | 206.6 | — | 337.3 | 248.8 | 359.1 | 365.0 |
| 145 | 220.6 | — | 358.9 | 265.2 | 381.1 | 390.0 |
| 150 | 234.6 | — | 381.5 | 282.3 | 403.9 | 420.0 |
| 155 | 249.9 | — | 405.2 | 300.1 | 427.8 | 450.0 |
| 160 | 265.12 | — | 429.8 | 318.7 | 452.6 | 490.0 |

Bold figures, inches mercury vacuum; light figures, psig.
*Source:* Alco Controls Division, Emerson Electric Co.

A compressor designed for heat pump applications differs from other compressors in several ways. For instance, it has a larger connecting rod and bearings. This reduces the bearing loads by as much as 20 to 41%. A low-foaming lubricant is used. It is a white oil. The older types of lubricating oil are not satisfactory for heat pump applications. Also included is an improved motor-winding protector. The pistons are designed with rings, not lapped, to improve cylinder and piston clearance. Many compressor designs use a nondirectional refrigerant flow to allow liquid and vapor separation and a reduced motor-winding temperature.

It is not good practice to use an air-conditioning compressor on a heat pump system. The old adage, "do not field convert," is extremely applicable to heat pump systems. Many have tried to field convert an air-conditioning system to a heat pump and have failed.

## REVERSING VALVE ————

The reversing valve changes the direction of the refrigerant flow in a heat pump system (Figure 3–2). It is an electrically actuated four-way solenoid valve. It mechanically changes the flow of refrigerant to place the system in either the heating or cooling cycle (Figure 3–3). In some applications, when the system calls for cooling, the solenoid coil is deenergized and the valve directs the compressor discharge vapor to the outdoor coil; the flow of low-pressure liquid–vapor mixture from the indoor coil is directed into the shell of the suction line accumulator, which may also incorporate a heat exchanger. In other words, during the cooling cycle the outdoor coil functions as the condenser and the indoor coil functions as the evaporator. When heating is required, the solenoid is electrically energized and the valve reserves its position. The compressor discharge vapor is now directed to the indoor coil, and the liquid–vapor mixture from the outdoor coil flows through the suction line accumulator. The outdoor coil now functions as the evaporator. The indoor coil functions as the condenser.

The reversing valve is made up of two pistons connected to a ported sliding block. It is actuated by a solenoid-operated plunger that opens and closes the vapor ports to cause a buildup in pressure, which moves the piston assembly either to the right or left, depending on whether the thermostat is calling for heating or cooling (Figure 3–4). During the cooling cycle the solenoid coil usually is de-energized, allowing the plunger to drop and close the lower vent line to the right side of the assembly. High-

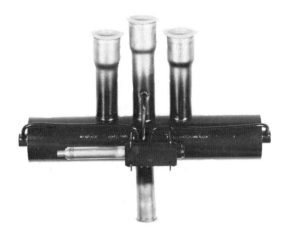

**Figure 3-2** Reversing Valve (Courtesy of Ranco Controls Division.)

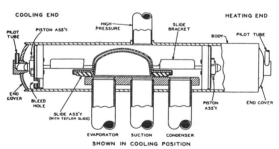

**Figure 3-3** Cutaway of Reversing Valve (Courtesy of Ranco Controls Division.)

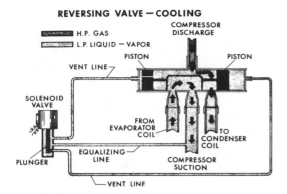

Figure 3-4 Reversing Valve in Cooling Position (Courtesy of Westinghouse Central Residential Air Conditioning Division.)

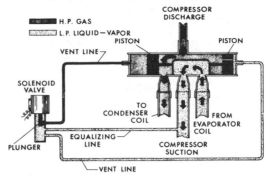

Figure 3-5 Reversing Valve in Heating Position (Courtesy of Westinghouse Central Residential Air Conditioning Division.)

pressure discharge vapor seeps through the bleed port in the right piston to build up pressure behind the piston. Since the left piston is equalized with the suction pressure, the piston assembly moves to the left, as shown. In this position, the compressor discharge vapor is directed to the outdoor coil, and the compressor reduces the pressure in the outdoor coil through the suction line accumulator. The outdoor coil now serves as the condenser, and the indoor coil serves as the evaporator.

In our example, when the thermostat calls for heating, the solenoid coil is energized, and the plunger is pulled up to cover the top vent line on the left side of the piston assembly (Figure 3–5). The discharge vapor pressure now seeps through the small port in the left piston, and vapor pressure builds up in back of the piston. With vapor pressure on the left piston and the right piston vented to the suction pressure, the piston assembly moves to the right. In this position, the valve directs the compressor discharge vapor to the indoor coil, and the compressor removes the vaporous refrigerant from the outdoor coil through the accumulator. In this situation, the outdoor coil serves as the evaporator and the indoor coil serves as the condenser.

Remember, some manufacturers energize the reversing valve solenoid coil during the heating season and others energize it during the cooling season. Therefore, it is necessary to check when the coil is energized if problems are encountered with this component.

## REFRIGERANT FLOW-CONTROL (METERING) DEVICES _____

The refrigerant flow is controlled, or metered, to the coil that is to absorb heat. During the heating cycle, the refrigerant is metered to the outdoor coil. During the cooling cycle, the refrigerant is metered to the indoor coil.

Many types of flow-control devices can be used. In some units the refrigerant flow is controlled by a single metering device that allows flow in either direction (Figure 3–6). This metering device is mounted in the liquid line. The refrigerant flow direction through this device reverses when the unit changes from one cycle to the other.

The most common method of controlling the flow of refrigerant is to use two separate flow-control devices and two check valves, one set for each coil (Figure 3–7). The flow-control

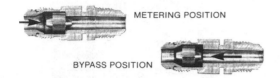

METERING POSITION

BYPASS POSITION

*Figure 3-6* Refrigerant Accurator (Courtesy of Carrier Air Conditioning.)

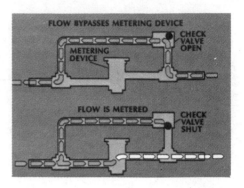

*Figure 3-8* Flow-control Device and Valve Schematic (Courtesy of Carrier Air Conditioning.)

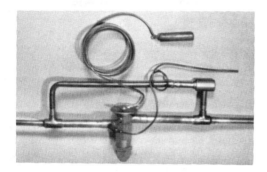

*Figure 3-7* Flow-control Device and Check Valve Arrangement (Courtesy of Carrier Air Conditioning.)

devices could be thermostatic expansion valves, capillary tubes, or a combination of the two. With this type of refrigerant flow control, the check valves are used to allow refrigerant to bypass the metering device not in use at that time. The liquid refrigerant is directed to the proper coil. The check valve will allow flow in only one direction (Figure 3–8).

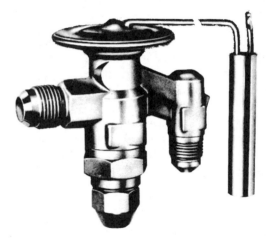

*Figure 3-9* Thermostatic Expansion Valve (Courtesy of Alco Controls Division, Emerson Electric Co.)

*Thermostatic Expansion Valve:* The most common flow-control device used on heat pump systems to meter the refrigerant flow into the coil is the thermostatic expansion valve (Figure 3-9). An orifice in the valve meters the flow into the coil. The rate of flow is modulated as required by a needle-type plunger and seat that varies the orifice opening.

The thermostatic expansion valve is a precision device that is designed to meter the flow of refrigerant into an evaporator in exact pro-portion to the rate of evaporation of the liquid refrigerant in the evaporator, thereby preventing the return of liquid refrigerant to the compressor. (For simplicity, we will use evaporator rather than indoor and outdoor coil in this discussion.) By being responsive to the temperature of the refrigerant vapor leaving the evaporator and the pressure in the evaporator, the thermostatic expansion valve can control the refrigerant vapor leaving the evaporator at a predetermined superheat.

A vapor is said to be superheated whenever its temperature is higher than the saturation temperature corresponding to its pressure. The amount of superheat is the temperature increase above the saturation temperature of the existing pressure.

Consider a refrigeration evaporator operating with R-12 as the refrigerant at 37 psig (255.106 kPa) suction pressure (Figure 3–10). The R-12 saturation temperature at 37 psig is 40 °F (4.4 °C). As long as any liquid refrigerant exists at this pressure, the refrigerant temperature will remain at 40 °F.

As the refrigerant moves along in the evaporator, the liquid boils off into a vapor, and the amount of liquid decreases. At point A in Figure 3–10, all the liquid has evaporated owing to the absorption of a quantity of heat from the surrounding atmosphere that is equal to the latent heat of vaporization of the refrigerant. The refrigerant vapor continues along in the evaporator and remains at the same pressure (37 psig); however, its temperature increases owing to the continued absorption of heat from the surrounding atmosphere. By the time the refrigerant vapor reaches the end of the evaporator at point B, its temperature is 50 °F (10 °C). This refrigerant vapor is now superheated, and the amount of superheat is 50 °F − 40 °F or 10 °F (5.6 °C). The degree to which the refrigerant vapor is superheated is a function of the amount of refrigerant being fed into the evaporator and the load to which the evaporator is exposed.

In operation, three forces govern the operation of a thermostatic expansion valve (Figure 3–11): (1) the pressure created by the remote bulb and power assembly ($P_1$), (2) the evaporator pressure ($P_2$), and (3) the equivalent pressure of the superheat spring ($P_3$).

The remote bulb and power assembly is a closed system, and in the following discussion it is assumed that the remote bulb and power assembly charge is the same refrigerant as that used in the system. The pressure within the remote bulb and power assembly, $P_1$ in Figure 3–11, then corresponds to the saturation pressure of the refrigerant temperature leaving the evaporator and moves the valve pin in the opening direction. Opposed to this force, on the underside of the diaphragm, and acting in the closing direction, is the force exerted by the evaporator pressure ($P_2$), along with the pressure exerted by the superheat spring ($P_3$). The valve will assume a stable control position

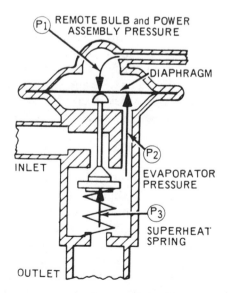

**Figure 3-11** Thermostatic Expansion Valve Basic Forces (Courtesy of Alco Controls Division, Emerson Electric Co.)

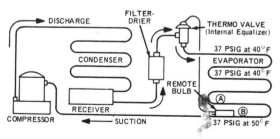

**Figure 3-10** Basic Refrigeration Schematic (Courtesy of Alco Controls Division, Emerson Electric Co.)

when these three forces are in equilibrium, that is, when $P_1 = P_2 + P_3$. As the temperature of the refrigerant vapor at the evaporator outlet increases above the saturation temperature that corresponds to the evaporator pressure, the vapor becomes superheated. The pressure thus generated in the remote bulb and power assembly increases above the combined pressures of the evaporator pressure and the superheat spring, causing the valve pin to move in the opening direction. Conversely, as the temperature of the refrigerant vapor leaving the evaporator decreases, the pressure in the remote bulb and power assembly also decreases, and the combined evaporator and the spring pressures cause the valve pin to move in the closing direction.

As the operating superheat setting is raised, the evaporator capacity decreases, since more of the evaporator surface is required to produce the superheat necessary to open the valve (Figure 3–12). Thus it is most important to adjust the operating superheat correctly. It is of vital importance that only a minimum change in superheat be required to move the valve pin to the full open position, because this provides savings in both initial evaporator cost and the cost of operation. Accurate and sensitive control of the liquid refrigerant flow into the evaporator is necessary to provide maximum evaporator capacity under all load conditions.

*Capillary Tubes:* The capillary tube is the simplest type of refrigerant flow-control device used on modern refrigeration systems. A capillary tube is a small-diameter tube through which the refrigerant flows to the evaporator (Figure 3–13). The capillary tube is not a true valve since it is not adjustable and cannot be readily regulated. It is used only on flooded systems and allows the liquid refrigerant to flow into the evaporator at a predetermined rate, which is determined by the size of the refrigeration unit and the load it must carry. The

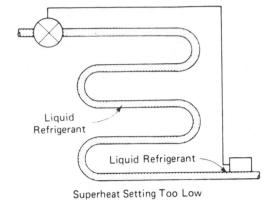

Liquid Refrigerant

Liquid Refrigerant

Superheat Setting Too Low

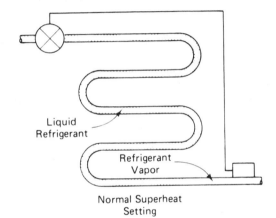

Liquid Refrigerant

Refrigerant Vapor

Normal Superheat Setting

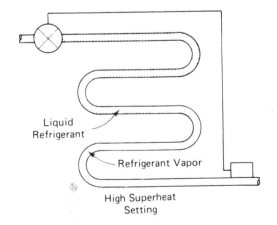

Liquid Refrigerant

Refrigerant Vapor

High Superheat Setting

**Figure 3-12** Superheat Setting and Evaporator Capacity

Figure 3-14   Capillary Tube and Strainer
(Courtesy of Jarrow Brinda, Inc.)

Figure 3-13   Capillary Tubing

capillary tube acts in exactly the same manner as a small-diameter water pipe that holds back water, causing a higher pressure to be built up behind the water column with only a small rate of flow. In the same manner, the small-diameter capillary tube holds back the liquid refrigerant, causing a high pressure to be build up in the condensing coil during operation of the unit and, at the same time, permitting the liquid refrigerant to flow slowly into the evaporator.

Since the orifice is fixed, the rate of feed is relatively inflexible. Under conditions of constant load and constant discharge and suction pressures, the capillary tube performs very satisfactorily. However, changes in the evaporator load or fluctuations in head pressure can result in under- or overfeeding of the coil with refrigerant.

When the condensing unit stops, the condenser pressure and the evaporator pressure gradually equalize as the liquid refrigerant continues to flow through the capillary tube. Eventually, the compressor is able to start under balanced pressure conditions. This allows the use of low-starting-torque motors, a big advantage of capillary tubes.

Because of the small bore of the capillary tube, it is essential that the system be kept free from dirt and foreign matter. Usually a filter is placed before the capillary tube to prevent dirt from plugging the small bore (Figure 3–14). If the capillary tube becomes plugged, the evaporator will defrost, the unit will run continuously, or the thermal overload may cut out. The discharge pressure will become very high unless the condenser has capacity enough to take the entire charge of refrigerant. The capillary tube is connected directly to the outlet end of the condenser. Thus most of the refrigerant is in the evaporator, and it is evident that a plugged capillary tube will result in an excessively high head pressure.

The refrigerant charge is critical in capillary tube systems because normally there is no receiver to store the excess refrigerant. Too much refrigerant will cause high head pressures, compressor motor overloading, and possible liquid floodback to the compressor during the off cycle. Too little refrigerant will allow vapor to enter the capillary tube, which will cause a loss in system capacity.

Due to its basic simplicity and the low-starting-torque motor requirement, a capillary tube system is the least expensive of all flow-control systems. The proper size of capillary tube is difficult to calculate accurately and can best be determined by actual test on the system. Once the size is determined, the proper capillary tube can be applied to identical systems, so it is well adapted to production-type units.

*Subcooling Control Valve:*   This flow-control device is used on Westinghouse Hi/Re/Li heat pump units. The same valve controls refrigerant flow during both the heating and cooling cycles (Figure 3–15), thus taking the place of both flow-control devices required on a conventional

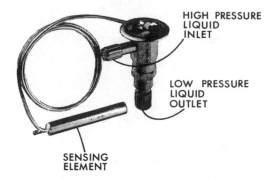

HIGH PRESSURE LIQUID INLET

LOW PRESSURE LIQUID OUTLET

SENSING ELEMENT

*Figure 3-15* Hi/Re/Li Subcooling Control Valve (Courtesy of Westinghouse Central Residential Air Conditioning Division.)

a remote sensing element that is pressure sensitive to the temperature sensed by the bulb. The valve is installed in the high-pressure liquid line, with its sensing bulb clamped to the liquid line leaving the condensing coil, instead of to the suction line as in a thermostatic expansion valve (Figure 3-16). To control liquid subcooling, in this location the subcooling control valve is actuated by the temperature of the liquid refrigerant leaving the condensing coil instead of by the temperature of the suction vapor, which actuates the thermostatic expansion valve to control the evaporator superheat.

The subcooling control valve consists of a power element and the internally equalized, spring-balanced valve. The power element contains a diaphragm that flexes to close or open the control valve as more or less pressure is exerted upon it. The upper surface of the diaphragm is exposed to the pressure exerted by the remote sensing element, while the underside is exposed to the condenser pressure through

heat pump. Although the subcooling control valve resembles a thermostatic expansion valve, its location, function, and operation are entirely different. To understand the principle of the Hi/Re/Li system, it is important to have a thorough knowledge of this valve assembly, which consists of a spring-balanced valve and

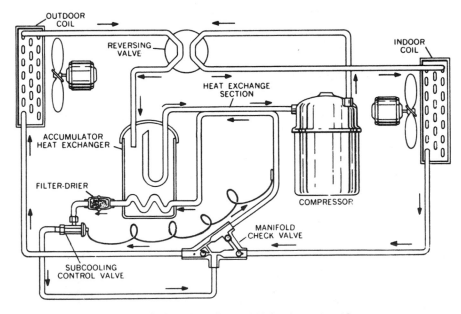

OUTDOOR COIL

REVERSING VALVE

HEAT EXCHANGE SECTION

INDOOR COIL

ACCUMULATOR HEAT EXCHANGER

FILTER-DRIER

COMPRESSOR

MANIFOLD CHECK VALVE

SUBCOOLING CONTROL VALVE

*Figure 3-16* Subcooling Control Valve Location (Courtesy of Westinghouse Central Residential Air Conditioning Division.)

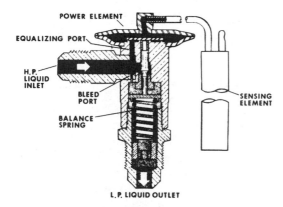

POWER ELEMENT

EQUALIZING PORT

H. P. LIQUID INLET

BLEED PORT

BALANCE SPRING

SENSING ELEMENT

L. P. LIQUID OUTLET

*Figure 3-17* Subcooling Control Valve Cutaway (Courtesy of Westinghouse Central Residential Air Conditioning Division.)

an internal equalizing port in the valve body (Figure 3-17). The valve is factory set to maintain 10 to 15 °F (5.6 to 8.4 °C) of subcooling in the liquid leaving the condensing coil. During operation at any condensing pressure, if the liquid temperature rises, the control valve bulb will sense the warmer temperature, increase the pressure on the upper side of the diaphragm, and cause the valve to move toward the closed position. As the valve closes, the liquid refrigerant backs up slightly in the condensing coil until the liquid temperature is reduced to correspond to 10 to 15 °F of subcooling or 10 to 15 °F below saturation.

On the other hand, if the liquid refrigerant leaving the condensing coil is subcooled more than 10 to 15 °F, the sensing bulb of the control valve senses the drop in temperature and exerts less pressure on the top of the diaphragm. Thus, the condenser pressure under the diaphragm, assisted by the balance spring, is allowed to move the valve toward the open position, thereby increasing the flow of liquid refrigerant from the condensing coil and reducing the degree of subcooling.

Unlike the thermostatic expansion valve, which opens when its sensing bulb feels a

warmer temperature, the subcooling control valve closes as its bulb temperature rises. If the diaphragm should rupture or the sensing element loses its charge, liquid refrigerant simply flows through the evaporating coil into the suction line accumulator, which is large enough to hold the refrigerant charge and protect the compressor from liquid flood-back.

The subcooling control valve contains a small bleed port that bypasses the valve between the liquid inlet from the condensing coil and the outlet to the evaporating coil. This tiny port is not only essential to valve operation; it is extremely important to the success of the overall system performance in three different ways (Figure 3-18). First, the bleed port assures quick equalization of the suction and discharge pressures after shutdown on the cooling cycle. On the heating cycle, pressure equalization after shutdown is accomplished by simply deenergizing the reversing valve and allowing it to return to the cooling position. Pressure equalization is important because the compressor is not required to start against a high pressure differential.

Second, the bleed port provides additional assurance of proper oil return during all types of operating conditions (Figure 3-19). Even when the liquid subcooling valve is closed, a

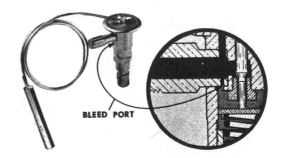

BLEED PORT

*Figure 3-18* Subcooling Control Valve Bleed Port for Pressure Equalization (Courtesy of Westinghouse Central Residential Air Conditioning Division.)

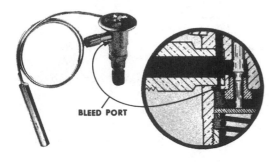

*Figure 3-19* Subcooling Control Valve Bleed Port for Oil Return (Courtesy of Westinghouse Central Residential Air Conditioning Division.)

small but sufficient amount of refrigerant is allowed to bypass the control valve to assure adequate oil return during both the heating and cooling cycles.

Third, the bleed port assures that the valve responds to changing system conditions. For example, at the instant the system is reversed from heating to cooling for defrosting the outdoor coil, the high side pressure tends to build up slowly under the valve diaphragm, and the valve would be slow in opening if it were not for the bleed port (Figure 3–20). The bleed port permits enough refrigerant to bypass the valve and flow into the system during this period, which causes the pressure to build up much

more rapidly and allows the control valve to recover full control in a very short time.

## CHECK VALVES ————————

In heat pump systems, check valves are used for two purposes: (1) to cause the refrigerant to flow through the flow-control device, or (2) to allow the refrigerant to bypass the flow-control device (Figure 3–21). They are installed in a loop that bypasses the flow-control device. Check valves will only open when pressure is exerted in the right direction; therefore, they should be installed with the arrow pointing in the proper direction of refrigerant flow at the point of installation.

In operation, the refrigerant either pushes against the valve seat to close it tighter or against its face to cause it to open and allow refrigerant to pass through. These valves are usually spring loaded and will open when the pressure difference on the seat reaches 15 to 20 psig (103.421 to 137.895 kPa). However, some types use a ball-type check valve, which refrigerant force causes to open or close, depending on the direction of flow.

*Manifold Check Valve:*  This is a fairly simple device that serves a vital function in the Westinghouse Hi/Re/Li heat pump system. It

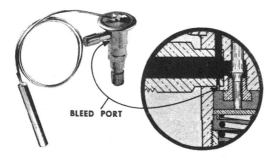

*Figure 3-20* Subcooling Control Valve Bleed Port for Quick Recovery (Courtesy of Westinghouse Central Residential Air Conditioning Division.)

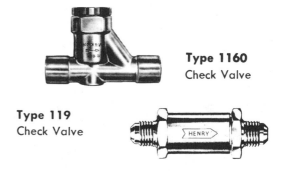

**Type 1160**
Check Valve

**Type 119**
Check Valve

*Figure 3-21* Check Valve (Courtesy of Henry Valve Company.)

is a combination of four separate ball check positions in a single housing (Figure 3-22). The balls are positioned by changes in the system pressure created when the electrically operated reversing valve changes its position from heating to cooling or cooling to heating. The valve's purpose is to direct the flow of liquid refrigerant from the condensing coil through the coil of the accumulator-heat exchanger and the subcooling control valve to the evaporating coil during both the heating and cooling cycles. The manifold check valve is simple and requires no adjustment.

When the reversing valve moves to the cooling position, the outdoor coil becomes the condensing coil and the indoor coil becomes the evaporating coil. System pressures move the balls to cover the check valve ports (Figure 3-23). The high-pressure liquid from the outdoor coil is directed to the coil of the accumulator-heat exchanger, and the low-pressure liquid is directed to the indoor coil.

When the thermostat calls for heat, the reversing valve moves to reverse the flow of refrigerant vapor. The indoor coil is now the condensing coil, and the outdoor coil is the evaporating coil. The system pressures are reversed, as are the positions of the balls in the manifold check valve. However, the flow of liquid remains the same with respect to the condensing and evaporating coils. The high-

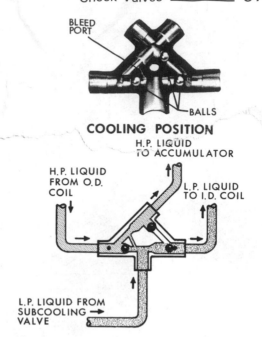

**COOLING POSITION**

**Figure 3-23** Manifold Check Valve in Cooling Position (Courtesy of Westinghouse Central Residential Air Conditioning Division.)

pressure liquid from the condensing coil still flows to the coil of the accumulator-heat exchanger; the low-pressure liquid from the subcooling valve still flows through the evaporating coil. Only the path of liquid through the check valve is changed. The actual liquid refrigerant flow remains the same.

The small bleed port between the check valve passages is used only during the cooling cycle and serves two important purposes (Figure 3-24):

1. It assists the bleed port in the liquid subcooling valve to quickly equalize the system pressures on the shutdown.

2. It helps to provide a minimal flow of refrigerant during operation at extremely low load conditions, an important factor in maintaining a cool-running

**Figure 3-22** Manifold Check Valve (Courtesy of Westinghouse Central Residential Air Conditioning Division.)

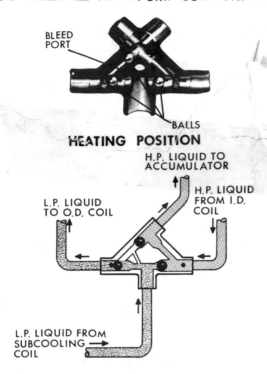

BLEED PORT

BALLS

**HEATING POSITION**

H.P. LIQUID TO ACCUMULATOR

L.P. LIQUID TO O.D. COIL

H.P. LIQUID FROM I.D. COIL

L.P. LIQUID FROM SUBCOOLING COIL

*Figure 3-24* Manifold Check Valve in Heating Position (Courtesy of Westinghouse Central Residential Air Conditioning Division.)

compressor and assuring adequate oil return for proper compressor lubrication.

## ACCUMULATORS _____

Compressors are designed to compress vapors, not liquids. Many systems, especially heat pump and low-temperature systems, are subject to the return of excessive quantities of liquid refrigerant to the compressor. Liquid refrigerant returning to the compressor dilutes the oil, washes out the bearings, and in some cases causes complete loss of oil in the compressor crankcase. This condition is known as *oil pumping* or *slugging* and results in broken valve reeds, pistons, rods, crankshafts, and the

like. The purpose of the accumulator is to act as a reservoir to temporarily hold the excess oil-refrigerant mixture and to return it at a rate that the compressor can safely handle. Some accumulators include a heat-exchanger coil to aid in boiling off the liquid refrigerant while subcooling the refrigerant in the liquid line (Figure 3-25), thus helping the system to operate more efficiently.

Proper installation of a suction accumulator in the suction line just after the reversing valve and before the compressor helps to eliminate damage (Figure 3-26). If the accu-

*Figure 3-25* Suction Line Accumulator With a Heat Exchanger (Courtesy of Refrigeration Research, Inc.)

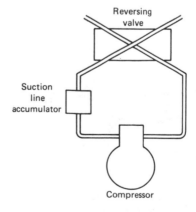

Reversing valve

Suction line accumulator

Compressor

*Figure 3-26* Accumulator Installation

mulator is correctly sized, relatively large quantities of liquid refrigerant may return through the suction line, and the suction accumulator will prevent damage to the compressor. The liquid refrigerant is temporarily held in the suction accumulator and metered back to the compressor, along with any oil, at a controlled rate through the metering orifice (Figure 3–27). Therefore, damage to the compressor is prevented and the compressor immediately and quietly goes to work.

*Accumulator Soft Plug:* Some manufacturers include a "soft plug" in the top of the accumulator used on their equipment. Soft plugs provide over-temperature protection for the equipment should something happen to the outdoor fan or the outdoor coil become clogged, or any other condition that would cause an over-temperature condition, causing the discharge refrigerant temperature to rise to a dangerous

point. These plugs generally are not replaceable in the field. Therefore, when a leak develops here, the accumulator must be replaced.

*Accumulator-Heat Exchanger:* This component serves three important functions:

1. It adds subcooling to the high-pressure liquid on its way to the evaporating coil of approximately 35 to 45 °F (19.6 to 25.2 °C) during the cooling cycle and approximately 50 to 60 °F (28 to 33.6 °C) during the heating cycle.
2. It provides a positive separation of the low-pressure liquid and vapor from the evaporating coil so that only dry, nearly unsaturated, vapor reaches the compressor suction.
3. It assures positive oil return to the compressor at all times during operation.

The accumulator-heat exchanger assembly consists of a steel shell containing a circular liquid subcooling coil and a U-shaped suction connection with a filtered oil return orifice.

During operation in either the cooling or heating cycle, the mixture of low-pressure vapor and unevaporated liquid from the evaporating coil dumps into the accumulator-heat exchanger shell (Figure 3–28). The saturated vapor separates in the upper part of the shell, ready for return to the compressor. The cold liquid refrigerant accumulates in the bottom of the shell around the liquid subcooling coil. The warm, high-pressure liquid from the condensing coil flows through the subcooling coil on its way to the flow-control device and the evaporating coil. As the liquid refrigerant passes through the coil, it is subcooled approximately 35 to 45 °F (19.6 to 25.2 °C) during the cooling cycle and 50 to 60 °F (28 to 33.6 °C) during the heating cycle. In addition to subcooling the high-pressure liquid, this heat exchange also boils off some of the low-pressure liquid

**Figure 3-27** Internal View of a Suction Accumulator (Courtesy of Refrigeration Research, Inc.)

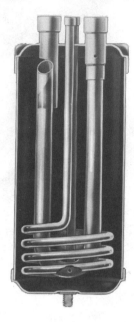

**Figure 3-28** Accumulator-Heat Exchanger Cutaway (Courtesy of Refrigeration Research, Inc.)

in the accumulator shell. The compressor removes the dry saturated vapor in the upper part of the shell through the U-shaped suction connection. As the suction vapor passes through the bottom of the U-tube at a high velocity, a low-pressure area is created, and a mixture of refrigerant and oil is constantly drawn through the filtered orifice to assure a positive and continuous oil return to the compressor. Any traces of liquid refrigerant in the oil are dried up before it enters the compressor. To obtain the best possible heat-exchange efficiency for maximum subcooling effect and to prevent the formation of condensation and frost, the entire accumulator-heat exchanger assembly is completely insulated.

## COILS

The indoor and outdoor coils used on heat pump units have a large face area and are

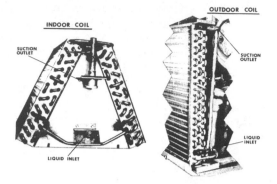

**Figure 3-29** Heat Pump Coils (Courtesy of Westinghouse Central Residential Air Conditioning Division.)

engineered for both condensing and evaporating functions (Figure 3–29). The staggered tube arrangements, as well as the fin configuration and spacing, are precisely engineered for uniform distribution of air flow through the coil for best heat-transfer characteristics and high capacity. Both coils are carefully sized to provide balanced system performance and to obtain maximum efficiency during both heating and cooling cycles. In the conventional heat pump system, refrigerant distribution to the indoor coil presents serious difficulties, which many elaborate distributors have been developed to overcome. This is another reason never to field convert a regular air-conditioning unit to a heat pump system.

## AUXILIARY HEATING ELEMENTS

Electric resistance heating elements are used to aid the heat pump when the outdoor ambient temperature falls below the balance point (the outdoor temperature at which the heat pump does not have enough capacity to heat the structure) (Figure 3–30). These elements also temper the indoor air supply during the defrost cycle. They may be duct heaters mounted in the dis-

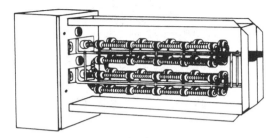

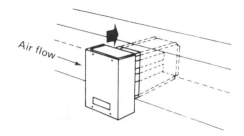

**Figure 3-30** Auxiliary Heater Coils and Rack (Courtesy of Tutco, Inc.)

charge air plenum or specially designed to be installed in the unit (Figure 3-31).

The resistance heaters used on heat pumps are factory assembled units consisting of a steel frame, open coil heating elements, and an integral control compartment. These matched combinations are fabricated to order in a wide range of standard and custom sizes, heating capacities, and control modes. They are prewired at the factory, inspected, and ready for installation at the job site.

The open coil heating elements are a high nickel-chromium alloy wire (usually 80% nickel and 20% chromium) physically designed to meet the application requirements of auxiliary heaters (Figure 3-32). Essentially 100% of the electrical energy input to the coil is converted to heat energy, regardless of the temperature of the surrounding air or the velocity with which it passes over the heating element.

Ceramic bushings are used to insulate the coil from surrounding sheet metal; the coil can float freely in the embossed openings. Also, the bushings prevent binding and cracking as the heater cycles (Figure 3-33). The bushings are held in place by curved metal tabs. Their extra heavy body enables the bushings to withstand high-humidity conditions and a 2000-V dielectric test.

Stainless-steel terminals have threads for securing electrical connections (Figure 3-34). The coil is mechanically crimped into its terminal with closely adjusted tools to ensure cool,

**Figure 3-31** Auxiliary Heater Locations (Courtesy of Square D Company.)

minimum-resistance connections. High-temperature molded phenolic terminal insulators are used that will not crack or chip during normal use.

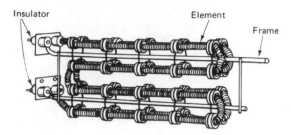

**Figure 3-32** Open Wire Heating Element (Courtesy of Tutco, Inc.)

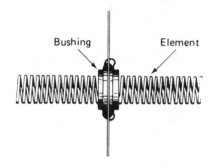

**Figure 3-33** Coil and Bushing (Courtesy of Industrial Engineering and Equipment Co.)

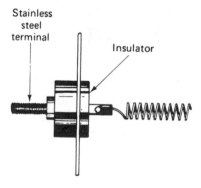

**Figure 3-34** Heater Element Terminal (Courtesy of Industrial Engineering and Equipment Co.)

Steel brackets are used to support the coils (Figure 3-35). They are spot welded in place and spaced to prevent coil sag. The brackets are reinforced (1) by ribbing the bends at each end

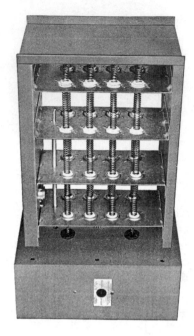

**Figure 3-35** Element Brackets (Courtesy of Industrial Engineering and Equipment Co.)

and (2) by grooving the edges facing the airstream.

## CRANKCASE HEATERS ————

The crankcase heater is a low-wattage resistance element. It may be energized continuously or only when the compressor is not running. It must be carefully selected to avoid overheating of the compressor lubricating oil. Compressor manufacturers generally have a list of recommended heaters for any particular model not equipped with one from the factory.

Because of the severe operating conditions required of a heat pump compressor, crankcase heaters are a must. This is especially true when a compressor is installed in a location where it will be exposed to ambient temperatures colder than the evaporator; in typical heat pump installation, refrigerant migration to the crankcase can be aggravated by the resulting pressure difference between the indoor coil and the compressor during the off cycle. To protect against the possibility of refrigerant migration, crankcase heaters are often used to keep the oil in the crankcase at a temperature high enough so that any liquid refrigerant that enters the crankcase will evaporate and create a pressure sufficient to prevent a large-scale refrigerant migration to the compressor.

If liquid refrigerant does migrate to the compressor, it will be absorbed by the oil, and a rapid reduction in suction pressure on compressor startup will cause the liquid to boil (evaporate). When this boiling action occurs, the oil-refrigerant mixture will foam. Some of this foam will leave the crankcase and pass into the cylinder and cause a condition known as slugging, which will damage the compressor valves. The liquid refrigerant also reduces the lubricating effectiveness of the oil.

Crankcase heaters are made in three types: (1) insert type (Figure 3–36), (2) wrap around (Figure 3–37), and (3) positive temperature coefficient resistor (PTCR).

The insert and the PTCR types of crankcase heaters are installed in a fitting in the compressor crankcase side. The heating elements extends into the lubricating oil to heat the oil

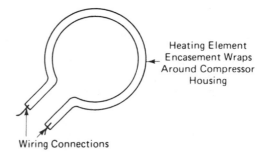

*Figure 3-37* Wrap-Around-Type Crankcase Heater

directly. The insert type applies a given amount of heat at all times, depending on the wattage rating of the heater. The PTCR heater uses a resistor inside the insert that senses the oil temperature. As the oil temperature increases the resistance of the resistor also increases, allowing less electricity to flow through the resistor. With the PTCR heater, the oil, therefore, is kept at a relatively constant temperature.

The wraparound type is placed around the compressor shell. The heater heats the shell, which in turn heats the lubricating oil. This type is very popular on hermetic-type compressors, which are not equipped with fittings in their side. It is also used on equipment already in the field.

## REFRIGERANT PIPING AND PIPE INSULATION ————

The refrigerant lines used on a heat pump deserve special consideration, because the lines serve different purposes during each cycle. The tubing should be run in as direct a route as possible. The line length and diameters must be sized in accordance with the equipment manufacturer's recommendations for the unit being designed to ensure that capacity and oil return are not reduced. Any piping used must be approved for ACR use and be kept dry and clean.

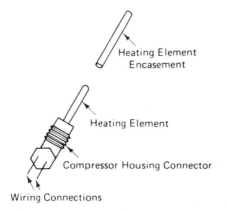

*Figure 3-36* Insert-type Crankcase Heater

All vapor lines, including those field piped inside the cabinet, should be insulated to prevent capacity loss and/or sweating. The insulation must act as a vapor barrier and be able to withstand both hot and cold temperatures because of the reversed cycle operation. The liquid and vapor line should not contact except as specified by the manufacturer of the equipment. The insulation usually recommended is $\frac{1}{2}$-in. (12.70-mm) thick armaflex.

## HEAT EXCHANGER _____

Any device that brings into contact two substances of different temperatures for the purpose of heating or cooling one of these substances is called a *heat exchanger*. In refrigeration and air conditioning, the heat exchanger brings the liquid refrigerant and the suction vapor into thermal contact with each other.

Refrigerant in the liquid line is ordinarily above room temperature. Suction vapor leaves the evaporating coil at a temperature near that maintained in the unit. Lowering the temperature of the liquid refrigerant before it reaches the flow-control device increases the overall efficiency of the unit by reducing the quantity of flash gas. During the heat-exchange process, the liquid is subcooled, and any liquid present in the vapor line is evaporated.

Although it is sometimes possible to solder together lengths of the liquid and vapor line to effect a heat exchange (Figure 3–38), a carefully designed heat exchanger allows complete and more reliable use of evaporating coil surface. There are three styles of highly efficient heat exchangers. The first is a fountain-type heat exchanger, which is a double-tube device in which liquid refrigerant passes through the space between the inner and outer tubes. A perforated internal tube causes the cool suction vapor to impinge at a high velocity on the walls of the

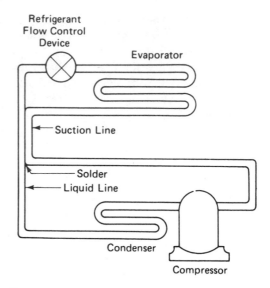

**Figure 3-38** Liquid and Vapor Lines Soldered Together

inside tube, which results in a high rate of heat exchange (Figure 3–39). A second type is a coil-type heat exchanger, which has a coil made from small-diameter tubing within a larger, copper tube shell (Figure 3–40). A third type is a suction line accumulator equipped with a

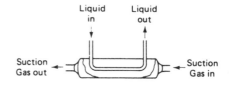

**Figure 3-39** Fountain-Type Heat Exchanger

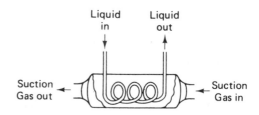

**Figure 3-40** Coil-Type Heat Exchanger

**Figure 3-43** Cleanable Angle-Type Filter-Strainer (Courtesy of Henry Valve Company.)

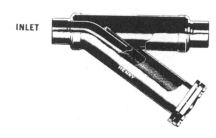

INLET

**Figure 3-44** Cleanable Y-Type Filter-Strainer (Courtesy of Henry Valve Company.)

(2) cleanable angle type (Figure 3–43), and the cleanable Y type (Figure 3–44).

**Figure 3-41** Heat Exchanger-Suction Accumulator (Courtesy of Refrigeration Research, Inc.)

heat-exchange coil in the bottom where the liquid refrigerant collects (Figure 3–41). This is the most popular type used on heat pump systems. It subcools the liquid refrigerant while assuring that no liquid refrigerant will return to the compressor.

## STRAINERS

Strainers remove foreign matter such as dirt and metal chips from the refrigerant lines. If left in the system, foreign matter could clog the small orifices of the flow-control devices and check valves and also enter the compressor.

There are three general types of strainers: (1) straight-through sealed type (Figure 3–42),

**Figure 3-42** Straight-Through Sealed-Type Filter-Strainer (Courtesy of Henry Valve Company.)

## DRIERS

Most authorities agree that moisture is the single most detrimental factor in a refrigeration system. A unit can stand only a very small amount of moisture. For this reason, the majority of both field- and factory-assembled refrigeration and air-conditioning systems are equipped with driers (Figure 3–45).

The following factors influence the selection of the correct size of drier:

1. Type and amount of refrigerant
2. Refrigeration system tonnage
3. Line size
4. Allowable pressure drop

**Figure 3-45** Liquid Line Drier (Courtesy of Alco Controls Division, Emerson Electric Co.)

When the refrigerant type, line size, and equipment application are known, the drier is generally selected on the basis of recommended capacities, which take into account both drying and refrigerant flow capacity. The flow and moisture capacity information is published by the drier manufacturer in tabular form for all popular refrigerants (Table 3-2).

Several designs of heat pump units use two driers, one for refrigerant flow during each cycle. However, another type of liquid line drier has been especially developed for use on heat pump systems. This filter-drier provides system protection in both heating and cooling cycles with one filter-drier. These components are equipped with internal check valves that allow refrigerant flow and filtration in either direction. Thus, the external check valves normally used on heat pump systems using capillary tube flow-control devices are eliminated (Figure 3-46). These units reduce system complexity and cost, while providing effective removal of moisture, acid, and solid contaminants.

Liquid line driers have been used in refrigeration systems for many years. Their primary function, as their name implies, is to remove moisture from the refrigerant. Moisture in a refrigeration system is a contributing factor in the formation of acids, sludge, and corrosion. Hydrochloric and hydrofluoric acids are formed by the interaction of halocarbon refrigerants and small amounts of moisture. Moisture in a refrigeration system can cause a multitude of difficulties, varying from flow-control device freeze-up to possible hermetic motor-compressor burnout. In large amounts, moisture can directly produce mechanical malfunctions. It is important to remove quickly and completely any moisture and any acid or sludge that may have formed in a refrigeration system.

Over the years, refrigeration systems have become more complex and driers are required to perform functions other than drying. Modern-day liquid line driers usually perform such functions as moisture removal, filtering, and acid removal.

Two general types of liquid line driers are in use. The straight-through sealed type (Figure 3-47), and the angle-replaceable core (replaceable cartridge) type (Figure 3-48).

*Applying Filter-Driers on Heat Pump Systems:* To obtain a long-term system life it is important to keep the level of contaminants at a minimum. This is particularly true of a heavy-duty application such as a heat pump.

**Figure 3-46** Heat Pump Liquid Line Drier (Courtesy of Watsco, Inc.)

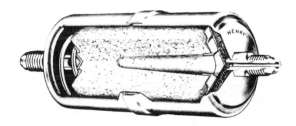

**Figure 3-47** Straight-Through Sealed-Liquid Line Drier (Courtesy of Henry Valve Company.)

**TABLE 3–2**

Filter Drier Selection

| FILTER-DRIER SELECTION SERIES "H" DRI-COR | | | | | | | | RECOMMENDED TONNAGE BASED ON BOTH DRYING AND FLOW CAPACITY | | CAPACITY RATINGS ACCORDING TO A.R.I. STANDARD 710 | | | | | | |
|---|---|---|---|---|---|---|---|---|---|---|---|---|---|---|---|---|
| CATALOG NUMBERS | | SIZE CONN. | CORE FILTER AREA SQ. IN. | DIMENSIONS INCHES | | | WEIGHT POUNDS | REFRIGERANTS | | DROPS OF WATER | | | | FLOW CAPACITY @ 2 P.S.I. IN TONS | | |
| | | | | SHELL DIA. | OVER-ALL LENGTH | | | | | R-12 (15 P.P.M.) | | R-22 (60 P.P.M.) | | | | |
| | | | | | | | | | | LIQUID LINE TEMP. °F. | | | | | | |
| FLARE | O.D. SOLDER | | | | FLARE | O.D.S. | | R-12 | R-22 | 75° | 125° | 75° | 125° | R-12 | R-22 | R-502 |
| H032 | H032-S | 1/4" | 11 | 1 5/8 | 4 1/4 | 3 13/16 | 3/8 | 3/4 | 3/4 | 46 | 33 | 31 | 20 | 2.3 | 3.0 | 2.0 |
| H052 | — | 1/4" | 17 | 2 5/16 | 4 11/16 | | 3/4 | 1 | 1 | 92 | 66 | 62 | 40 | 2.3 | 3.0 | 2.0 |
| H053 | — | 3/8" | | | 5 3/16 | | 7/8 | 1 | 1 | | | | | 4.0 | 5.2 | 3.5 |
| H082 | H082-S | 1/4" | 24 | 2 5/16 | 5 9/16 | 5 1/8 | 1 1/8 | 1 | 1 | 156 | 112 | 107 | 68 | 2.7 | 3.5 | 2.4 |
| H083 | H083-S | 3/8" | | | 6 1/16 | 5 1/4 | 1 1/8 | 2 | 2 | | | | | 5.3 | 6.8 | 4.7 |
| H084 | H084-S | 1/2" | | | 6 1/4 | 5 3/8 | 1 1/4 | 2 | 2 | | | | | 8.2 | 10.6 | 7.2 |
| H162 | — | 1/4" | 36 | 2 7/8 | 6 5/16 | — | 1 1/2 | 2 | 2 | 282 | 202 | 192 | 122 | 2.7 | 3.5 | 2.4 |
| H163 | H163-S | 3/8" | | | 6 3/4 | 5 15/16 | 1 1/2 | 3 | 3 | | | | | 5.5 | 7.1 | 4.8 |
| H164 | H164-S | 1/2" | | | 7 | 6 1/16 | 1 3/4 | 4 | 4 | | | | | 8.7 | 11.2 | 7.7 |
| H165 | H165-S | 5/8" | | | 7 1/4 | 6 5/16 | 1 3/4 | 5 | 5 | | | | | 11.0 | 14.2 | 9.7 |
| H303 | — | 3/8" | 57 | 3 | 9 11/16 | — | 3 3/8 | 4 | 5 | 490 | 352 | 335 | 212 | 5.8 | 7.5 | 5.1 |
| H304 | H304-S | 1/2" | | | 9 7/8 | 9 | 3 3/8 | 7 1/2 | 7 1/2 | | | | | 11.8 | 15.2 | 10.4 |
| H305 | H305-S | 5/8" | | | 10 3/16 | 9 1/4 | 3 3/8 | 10 | 10 | | | | | 15.3 | 19.7 | 13.5 |
| — | H307-S | 7/8" | | | — | 9 7/8 | 3 1/4 | 10 | 15 | | | | | 24.9 | 32.1 | 21.9 |
| H414 | | 1/2" | 71 | 3 1/2 | 9 15/16 | — | 4 3/8 | 10 | 10 | 710 | 506 | 482 | 305 | 12.1 | 15.6 | 10.6 |
| H415 | H415-S | 5/8" | | | 10 1/4 | 9 5/16 | 4 1/2 | 10 | 15 | | | | | 16.0 | 20.6 | 14.1 |
| — | H417-S | 7/8" | | | — | 9 15/16 | 4 1/2 | 15 | 20 | | | | | 25.9 | 33.4 | 22.8 |
| — | H419-S | 1 1/8" | | | — | 9 7/8 | 4 1/2 | 15 | 20 | | | | | 31.0 | 40.0 | 27.3 |
| — | H607-S | 7/8" | 106 | 3 | — | 16 9/16 | 6 | 20 | 25 | 1158 | 579 | 562 | 432 | 31.0 | 40.0 | 27.3 |
| — | H609-S | 1 1/8" | | | | | | 25 | 30 | | | | | 34.0 | 45.0 | 30.0 |
| — | H755-S | 5/8" | 123 | 3 1/2 | — | 15 1/16 | 8 1/4 | 15 | 20 | 1320 | 950 | 905 | 570 | 17.5 | 22.6 | 15.4 |
| — | H757-S | 7/8" | | | — | 15 11/16 | | 25 | 30 | | | | | 28.2 | 36.4 | 24.8 |

Recommended Tonnage Rating based on both Drying and Flow Capacity is shown in table. For Refrigerants 500 and 502 use data shown for Refrigerant 12.
*Source:* Henry Valve Company.

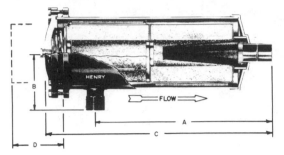

Dimension "D" is the minimum space required to remove the filter-drier core from the shell.

*Figure 3-48* Angle-Replaceable Core Liquid Line Drier (Courtesy of Henry Valve Company.)

Therefore, all heat pumps should have at least one filter-drier. Two standard driers are sometimes preferred, but where this creates a piping problem, adequate protection can be obtained with a single reversing drier.

*Using two standard driers.* Many original-equipment manufacturers prefer to use two standard driers rather than a single reversible drier. This has several advantages: more desiccant in the system, less complicated drier parts, and lower cost. Field service people

should follow the recommendations of the equipment manufacturer and use driers in the manner recommended by their design. The use of two driers will give protection equal to, or greater than, the use of a single reversible filter-drier (Figure 3–49).

Standard driers are often installed directly ahead of the expansion device—one in the outdoor section and one in the indoor section. Another common design is to locate both driers in the outdoor section, where they are easier to service. In this design one drier is ahead of the expansion device, and the other drier is ahead of the check valve in the outdoor section. When installed in these locations, the flow through each drier is always in the same direction (Figure 3–50). Standard filter-driers will not tolerate flow in the reverse direction. Reverse flow washes out the dirt previously collected, and also tends to result in a high-pressure drop.

In servicing units in the field it is advisable to replace the originally installed filter-drier with the next larger size, or the size recommended by the equipment manufacturer.

*Drier-check valve combination.* Some heat pumps use a drier with a check valve built

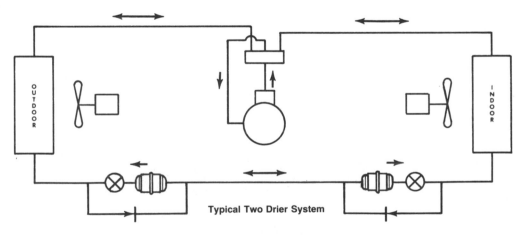

*Figure 3-49* Typical Two Drier System (Courtesy of Sporlan Valve Company.)

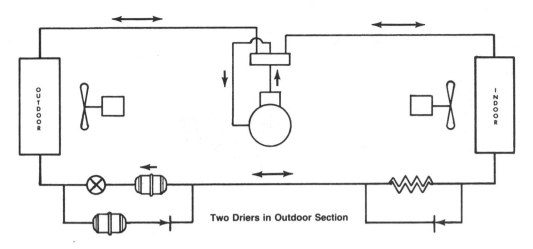

**Figure 3-50** Two Driers in Outdoor Section (Courtesy of Sporlan Valve Company.)

into the outlet fitting of the filter-drier. Filter-driers of this type must be replaced with an identical replacement drier obtained from the equipment manufacturer. If an identical replacement is not available, repair of the unit would require replacing the drier with a standard filter-drier and a separate check valve. Obviously, this would involve extensive re-piping (Figure 3–51).

*Using reversible filter-driers.* The simplest way to install a drier on a heat pump system

is to install a reversible filter-drier. These driers are installed in the reversing liquid line that runs between the indoor and the outdoor sections of the unit. Reversible driers should never be installed in the reversing gas line that runs between the indoor coil and the four-way valve, or the reversing gas line that runs between the outdoor coil and the four-way valve. Installation in this location will not give protection to the close-tolerance parts of the system, and may result in an excessive pressure drop. If the reversible drier is to be used on a highly con-

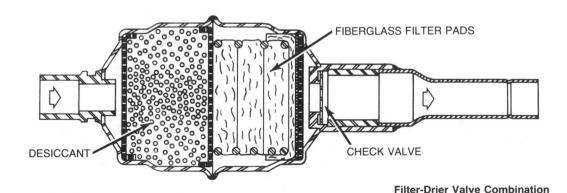

**Filter-Drier Valve Combination**

**Figure 3-51** (Courtesy of Sporlan Valve Company.)

**Reversible HPC Filter-Drier**

*Figure 3-52*   (Courtesy of Sporlan Valve Company.)

taminated system, such as after a hermetic motor burnout, it is essential that the old filter-driers be removed (Figures 3–52 and 3–53).

Some heat pumps will have only one expansion device—a bidirectional thermostatic expansion valve, or special short-tube flow restrictor. These systems are properly protected by using a reversible filter-drier in the line between the expansion device and the outdoor coil. In this location the drier will receive solid liquid on the cooling cycle and a mixture of liquid and vapor on the heating cycle.

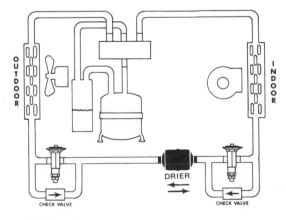

*Figure 3-53*   (Courtesy of Sporlan Valve Company.)

*Cleanup after burnout.*   Typical recommendations for cleanup after a hermetic compressor motor burnout are as follows:

1.  Install the same-size drier (or an oversized drier) in the liquid line ahead of each expansion device, and a drier in the common suction line. Of course, units with two liquid line driers should have both changed.
2.  Install an oversized drier in the common liquid line and, if possible, a drier in the common suction line. Run the unit in one mode of operation for a day. Then replace the liquid line drier with a reversible filter-drier.
3.  Install a reversible filter-drier in the common liquid line. If possible, install a drier in the common suction line.

*Suction line drier location.*   A filter-drier should be installed in the suction line to clean up a heat pump after a severe hermetic motor burnout. First, make sure that the burnout is "severe" by testing the acidity of the oil from the burned-out compressor using an acid test kit. If this test or other contamination indicates that the burnout was severe, install a standard drier in the common suction line. This drier can be installed either before or after the accumulator, but always between the four-way valve and the compressor. If some contaminants should remain in the accumulator, the preferred location is between the accumulator and the compressor. This is usually a crowded location on the unit, and installing the drier may be difficult. In some cases it may be necessary to re-pipe the suction line so that the drier is located outside the cabinet.

Most heat pumps are compact, unitary systems with cramped piping, making it difficult to replace or install filter-driers. Here are some suggestions indicating the principles involved in solving these practical problems.

1. The reversible filter-drier is usually the simplest to install because it can be installed anywhere in the reversing liquid line leading from the outdoor section to the indoor section. Even when using the reversible filter-drier, it is important to remove any old filter-driers that are installed on the unit, since they may be plugged if the unit is severely contaminated. If it is not possible to replace the previous filter-driers directly, the best alternative is to remove them. Reattach the line at the drier location, and use a reversible filter-drier to protect the system in the future. The old driers must be removed; they may be severely plugged.

2. Most filter-drier manufacturers make compact filter-driers that can be used in most units with cramped locations.

3. The installation of two standard driers will usually be easier if both are installed in the outdoor section. One drier is installed directly ahead of the expansion device in the outdoor coil, and the other drier is installed directly ahead of the check valve that leads to the expansion device on the indoor coil. Each of

these locations has refrigerant flow in only one direction at all times. If it seems there is not sufficient space for two standard driers on the particular unit involved, perhaps satisfactory contaminant protection can be obtained by using only one drier, particularly if this drier will be functional when the unit is started up. When there is no room inside the cabinet to install or replace the drier, the serviceman has no choice but to mount the drier outside the cabinet and do the necessary repiping.

4. The suction line connection on the four-way valve is usually the middle connection of the three connections on one side. The single connection on the opposite side of the valve is the discharge line connection.

5. Whenever a drier is added to the liquid line of a system that did not originally have a drier, some additional refrigerant charge should be added to the unit to fill the interior space of the drier. No additional charge is necessary for a drier installed in the suction line.

6. In some cases it may be desirable to install a drier in the discharge line of a heat pump. A standard filter drier can be installed in the discharge line between the compressor and the four-way valve. Only a core-type filter drier should be installed in the discharge line location. It is believed that the pulsations will cause the desiccant to break down in a granular desiccant style filter-drier. The discharge line is not normally chosen for drier location because the water capacity is reduced about 30%. However, on a heat pump this location has the advantage of protecting the four-way valve in case of a hermetic motor burnout. Also, in some instances, the field service engineer may

feel it is necessary to apply a drier in the discharge line because of cramped piping considerations.

## DISCHARGE MUFFLERS _____

On systems where noise transmission must be reduced to a minimum or where compressor pulsation might create vibration problems, a discharge muffler is frequently installed to dampen and reduce compressor discharge noise. The muffler is basically a shell with baffle plates; the required internal volume is primarily dependent on the compressor displacement, although the frequency and intensity of the sound waves are also factors in muffler design (Figure 3-54). The purpose of the muffler is to dampen or remove the hot vapor pulsations set up by a reciprocating compressor.

Every reciprocating compressor will create some hot vapor pulsations. Although great effort is made to minimize pulsations in the system and compressor design, vapor pulsations can be severe enough to create two closely related problems: (1) noise, which, although irritating to equipment users, does not necessarily have a harmful effect on the system, and (2) vibration, which can result in refrigerant line breakage. Frequently, these two problems appear simultaneously.

The discharge muffler is installed in the discharge line as close to the compressor as

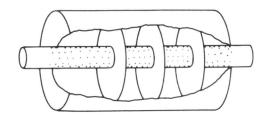

**Figure 3-54** Discharge Muffler

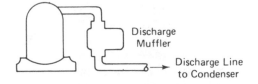

Discharge Muffler

Discharge Line to Condenser

**Figure 3-55** Discharge Muffler Location

possible. In the case of hermetic compressors, the muffler is frequently manufactured in the compressor shell itself.

Because the normal construction of a muffler is within a shell, a natural trap is formed. Mufflers will trap oil easily and may trap liquid refrigerant. If a muffler is used, it should be installed in either a downward or horizontal line (Figure 3-55).

The selection of a discharge muffler can be a difficult engineering problem. Improper field selection of a muffler will sometimes lead to increased rather than decreased vibration.

## AIR FILTERS _____

The air filters used in any heating and/or cooling system should be kept clean, primarily so that the volume of air circulated through the unit will be at the design rate. This is especially necessary for heat pump systems, because any decrease in indoor air flow will cause a corresponding increase in compressor compression ratio.

The purpose of air filters is to trap dust and dirt before the air reaches the indoor coil and the dust is deposited on the coil surface. When filters get dirty, they are performing their designed function and should be cleaned or replaced, not removed and thrown away.

Filters are installed in the air stream ahead of the indoor coil (Figure 3-56). Here they protect the entire indoor unit from collecting dust and dirt.

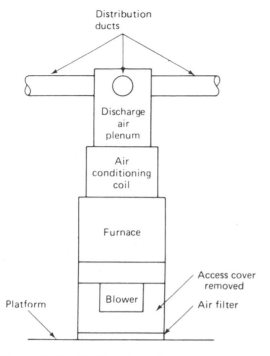

Figure 3-56  Air Filter Location

## CAPACITORS _____

Almost all single-phase motors used in the refrigeration and air-conditioning industry use capacitors to provide greater starting torque, to increase efficiency, or both. They accomplish this task by producing a second phase of current, which is needed for the motor to perform as designed.

Capacitors are rated in millionths of a farad, or in microfarads ($\mu$F). A large microfarad rating indicates greater storage capacity, whereas a small microfarad rating results in a smaller storage capacity. Therefore, a large phase shift will occur between the applied voltage and the load when a capacitor with a large microfarad rating is connected in series with a load. This high phase shift pro-

vides the high starting torque for single-phase motors. A low phase shift and low starting torque results when a capacitor with a low microfarad rating is used.

Two types of capacitors are used to produce the second phase current necessary to start single-phase motors (Figure 3-57). The physical dimensions have nothing to do with the microfarad rating of a capacitor. A run capacitor will usually have a low microfarad rating. It is physically larger because of the oil-type dielectric used. This oil dissipates the internal heat from the plates to the outside atmosphere. The run capacitor provides a low starting torque and a low current flow through the series load.

The starting capacitor usually has a higher microfarad rating than the run capacitor. The starting capacitor is switched out of the circuit a few seconds after the motor starts, because it is not designed to carry heavy current flow for an extended period of time. If the starting capacitor is left in the circuit for an extended period of time, permanent damage may be done to the capacitor.

When replacing capacitors, be sure to use the size recommended by the equipment manufacturer. Some will allow a certain percentage of variation in replacement capacitor microfarad rating, while others will not.

Run                    Start

Figure 3-57  Motor Capacitors

## ———————— SUMMARY ————————

The compressor on a heat pump system circulates the refrigerant through the system. Reciprocating compressors are the most popular type used on heat pump units.

The compression ratio (CR) is of great importance in heat pump systems. A high compression ratio could possibly cause compressor damage. Most hermetic compressors are not designed to operate with a CR greater than 7.5.

Lubrication is another problem with compressors in heat pump systems because liquid refrigerant often enters the compressor. Caution should be exercised to prevent a mixture of greater than 10% liquid in oil. Use a low-foaming oil in heat pump compressors.

There are several design differences between heat pump compressors and other compressors, such as larger connecting rods and bearings, improved motor winding protector, and pistons designed with rings, not lapped.

The reversing valve is used to change the direction of the refrigerant from the cooling to the heating cycle. It mechanically changes the flow of refrigerant. The valve is actuated by a solenoid-operated plunger that opens and closes the vapor ports to cause a buildup of pressure to move the piston assembly.

The most popular types of refrigerant flow control devices used on heat pump systems are the thermostatic expansion valve, the capillary tube, and a combination of these two types. Check valves are used to divert the flow of refrigerant around the flow-control device not in use during that cycle.

The refrigerant charge is critical in capillary tube systems because normally there is no receiver to store excess refrigerant.

The subcooling control valve is the type of flow-control device used on Westinghouse Hi/Re/Li heat pump units. The same valve controls the refrigerant flow during both cycles. This device controls the amount that the refrigerant is subcooled as it leaves the condensing coil.

The bleed port in a subcooling control valve does three things: (1) it assures quick equalization of the suction and discharge pressures after shutdown on the cooling cycle, (2) it assures proper oil return during all types of operating conditions, and (3) it assures that the valve quickly responds to changing system conditions.

Check valves are used in heat pump systems for two reasons: (1) to cause the refrigerant to flow through the flow-control device, or (2) to allow the refrigerant to bypass the flow-control device.

The manifold check valve used on Westinghouse Hi/Re/Li heat pump systems is a combination of four separate ball check positions in a single housing. The balls are positioned by changes in system pressures created when the reversing valve changes its position. A small bleed port between the check valve passages serves two purposes: (1) it assists the bleed port in the liquid subcooling valve to quickly equalize the system pressures on shutdown, and (2) it helps to provide a minimal flow of refrigerant during operation under extremely low load conditions to aid in maintaining a cool running compressor and oil return.

Accumulators are installed in the suction line of compressors to trap and evaporate liquid refrigerant, thus preventing dilution of the lubricating oil and possible damage to the compressor.

The purpose of the "soft plug" in an

accumulator is to provide over-temperature protection for the equipment.

The accumulator–heat exchanger serves three important functions: (1) it adds subcooling to the hot high-pressure liquid on its way to the evaporating coil, (2) it provides a positive separation of the low-pressure liquid and vapor from the evaporating coil, and (3) it assures positive oil return to the compressor at all times during operation.

Both the indoor and outdoor coils used on heat pump units have a large face area and are engineered for both condensing and evaporating functions. Both coils are carefully sized to provide balanced system performance and to maintain maxiumum efficiency during both heating and cooling cycles.

Auxiliary heating elements are used to aid the heat pump when the outdoor ambient temperature falls below the balance point. They are also used to temper the indoor air during the defrost period.

Crankcase heaters are placed on compressors to prevent liquid migration to the compressor during the off cycle. They are low-wattage resistance elements and may be energized continuously or only when the compressor is not running.

The refrigerant piping in heat pump systems should be sized in accordance with the equipment manufacturer's recommendations for the unit being designed. All vapor lines should be insulated to prevent capacity loss and/or sweating.

Any device that brings into contact two substances of different temperatures for the purpose of heating or cooling one of these substances is called a heat exchanger. Lowering the temperature of the liquid refrigerant before it reaches the flow-control device increases the overall efficiency of the unit by reducing the quantity of flash gas.

Strainers remove foreign matter such as dirt and metal chips from the refrigerant lines. The three general types of strainers are (1) straight-through sealed type, (2) cleanable angle type, and (3) cleanable Y type.

The primary function of a drier is to remove moisture from the refrigeration system. Moisture in a refrigeration system can cause a multitude of difficulties, varying from flow-control device freeze-up to possible hermetic motor-compressor burnout.

On systems where noise transmission must be reduced to a minimum or where compressor pulsation might create vibration problems, a discharge muffler is frequently installed to dampen and reduce compressor discharge noise. The muffler is basically a shell with baffle plates; the required internal volume is primarily dependent on the compressor displacement, although the frequency and intensity of the sound waves are also factors in muffler design.

The air filters used in any heating and/or cooling system should be kept clean, primarily so that the volume of the air circulated through the unit will be at the design rate. This is especially necessary for heat pump systems because any decrease in indoor air flow will cause a corresponding increase in compressor compression ratio.

Almost all single-phase motors used in the refrigeration and air-conditioning industry use capacitors to provide greater starting torque, to increase motor efficiency, or both. They accomplish this task by producing a second phase of current, which is needed for the motor to perform as designed.

## REVIEW QUESTIONS

1. What is the primary function of a compressor in a refrigeration system?
2. What is the maximum compression ratio for

which hermetic compressors are designed to operate?

3. What will cause an increase in compression ratio when the heat pump is operating in the heating cycle?

4. Why should liquid refrigerant be kept from the compressor crankcase?

5. What is the maximum percentage allowed of liquid refrigerant in oil?

6. Why do heat pump compressor motors run hotter than air-conditioning compressors?

7. What percentage of reduction in bearing load do the larger connecting rods and bearings provide in a heat pump compressor?

8. Is it recommended to field convert an air-conditioning unit to a heat pump?

9. What is the purpose of the reversing valve?

10. What causes the piston assembly to move in a reversing valve?

11. What are the names of the coils used on heat pump systems?

12. What component is placed in the refrigerant line between the compressor and the reversing valve on a heat pump system?

13. Name the refrigerant flow-control devices most commonly used on heat pump systems.

14. Does a subcooling control valve operate like a thermostatic expansion valve?

15. What device directs the refrigerant either through or around the flow-control device on a heat pump system?

16. Of what is the manifold check valve a combination?

17. What two functions do the indoor and outdoor coils have on a heat pump system?

18. What are the two purposes of the auxiliary heat strips used on heat pump units?

19. What prevents liquid refrigerant from collecting in the compressor crankcase during the off cycle?

20. What does PTCR mean?

21. What should be done to the vapor line during installation of the unit?

22. What device brings into contact two substances for the purpose of cooling or heating one of the substances?

23. What is the purpose of the strainers used in refrigeration systems?

24. What device is used to remove moisture from a refrigeration system?

25. What effect will a dirty air filter have on the compression ratio of a heat pump compressor?

# 4

# Heat Pump Controls

## INTRODUCTION

The controls and their connecting wiring are the devices that signal the heat pump components when their function is desired. It is necessary that design, service, and installation people be familiar with how each control functions so that the proper control can be used in the design and in servicing the installation after completion. The more a person in this industry knows about controls, the easier the job will be.

## *TRANSFORMER*

A transformer is a device used to transfer electrical energy from one circuit to another. In airconditioning and refrigeration work the transformer reduces line voltage (240 or 120 V) to 24 V, which is sometimes referred to as low voltage or the control circuit voltage.

A transformer is a very basic device. It consists of two or more electric coils wound around one laminated core so that all the lines of flux from one coil will cut across all the turns of the other coil (known as *unity coupling*). A transformer has no moving parts and, therefore, requires very little maintenance. Transformers are simple, rugged, and efficient (Figure 4–1).

In operation, the primary coil is connected to and draws power from the electrical power source. The secondary coil is connected to and delivers electrical power to the load in the control circuit. The amount of power being transferred from the primary winding to the secondary winding is determined by the amount of electric current flowing in the secondary control circuit. The amount of current flow in the secondary circuit is dependent on the amount of power required by the load. If the load has a low resistance and requires a great amount of power, a high current will flow in the control circuit. This high current flow causes a decrease in the electromotive force (EMF) of the magnetic field, which is necessary for the

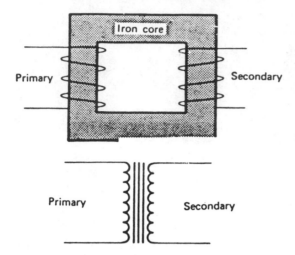

*Figure 4-1* Symbolic Sketch and Schematic Symbol of a Transformer

high current flow in the secondary circuit. Thus the transformer regulates the transfer of electrical power from the source to the load in response to the load requirements.

When selecting a transformer to power a low-voltage control system, careful consideration should be exercised. For example, inductive devices like contactors, relays, and solenoid valves require more power on starting than during steady operation. A transformer delivers the maximum possible inrush of electrical current to the load when the transformer impedance equals the impedance of the load. Therefore, the performance of a less expensive transformer that is properly matched to the load can be equal to that of a more expensive, unmatched transformer.

## THERMOSTATS ————

Thermostats signal the equipment to start or to stop in response to the temperature requirements of the building. Indoor and outdoor thermostats are used on heat pump systems. Indoor thermostats generally use a bimetal control to actuate the contacts or a mercury bulb; outdoor thermostats generally use a remote bulb for this purpose.

*Indoor Thermostats:* Indoor thermostats are installed in the conditioned space and sense the temperature there. These controls signal the equipment to provide either cooling, heating, or ventilation as indicated by the space requirements (Figure 4–2).

There are many types of thermostats; however, the most popular are single-stage heating single-stage cooling, double-stage heating and single-stage cooling, and double-stage heating and double-stage cooling (Figure 4–3).

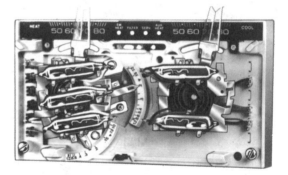

*Figure 4-2* Heat Pump Thermostat (Courtesy of Honeywell Inc.)

*Figure 4-3* Thermostat with Different Stages (Courtesy of Honeywell Inc.)

Some thermostats are equipped with automatic changeover. The first stage on a thermostat usually controls the heat pump, and the second stage controls the auxiliary heating equipment. The first stage is also used occasionally to energize the reversing valve solenoid. Other manufacturers energize the reversing valve solenoid when the system changeover switch is switched. Remember, the reversing valve may be energized in either mode of operation depending on equipment design. When the first stage energizes the reversing valve solenoid, the second stage controls the equipment. The auxiliary heating equipment is then controlled by the outdoor thermostats.

Some manufacturers incorporate an emergency heat switch that turns off all the refrigeration equipment and energizes the auxiliary heating equipment. This is for use should the refrigeration part of the unit malfunction (Figure 4-4). A red light on the thermostat indicates when the emergency heat is energized. This is to warn the user that it has been switched.

The second stage on the thermostat usually controls the auxiliary heating equipment in conjunction with the outdoor thermostat, if used. This stage usually energizes the circuit after the indoor temperature has fallen approximately 3 °F (1.68 °C) below the setting for the first stage, thus indicating that an additional amount of heat is needed to maintain the desired indoor temperature.

As the preceding discussion indicates, there are many types of thermostats available for heat pump systems and many more ways of wiring them to obtain the desired operation of the equipment. Therefore, the proper thermostat and wiring connections must be used for the particular brand and model of heat pump in consideration.

*Outdoor Thermostats:* These thermostats are installed either inside the outdoor cabinet or under the roof eave. They sense the outdoor temperature and prevent unnecessary operation of the auxiliary heating equipment. Outdoor thermostats are electrically connected in series with the second stage of the indoor thermostat. Therefore, both thermostats must demand operation before any auxiliary heat is used. However, either one can de-energize the heat strips.

Outdoor thermostats are actuated by a remote bulb placed to sense the outdoor air temperature (Figure 4-5). If one outdoor thermostat is used, it is set to energize all the auxiliary heat strips. However, more than one outdoor thermostat can be used to provide even more economy and comfort. Each thermostat

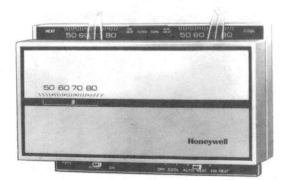

***Figure 4-4*** Thermostat with Emergency Heat Switch (Courtesy of Honeywell Inc.)

***Figure 4-5*** Outdoor Thermostat (Courtesy of Lennox Industries, Inc.)

is set at a balance point equal to the heat loss of all the equipment operating. For example, the first outdoor thermostat is set to bring on one or more heat strips at a temperature corresponding to the balance point of the heat pump only. The second outdoor thermostat is set to energize additional heat strips at a temperature corresponding to the balance point of the heat pump plus the energized heat strips. Each thermostat is set this way until all heating equipment is operating. The outdoor thermostats have no effect on operation of the equipment during the cooling mode.

Outdoor thermostats sometimes provide a dual function in equipment operation. Some heat pump manufacturers prefer that their refrigeration equipment not operate below certain outdoor temperatures. In this case, an outdoor thermostat may be used to deenergize the compressor and energize the auxiliary heat strips when this temperature is reached. The entire heating requirements are then provided by the heat strips. This control is sometimes called a *low-ambient switch* (Figure 4-6).

## HEAT ANTICIPATOR ————————

Heat anticipators are electric resistors used in indoor thermostats to provide a false heat inside the thermostat and thus to cause the thermostat to provide better control of the equipment (Figure 4-7). They are termed *heating anticipator* and *cooling anticipator*.

The heating anticipator is connected electrically in series with the equipment. The electric current flows through the resistor, which causes it to heat slightly. The adjustment is set to correspond with the amperage draw of the controlled equipment. The anticipator provides a false heat to the thermostat, which prevents temperature override inside the building. Each stage will have its own heating anticipator.

The most accurate method to determine the heat anticipator setting is to actually measure the current flow in that circuit with the equipment operating. Since this is usually a very small current flow, an energizer can be used, or wrap several turns of the wire around one tong of the ammeter (Figure 4-8). The amperage reading is then divided to determine the actual amperage flow.

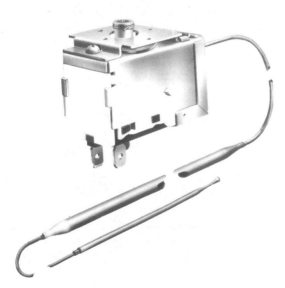

**Figure 4-6** Low-ambient Seitch (Courtesy of Ranco Controls Division)

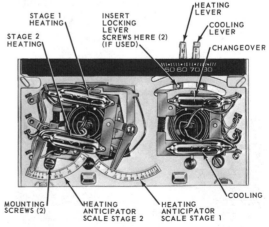

**Figure 4-7** Thermostat Heat Anticipators (Courtesy of Honeywell Inc.)

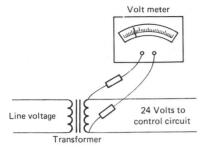

**Figure 4-8** Checking Temperature-control Circuit Voltage

The cooling anticipator is an electrical resistor placed in parallel with the cooling control circuit. It provides heat to the thermostat during the off cycle of the equipment. It is matched to the cooling circuit current flow and is nonadjustable. Cooling anticipators reduce

the length of the equipment off cycle (Figure 4-9).

## DEFROST CONTROLS ——————

Heat pump defrost controls are used to detect ice and frost on the outdoor coil during the heating cycle. Several methods are used to detect, initiate, and terminate the defrost cycle. Some of the more popular defrost controls are defrost thermostats, either single or double bulb (element), time–temperature, air pressure–temperature, solid state, and air-pressure differential.

*Defrost Thermostat:* Defrost thermostats may be used to control either a part of or the complete defrost cycle. They are generally of two

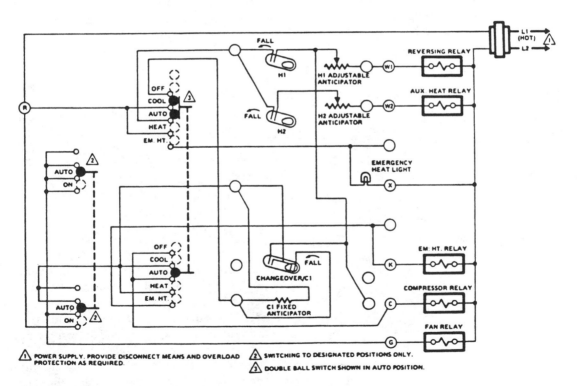

**Figure 4-9** Thermostat Schematic Showing Anticipators (Courtesy of Honeywell Inc.)

types, the single bulb (element), and the dual bulb (element) (Figure 4–10).

Single-bulb types are the most popular and are used in conjunction with a time clock to control the defrost cycle. The thermostat is responsible for terminating the defrost cycle, whereas the time clock, in combination with the thermostat, initiates the defrost cycle. When this type of thermostat is used, the bulb is attached to the outdoor coil at the point where the frost is melted last. Thus the unit will be kept in the defrost cycle by the defrost thermostat until the outdoor coil is completely defrosted.

In operation, initiation of the defrost cycle can occur only when the outdoor coil temperature is below the temperature of 26 °F ($-3.33$ °C) and during the first 60 seconds that the defrost timer attempts a defrost; then the cycle is skipped. If the outdoor coil temperature should drop to 26 °F or below during the first 60 seconds while the timer is trying to defrost, a defrost cycle will occur. The termination temperature can be set to a mimimum temperature to assure a completely defrosted

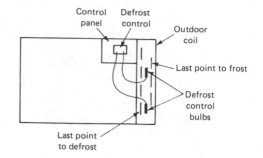

**Figure 4-11** Sensing Bulb Locations of Double Bulb Defrost Thermostat

coil. This temperature will usually be around 60 °F (15.56 °C).

The double-bulb type of defrost thermostat both initiates and terminates the defrost cycle. The initiation bulb is attached to the outdoor coil at a point where the frost leaves the outdoor coil last during the defrost cycle. The termination bulb is placed where it will sense the outdoor coil temperature at a point where the unit efficiency will be reduced rapidly should the coil frost over beyond that point (Figure 4–11).

This type of defrost thermostat is not very popular because it is not as effective as some of the other defrost methods. Should the cold outdoor air happen to blow on the sensing bulb, the thermostat may put the unit into defrost needlessly or cause it to remain in the defrost cycle too long, thus decreasing overall unit efficiency.

*Time-Temperature Defrost:* This very popular method of defrost control uses both time and temperature to cause a defrost cycle. Therefore, both the time and temperature sections of the control must demand defrost before the defrost cycle can be initiated (Figure 4–12).

When the defrost cycle is initiated, the reversing valve changes position and the outdoor fan motor stops running. The clock motor operates only when the compressor is running.

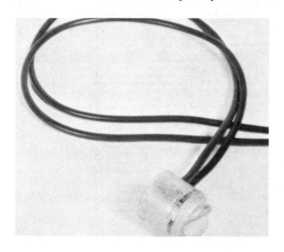

**Figure 4-10** Defrost Thermostat (Courtesy of Control Products Division, Texas Instruments, Incorporated)

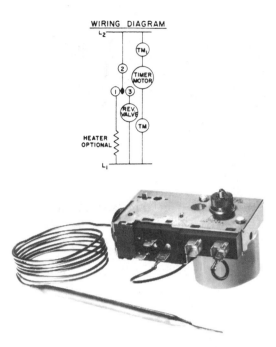

WIRING DIAGRAM

*Figure 4-12* Time-temperature Defrost Control and Schematic (Courtesy of Ranco Controls Division)

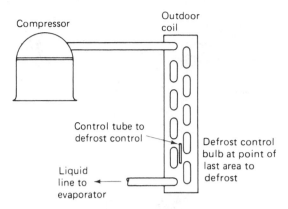

*Figure 4-13* Installation of Temperature Bulb for a Defrost Control

The timer can be set to initiate the defrost cycle on 30-, 60-, and 90-minute intervals. If it is found that the 90-minute cycle allows too much frost to accumulate on the outdoor coil, the timer can be adjusted to allow less frost to accumulate before defrost initiation. The defrost cycle can be initiated only by demand from both the timer and the temperature sections of the defrost control. The defrost control temperature will demand a defrost cycle at approximately 30°F ($-1.1$°C) and will terminate the defrost cycle at approximately 60°F (15.6°C) outdoor temperature.

The timer motor is operated by 240 V and operates only when the compressor is operating. The bulb of the temperature section of the defrost control is attached to a return bend on the outdoor coil (Figure 4-13). The bulb is usually insulated to ensure that the ambient air does not affect the bulb.

*Air Pressure-Temperature Defrost:* When this control is used, the complete process of defrost initiation and termination is fully automatic and is accomplished by the use of a single control (Figure 4-14). The control monitors any frost buildup or blockage due to leaves and other debris by sensing an air pressure increase or decrease at the outdoor coil. (This control may be used on push-through or pull-through coils.) The control initiates the defrost cycle by turning off the outdoor fan motor and changing the reversing valve to reverse the heat pump cycle. The defrost sensor (thermostat) terminates the defrost cycle by sensing the outdoor coil temperature.

In operation, two requirements must be met before the defrost cycle can be initiated: (1) the saturated vapor temperature of the refrigerant entering the outdoor coil through the flow-control device must be 32°F (0°C) or less, and (2) the air flow through the outdoor coil must be restricted from 70 to 90% with frost buildup. Most areas have the best results at about 75% blockage. Dirty outdoor coils will cause unnecessary defrost cycles, even when the system is properly adjusted, and increases the operating costs of the unit.

During normal operation of this control,

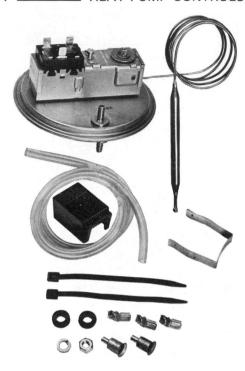

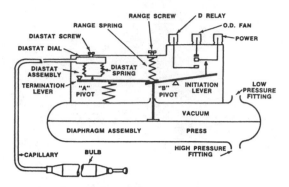

**Figure 4-15** Air-pressure Defrost Control in Normal Operating Condition (Courtesy of Robertshaw Controls Company, Uni-Line Division)

**Figure 4-14** Air Pressure-temperature Defrost Control Kit (Courtesy of Robertshaw Controls Company, Uni-Line Division)

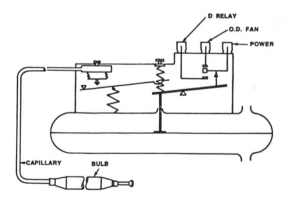

**Figure 4-16** Termination Lever in Up Position (Courtesy of Robertshaw Controls Company, Uni-Line Division)

the switch levers and contacts are positioned as in Figure 4–15. During operation, when the refrigerant entering the outdoor coil has fallen to 32 °F (0 °C) or less, the terminate lever moves to the up position (Figure 4–16). The terminate lever is moved up or down by the contracting or expanding of the alcohol charge stored in the sensing bulb, capillary, and diastat assembly.

After the terminate lever has moved up and the outdoor coil has a sufficient accumulation of frost to cause a vacuum effect at the outdoor coil to draw the diaphragm upward, the initiate lever will move to the down position (Figure 4–17). After the vacuum effect of the outdoor fan causes the diaphragm to initiate a defrost cycle, it is turned off by the switch, which allows the diaphragm to return to its original position. The initiate lever is then held

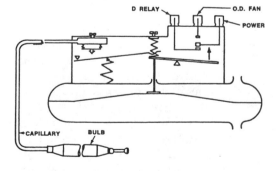

**Figure 4-17** Termination Lever in Down Position (Courtesy of Robertshaw Controls Company, Uni-Line Division)

in the down position by the switch spring (Figure 4–18).

When the temperature of the liquid refrigerant leaving the outdoor coil has risen to 55 to 58 °F (12.78 to 14.44 °C), the pressure of the alcohol in the sensing bulb will move the terminate lever, which places the switch in its position for normal operation, as shown in Figure 4–15.

*Solid-State Defrost:* This is another type of defrost control that is becoming more popular on heat pump systems. It is equipped with two temperature sensors (Figure 4–19). One sensor monitors the outdoor air temperature, while the other sensor monitors the outdoor coil temperature. This control operates on a temperature difference between the outdoor air temperature and the temperature of the outdoor coil. A defrost cycle is initiated when the outdoor temperature is below 45 °F (7.22 °C) and a temperature difference of 15 to 25 °F (−9.44 to −3.89 °C) exists between the outdoor coil sensor and the outdoor air sensor.

*Air-Pressure Differential (Dwyer) Defrost:* This control is mounted in the outdoor unit and senses the air pressure on both the air inlet and the air outlet sides of the outdoor coil (Figure 4–20). It is used in combination with a temperature-sensitive defrost termination control. Before a defrost cycle can be initiated with this type of system, two requirements must be met: (1) the saturated vapor temperature of the refrigerant between the outdoor coil and the refrigerant flow-control device must be 32 °F

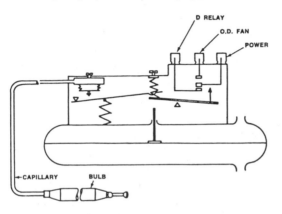

*Figure 4-18* Defrost Termination (Courtesy of Robertshaw Controls Company, Uni-Line Division)

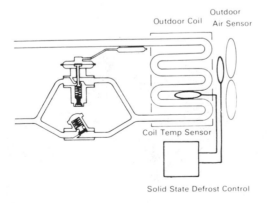

*Figure 4-19* Solid-state Control Installation (Courtesy of Lennox Industries, Inc.)

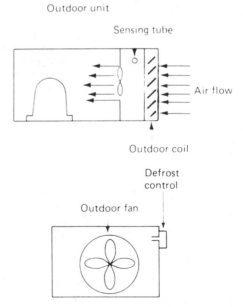

*Figure 4-20* Air-pressure Differential (Dwyer) Defrost Control Location

(0 °C) or less, and (2) the outdoor coil must be 70 to 90% restricted by the frost buildup.

In operation the vacuum-operated switch contacts are open. See Figure 4–21, in which the outdoor coil is not frosted enough to cause a vacuum effect of the outdoor fan to move the control diaphragm.

The defrost termination control contacts are normally open until the outdoor coil temperature falls to 32 °F (0 °C) or less (Figure 4–22). At this temperature the contacts will close and remain closed until the outdoor coil temperature has increased to 55 °F (12.78 °C).

When the outdoor coil has enough frost buildup for the outdoor fan to cause a vacuum on the control to move the diaphragm inward, the control switch contacts close (Figure 4–23). The diaphragm will remain in this position until the pressure differential across the outdoor

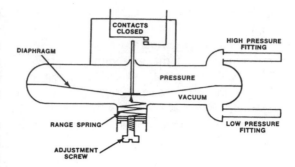

**Figure 4-23** Dwyer Defrost Control in Defrost Position (Courtesy of Ranco Controls Division)

coil is reduced enough to allow the diaphragm to move back into the normal operating position. The defrost cycle will remain in effect until the defrost termination control returns to its normal operating position.

## DEFROST RELAY _____

The defrost relay is a double-pole, double-throw relay with a 240-V coil (Figure 4–24). The defrost relay coil is energized when the defrost control demands a defrost cycle. This relay has several functions during the defrost period.

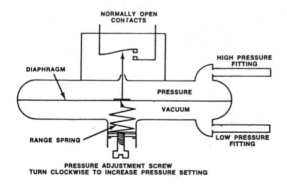

**Figure 4-21** Dwyer Defrost Control in Normal Operating Position (Courtesy of Ranco Controls Division)

**Figure 4-22** Defrost Termination Control Contacts Above 32 °F (Courtesy of Ranco Controls Division)

**Figure 4-24** Defrost Relay (Courtesy of Lennox Industries, Inc.)

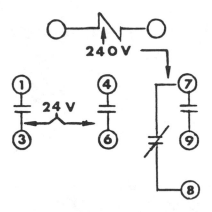

**Figure 4-25** Defrost Relay Schematic (Courtesy of Lennox Industries, Inc.)

trol circuit and is energized from the thermostat. It can be energized in two ways: (1) when the thermostat fan switch is switched to on, or (2) when the thermostat demands either heating or cooling. The coil terminals are numbers 4 and 5. The contact terminals are line voltage, either 120 or 240 V, with contacts 1 and 2 normally closed, while 2 and 6 are normally open (Figure 4-27). When color-coded wires are used, the blue wires are the coil connections, while red and yellow are normally closed and red and black are normally open (Figure 4-28).

When energized, it will deenergize the outdoor fan motor. It also changes the reversing valve back to the cooling position and energizes the auxiliary heat relay. Terminals 7–8 and 7–9 are in the 240-V circuit and terminals 1–3 and 4–6 are in the low (24-V) voltage circuit (Figure 4–25).

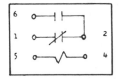

**Figure 4-27** Indoor Fan Relay Schematic

## INDOOR FAN RELAY _____

This relay is mounted in the control panel of the indoor unit. It is equipped with a 24-V coil and a set of single-pole, double-throw contacts (Figure 4-26). The coil is wired into the con-

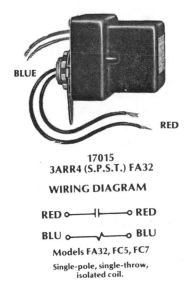

**17015**
**3ARR4 (S.P.S.T.) FA32**

**WIRING DIAGRAM**

RED o———||———o RED

BLU o————√————o BLU

Models FA32, FC5, FC7

Single-pole, single-throw, isolated coil.

3ARR4
Screw On Type

3ARR4
Push On Type

**Figure 4-26** Indoor Fan Relay (Courtesy of General Electric Co.)

**Figure 4-28** Indoor Fan Relay with Wiring Connections (Courtesy of General Electric Co.)

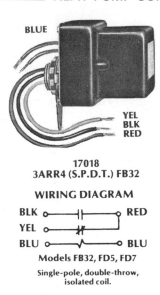

BLUE

YEL
BLK
RED

**17018**
**3ARR4 (S.P.D.T.) FB32**

**WIRING DIAGRAM**

BLK o———|⊢———o RED
YEL o———⊬
BLU o———⌇———o BLU

Models FB32, FD5, FD7

Single-pole, double-throw,
isolated coil.

*Figure 4-28* Continued

## EMERGENCY HEAT RELAY AND SUBBASE _____

The emergency heat relay is mounted in the control panel of the indoor unit and the subbase is mounted behind the thermostat (Figure 4–29). When the compressor fails to operate, the thermostat system switch should be moved to the emergency heat position. In this position, when the first-stage bulb demands heat, an emergency light is lighted on the thermostat. The light is to remind the operator that the compressor is not operating and that all the heating is being provided by the auxiliary heaters. Terminal E on the thermostat should be wired to provide operation of the fan, because intermittent fan operation is not possible from the first stage during the emergency heat cycle. Terminal E should also be wired to bypass all the outdoor thermostats so that the complete second-stage heating can be used.

The relay is energized by the manual switch on the thermostat to energize all the auxiliary heat if the compressor should fail. It de-

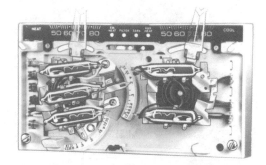

*Figure 4-29* Emergency Heat Thermostat and Subbase (Courtesy of Honeywell Inc.)

energizes the compressor and bypasses all the outdoor thermostats (Figure 4–30). In some cases it will also bypass the second-state heating bulb (H2), and as long as the first stage (H1) is demanding heat, it will control the auxiliary heating equipment (Figure 4–29).

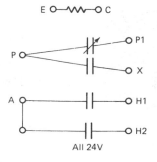

E o—⋀⋀⋀—o C

P o

P1
X

A o

H1
H2

All 24V

*Figure 4-30* Emergency Heat Relay Schematic

## CONTACTORS ————

The purpose of a contactor is to energize electrical circuits operating with high voltage, high current, or both. They are used on heat pump systems to energize the compressor and the emergency heat coils (Figure 4–31). The coil is operated by low (24-V) voltage, and the contacts are in the 240-V circuit. The electric wire (power) is brought from the disconnect switch or power source and connected to terminals L1

and L2 on the contactor (Figure 4–32). The load (the compressor or the electric heating elements) is then connected to terminals T1 and T2 on the contactor. Thus, when the contactor coil is energized, the contacts are closed, which completes the circuit between the power source and the load.

## OVERLOAD PROTECTORS ——

Basically two types of motor overload protectors are used on electric motors: (1) internal (thermostat), and (2) external. Almost all electric motors use one of these types, and some compressor manufacturers use both types to protect their equipment against faulty electric power or mechanical failure. These devices should be bypassed only when checking their condition. When an overload opens, there is a reason, and the problem must be found and corrected before permanent damage is caused.

*Internal (Thermostat) Overload:* This device is located in the motor winding at the hottest spot where it senses only temperature (Figure 4–33). The internal overload may be electrically connected into the circuit in many ways. First, it can be connected in the wiring between the common terminal and the motor winding (Figure 4–34). When placed this way, there are no external terminals for the overload. Second, it may be connected to external terminals (Figure 4–35). When the external terminals are present, the overload may be connected in the control (24-V) voltage circuit so that it will control the motor contactor, or it may be connected into the electric power to the motor. When connected in this manner, care should be exercised to make certain that all the power is turned off by the overload. Be sure to follow the manufacturer's recommendations when making electrical connections to ensure proper protection for the equipment.

*Figure 4-31* Contactor (Courtesy of General Electric Co.)

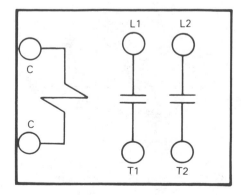

*Figure 4-32* Contactor Schematic

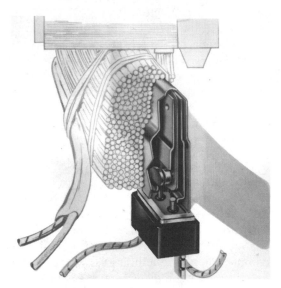

**Figure 4-33** Internal (Thermostat) Overload (Courtesy of Tecumseh Products Company)

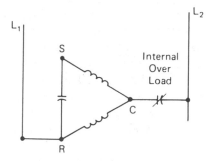

**Figure 4-34** Internal (Thermostat) Overload in Common Line

*External Overload:* These overloads are manufactured in three configurations: (1) two terminal, (2) three terminal, and (3) four terminal (Figure 4–36). They are mounted on the external surface of the compressor and may be replaced when found faulty. They are mounted in direct contact with the motor housing to sense the temperature at the point of contact.

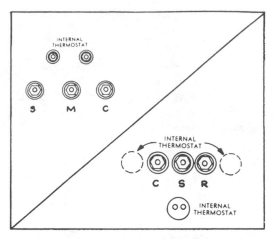

**Figure 4-35** Internal (Thermostat) Overload with External Terminals (Courtesy of Lennox Industries, Inc.)

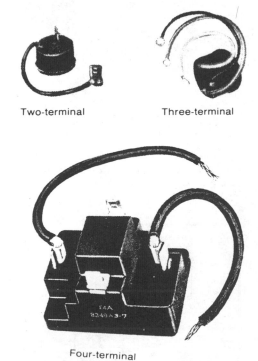

Two-terminal         Three-terminal

Four-terminal

**Figure 4-36** Two-, Three-, and Four-terminal Compressor Overloads (Courtesy of Control Products Division, Texas Instruments, Incorporated)

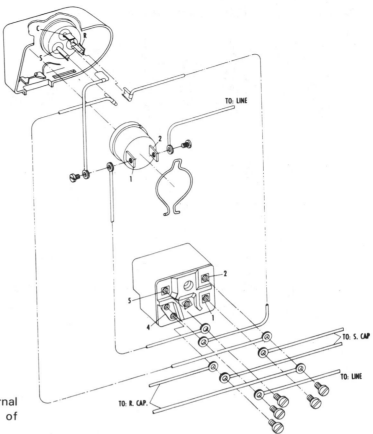

TO: LINE

TO: S. CAP

TO: LINE

TO: R. CAP.

*Figure 4-37* Location of External Overload Device (Courtesy of Tecumseh Products Company)

External overloads may be of the line break or thermostat type. These devices must be placed where the manufacturer has designated (Figure 4–37). An exact replacement must be used or proper protection of the motor will not be provided.

The line-break external thermostat is operated by a bimetal disk (Figure 4–38). The switch contacts are normally closed, and when an overload condition occurs the contacts open (Figure 4–39). They are electrically connected into the power circuit to the common terminal of the motor winding (Figure 4–40). When the resistance heater has provided sufficient heat, indicating an overload condition, the contacts open and interrupt the electrical circuit to the motor. The motor then stops until the overload device has cooled sufficiently to allow the con-

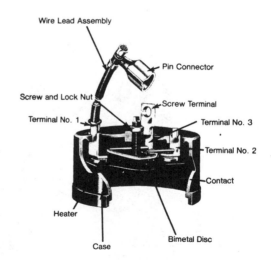

Wire Lead Assembly

Pin Connector

Screw and Lock Nut

Screw Terminal

Terminal No. 1

Terminal No. 3

Terminal No. 2

Contact

Heater

Bimetal Disc

Case

*Figure 4-38* External Line Break Overload (Courtesy of Tecumseh Products Company)

71

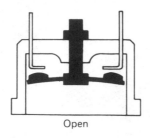 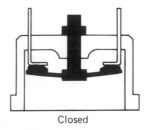

Open                    Closed

*Figure 4-39*  External Overload Contacts

### RESISTANCE START INDUCTION RUN

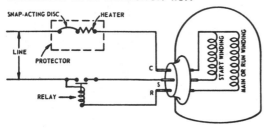

*Figure 4-40*  External Line Break Overload Connections

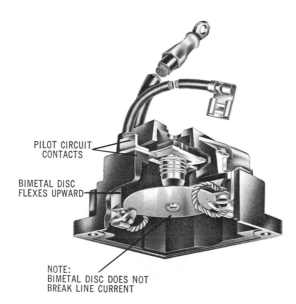

PILOT CIRCUIT CONTACTS

BIMETAL DISC FLEXES UPWARD

NOTE:
BIMETAL DISC DOES NOT
BREAK LINE CURRENT

*Figure 4-41*  External Thermostat Overload (Courtesy of Tecumseh Products Company)

tacts to close and complete the electrical circuit to the motor.

The thermostat type of external overload is mounted in the same way as the line-break type. However, the contacts do not open the main electrical power circuit (Figure 4–41). Instead, the control or pilot circuit is interrupted, which causes the contactor or starter to interrupt the main power supply (Figure 4–42).

## PRESSURE CONTROLS _____

Most heat pump systems are equipped with both high-and low-pressure control. These are in addition to the defrost termination pressure control used in combination with the solid-state defrost system.

*High-Pressure Control:*  These controls are mounted in the compressor discharge line between the compressor and the reversing valve (Figure 4–43). This control has a cutout point high enough so that normal operation of the system will not be affected by it. It may be either a manual or an automatic reset control (Figure 4–44). High-pressure controls are designed to protect the system against excessive compressor discharge pressures during both the heating and cooling cycles. The control contacts are in series electrically with the compressor contactor holding coil. The contacts are, therefore, in the 24-V electrical supply to the control system.

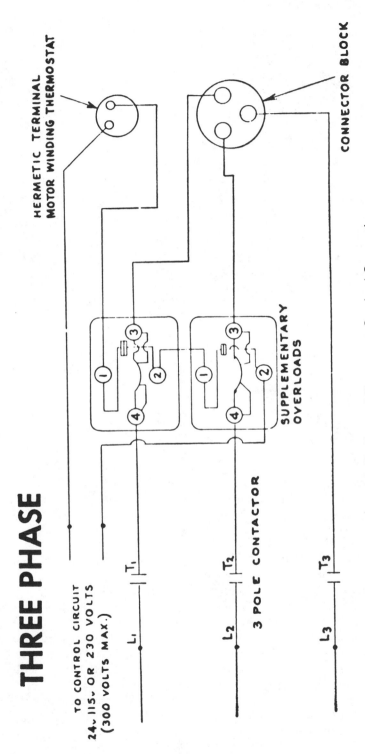

*Figure 4-42* External Thermostat Overload Connections

73

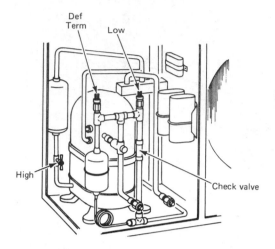

Figure 4-43 Location of High- and Low-pressure Controls (Courtesy of Lennox Industries, Inc.)

**Figure 4-44** Manual Reset High-pressure Control (Courtesy of Lennox Industries, Inc.)

*Low-Pressure Control (Loss of Charge Switch).* This control is located in the liquid line (Figure 4–43). It opens on a fall in pressure (Figure 4–45). The contacts are wired in series electrically with the compressor contactor holding coil. The control opens at a pressure low enough so that normal operation of the system will not be affected by it. The primary function of this control is to protect the system against a loss of refrigerant charge.

*Defrost Termination Pressure Switch:* This switch is located in the liquid line (Figure 4–43).

**Figure 4-45** Low-pressure Control (Loss-of-charge Switch) (Courtesy of Lennox Industries, Inc.)

**Figure 4-46** Defrost Termination Pressure Switch (Courtesy of Lennox Industries, Inc.)

It is used in combination with the solid-state defrost board (Figure 4–46). The defrost cycle is terminated by this high-pressure switch, which opens the electrical circuit to the control board when the refrigerant pressure in the outdoor coil increases to about 275 psi (1896.059 kPa). This pressure indicates that the outdoor coil temperature is about 124°F (51.11°C), which is high enough to ensure that all the frost has melted.

## LOW-AMBIENT SWITCH ————

This type of thermostat is used on some heat pump systems (Figure 4–47). It is generally located in the outdoor unit. When used, the equipment manufacturer will designate the temperature, usually around 10°F (−12.22°C), at which the thermostat will stop the compressor. Some manufacturers feel that operating the compressor at these low ambient temperatures saves very little energy, and, in fact, it increases wear on the compressor. This

*Figure 4-47* Low-ambient Switch (Courtesy of Lennox Industries, Inc.)

control is wired in series with the M terminal on the indoor thermostat and the compressor contactor holding coil.

## MILD-WEATHER CONTROL ——

In some installments it may be desirable to operate the unit in the heating mode when the outdoor temperature is mild (65 to 70°F or 18.33 to 21.11°C). When operating under these outdoor temperature conditions, an occasional

cutout by the high-pressure control might occur. The purpose of the mild-weather control is to cycle the outdoor fan motor to reduce the amount of heat taken from the outdoor air, thereby reducing the unit discharge pressure. The mild-weather control is located on the vapor line between the indoor coil and the reversing valve (Figure 4-48). The control is wired in electrical series with one lead of the outdoor fan motor (Figure 4-49).

*Figure 4-48* Location of Mild-weather Control (Courtesy of Lennox Industries, Inc.)

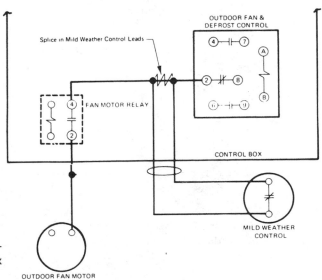

*Figure 4-49* Wiring Diagram of Mild-weather Control (Courtesy of Lennox Industries, Inc.)

## HEAT SEQUENCERS _____

Most manufacturers use a time-delay relay to bring on heating elements at different time intervals. They are generally used to prevent all the elements from coming on at one time and to prevent rapid cycling of the different components should the thermostat contacts chatter.

These relays have snap-acting contacts and may have many different contact arrangements. They may be either single stage or have many stages (Figure 4-50). They are designed to operate in any position. Some are resistance activated, while others have a solid-state positive temperature coefficient (PTC), which has some advantages over the other types.

These sequencers are actuated by the 24-V temperature control circuit. They may be wired in series according to the instructions to allow sequencing of virtually any number of heating elements and/or fans.

## STARTING COMPONENTS ____

All single-phase motors require some type of starting components. These may be in the form of a single running capacitor or a starting relay and starting capacitor in combination with the

*Figure 4-50* Heat Sequences (Courtesy of Control Products Division, Texas Instruments, Incorporated)

*Figure 4-50* Continued

running capacitor. Quite often the running capacitor will be used alone, and at times only the starting relay and starting capacitor are used without the running capacitor, depending on the equipment manufacturer's design and requirements. Capacitors were discussed in Chapter 3. Here we will discuss the starting relay.

*Starting Relay:* Starting relays are used to remove the starting circuit of a single-phase compressor from operation when the motor reaches approximately 75% of its normal running speed. Several types of relays are used for this purpose; here we will discuss only the potential relay and the solid-state starting relay.

*Potential (voltage) starting relays.* These relays operate on the electromagnetic principle. They incorporate a coil of very fine wire wound around a core. These starting relays are used on motors of almost any size. The contacts on this relay are normally closed and are caused to open when a plunger is pulled into the relay coil. These relays must have three connections to the inside in order for the relay to perform its function. These terminals are numbered 1, 2, and 5. Other terminals, numbered 4 and 6, are sometimes used as auxiliary terminals (Figures 4-51 and 4-52). The relay is installed

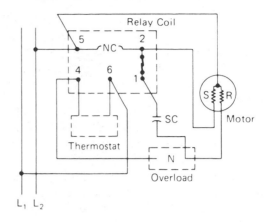

**Figure 4-52** Potential Starting Relay Connections

**Figure 4-51** Potential Starting Relay (Courtesy of General Electric Co.)

with terminal 5 connected to both the electrical line to the motor and to the common terminal on the motor. Terminal 2 is connected to the motor start winding. Terminal 1 is connected to the starting capacitor.

In operation, when the motor controller completes the electrical circuit to the motor, electricity is also supplied to the starting winding through the relay contacts between terminals 1 and 2 (Figure 4-52). As the motor reaches approximately 75% of its rated speed, the counter emf in the start winding has increased sufficiently to cause the relay to "pick up," which opens the contacts and removes the starting components from the circuit.

Remember, it is the voltage in the start winding that causes the relay to function. The relay is in the normal operating position. When the controller interrupts the electrical circuit to the motor, it begins slowing down. The counter emf in the start winding is reduced, and when a certain voltage is reached the relay "drops out," which closes the contacts. The relay is now in the motor starting position.

Potential relays are nonpositional. The sizing of these relays is not as critical as with some of the other types. A good way to determine the required relay is to manually start the motor and check the voltage between the start and common terminals while the motor is operating at full speed. Multiply the voltage obtained by 0.75; this will be the pickup voltage of the required relay.

*Solid-state starting devices.* These relays provide the additional starting torque necessary to solve permanent split capacitor (PSC) motor starting problems. This is done through the use of positive temperature coefficient (PTC) ceramic materials. This material increases its resistance with an increase in temperature. At a temperature known as its *anomaly* temperature, the resistance increase is very sharp (Figure 4–53).

One objective of a PTC start assist device is to provide a surge of current that lasts for a period of time sufficient to start the compressor. This additional current then decreases, permitting the motor to run in the normal PSC mode. When electrical power is applied to the PSC motor, current flows through the start winding and through the parallel combination of the run capacitor and the low-resistance PTC

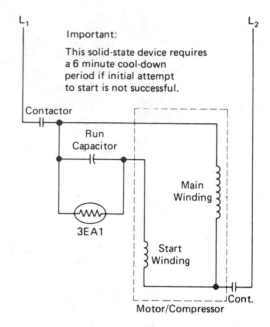

**Figure 4-54** Solid-state Start Component Connections (Courtesy of Control Products Division, Texas Instruments, Incorporated)

(Figure 4–54). The low resistance during starting not only increases the start winding current, but also reduces the angular displacement with the main winding current. In most PSC motors, this is advantageous, since the angle between these two currents is usually greater than 90%.

While the surge current increases motor starting torque, it is also flowing through the PTC material and heating it to its high-resistance region. The time it takes the PTC material to heat to its high resistance state is, unlike the operation of a potential relay, independent of when the motor starts. Instead, it is a function of the PTC material's mass, anomaly temperature, resistance, and the voltage applied to the material. When a 240-V compressor having a 9EA start assist is initially energized at nominal voltage, the 9EA switch-

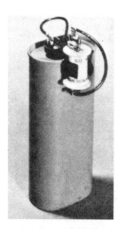

**Figure 4-53** Solid-state Starting Components (Courtesy of Control Products Division, Texas Instruments, Incorporated)

ing time is 16 electrical cycles. Starting the same compressor with 25% less voltage increases the 9EA switching time to 32 electrical cycles, which provides added assistance under these more difficult starting conditions.

The preceding switching times are ideal for starting PSC motors because the PTC material will only be in its low-resistance state long enough to overcome the initial inertia of the motor compressor. When the PTC material switching time is longer than normal motor start times, the low resistance PTC material effectively shunts out the run capacitor. The excessive on time retards the motor speed while the motor is trying to overcome the increasing load.

Once the PTC material heats up and reaches its anomaly temperature, its resistance increases dramatically to around 80,000 Ω, and it effectively switches itself out of the circuit without requiring an electromechanical relay. While the motor is running, the 9EA continues to draw only 6 mA of current, which is less than 1/1000 of the surge current provided by the 9EA during the motor starting period. This low current has no effect on the starting winding or on motor running performance.

When the motor is shut off, the PTC material is also deenergized and starts cooling to ambient temperature. Should the motor be restarted before the PTC material cools below

the anomaly temperature, the motor will try to start in the standard PSC mode. Cool-down times in excess of 1 minute will usually provide sufficient time for the PTC material to cool below its anomaly temperature and provide a start assist.

It is recommended that the 9EA1 be used on compressors up to 48,000 Btu or motors up to 4 hp. However, its use is not limited to this size range.

*Lockout Relay:* The purpose of the lockout relay is to prevent rapid cycling of the compressor should a malfunction occur in the refrigerant system. The purpose is to prevent damage to the compressor from cycling because of an overload condition. Any control in the control circuit can cause the lockout relay to activate and stop operation of the compressor by interrupting the electrical control circuit to the contactor coil (Figure 4–55).

It should be remembered that any time the lockout relay is activated an overload condition has occurred which must be located and corrected or the lockout relay will again be activated when the system is reenergized. To reset the lockout relay, turn the electrical power to the unit off and then back on. The unit should start. The system must then be checked and the problem corrected.

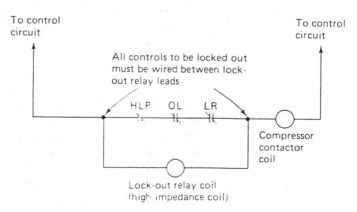

*Figure 4-55* Lockout Relay Connections.

## _____ SUMMARY _____

Controls and their connecting wiring are the devices that signal the heat pump components when their function is desired.

A transformer is a device used to transfer electrical energy from one circuit to another. It consists of two or more electric coils wound around one laminated core so that all the lines of flux from one coil will cut across all the turns of the other coil.

Thermostats are controls that signal the equipment to start or stop in response to the temperature requirements of a building.

Indoor thermostats are installed in the conditioned space and sense the temperature there. Outdoor thermostats are installed either inside the outdoor cabinet or under the eave of the building to sense the temperature there. Indoor thermostats are usually multistage devices.

Heat anticipators are electrical resistors used in indoor thermostats to provide a false heat inside the thermostat, which causes the thermostat to provide better control of the equipment. They are termed heating and cooling anticipators.

Defrost controls are used on heat pumps to detect ice and frost on the outdoor coil during the heating cycle. Their primary function is to detect frost and to initiate and terminate the defrost cycle.

Defrost thermostats may be either single- or double-bulb types of controls.

The defrost cycle can occur only when the outdoor coil temperature is below 26°F (−3.33°C) and during the first 60 seconds that the defrost timer attempts a defrost; otherwise, the cycle is skipped.

Some of the more popular types of defrost systems are defrost thermostats, either single or double bulb (element), time–temperature, air pressure–temperature, solid state, and air-pressure differential.

The time–temperature defrost system is a very popular method of defrost control. It uses both time and temperature to cause a defrost cycle.

The air pressure–temperature provides the complete process of defrost initiation and termination. It is fully automatic and uses a single control.

The solid-state defrost system has two sensors: one monitors the outdoor air temperature; the other monitors the outdoor coil temperature. The difference in these two temperatures is what initiates and terminates the defrost cycle.

The air-pressure differential (Dwyer) defrost system senses the air pressure on both the air inlet and the air outlet sides of the outdoor coil. It uses a temperature-sensitive defrost termination control.

The defrost relay is a double-pole, double-throw relay with a 240-V coil that is energized when the defrost control demands a defrost cycle. When energized, it stops the outdoor fan motor, switches the reversing valve into the cooling position, and energizes the auxiliary heat relay.

The indoor fan relay has a 24-V coil and a set of single-pole, double-throw contacts. It changes the speed of the indoor fan motor and allows for continuous fan operation from the thermostat.

The emergency heat relay and subbase provide for control of the auxiliary heat strips should something happen to stop the compressor. A light on the thermostat indicates that the auxiliary heat strips are on.

The two basic types of overloads used on electric motors are (1) internal (thermostat) and (2) external. The internal thermostat is located in the motor winding at the hottest spot, where is senses only temperature. External overloads are manufactured in three configurations: two terminal, three terminal, and four terminal. They are mounted on the external surface of the compressor and may be replaced when found faulty.

High-pressure controls are mounted in the compressor discharge line between the compressor and the reversing valve.

Low-pressure controls (loss-of-charge switch) are located in the liquid line. The contacts are wired electrically in series with the compressor contactor holding coil.

The defrost termination pressure switch is used in combination with the solid-state defrost board. It is located in the liquid line.

The low-ambient switch is a thermostat located in the outdoor unit (when used) to stop the compressor when the outdoor ambient temperature falls to a predetermined temperature.

The mild-weather control is used to cycle the outdoor fan motor when it is desirable to operate the unit in the heating mode in outdoor ambient temperatures of 65 to 70 °F (18.33 to 21.11 °C).

Heat sequencers are used to energize heating elements at different time intervals and to prevent rapid cycling of the different components should the thermostat contacts chatter.

All single-phase motors require some type of starting components. The most popular types of relays used on heat pump systems are (1) potential (voltage) and (2) the solid-state starting relay.

The purpose of the lockout relay is to prevent rapid cycling of the compressor should a malfunction occur in the refrigerant system.

## ——— REVIEW QUESTIONS ———

1. Define a transformer.
2. Name the control that signals the equipment to start and stop by temperature.
3. What is the purpose of the emergency heat switch on some thermostats?
4. What is the purpose of outdoor thermostats used on heat pump systems?
5. What is the purpose of a heat anticipator?
6. How is the heat anticipator connected electrically?
7. What are the purposes of the defrost controls used on heat pump systems?
8. Name two types of defrost thermostats.
9. Name the type of defrost system that uses both a time clock and thermostat to accomplish the defrost cycle.
10. At what point does the temperature sensor of the air pressure–temperature defrost control sense?
11. At what two points do the sensors on the solid-state defrost control system sense?
12. What causes the air-pressure differential (Dwyer) defrost system to initiate a defrost?
13. At what voltage is the coil rated in the defrost relay?
14. Name three functions of the defrost relay.
15. What type of contact configuration does the indoor fan relay use?
16. What energizes the emergency heat relay on heat pump systems?
17. What is the purpose of contactors?
18. Name two basic types of motor overloads.
19. What is the purpose of the high-pressure controls used on heat pump systems?
20. What is another name for the low-pressure controls used on heat pump systems?
21. At what place is the defrost termination pressure switch located?
22. What control prevents operation of the compressor during low outdoor ambient temperatures?
23. At what outdoor ambient temperatures is the mild weather control used?
24. What is the purpose of the heat sequencers used on heat pump systems?
25. Name the two types of starting relays most used on heat pump compressors.

# 5

# Water-Source Heat Pump Systems

Through research, development, and manufacture of materials that are capable of better heat transfer, water-source heat pumps are gaining popularity in the industry. These systems are generally referred to as closed-loop earth-coupled and open-loop (water well or groundwater) systems. When the closed-loop earth-coupled system is used, the water is circulated through special piping which is buried in the ground. This solution may also be a mixture of water and antifreeze. Therefore, there is no wastewater to be disposed of. The open-loop systems take water from the source and discharge it to an appropriate area. Sometimes disposing of the used water is a problem and is more expensive than the source of the water.

The basic operating principle of this system is that the heat is rejected to the water from the air to be cooled during the cooling cycle and heat is extracted from the water by the air to be heated during the heating cycle. Dehumidification is achieved in the same manner as with any cooling unit by its removal in the form of condensate. The heat-transfer medium is the refrigerant inside the refrigeration system. The refrigeration system components, with the exception of the water to refrigerant heat exchanger, are the same as those used on any other type of heat pump system. This type of system in most popular for heating and cooling residential structures, where an adequate source of water or ground area is available.

## CLOSED-LOOP EARTH-COUPLED SYSTEMS

In the closed-loop earth-coupled heat pump system a solution is circulated through special pipes buried in the ground. This solution may be water or a water and antifreeze mixture.

82

There are three ways used in installing these types of heat pump systems: (1) horizontal, (2) vertical, and (3) lake loops.

## TYPICAL CLOSED-LOOP EARTH-COUPLED METHODS __

These types of systems may use either vertical or horizontal tubing buried in the ground. The vertical system is the most popular when the site area is too small to accommodate the amount of piping needed for the horizontal type system. In these types of systems multiple boreholes are drilled into the ground of sufficient size and depth to allow the proper amount of piping to be buried for the size of system being installed

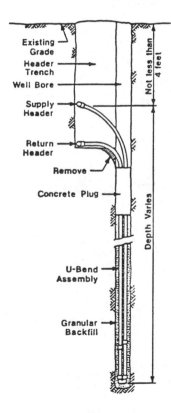

**FIGURE 5-1**  Vertical Piping Application (Courtesy of Command-Aire Corporation)

(Figure 5-1). Note that the style of letters used in Figures 5-1 and 5-2 matches the "corresponding feet per nominal ton" category in Table 5-1. The specifications and installation instructions for all methods shown are available from the manufacturer of the equipment being used. The earth-coupling-length information given here applies only to and has been specifically designed for Command-Aire Earth Energy Heat Pumps.

The horizontal system is used in applications where the building site is large enough to allow the proper amount of pipe to be buried in a horizontal manner (Figure 5-2). This type of system is especially advantageous when drilling is not desirable.

The choice of vertical, horizontal, or lake-loop earth coupling should be based on the characteristics of each application. Horizontal and vertical systems are designed to provide the same fluid temperature under a given set of conditions. The lake-loop system will provide slightly lower fluid temperatures, but the reduced installation cost should compensate for any minor reduction in performance. The three earth-coupling methods should be considered at each application, with the most cost-effective method chosen after all have been evaluated.

**TABLE 5-1**

Earth-Coupling Design Information[a]

| | Feet per nominal ton[b] | | | | |
|---|---|---|---|---|---|
| Zone | A | B | C | D | Antifreeze[c] |
| 1 | 150 | 375 | 450 | 600 | 20 |
| 2 | 160 | 400 | 475 | 650 | 15 |
| 3 | 160 | 425 | 500 | 675 | 15 |
| 4 | 170 | 450 | 525 | 700 | 15 |
| 5 | 180 | 475 | 550 | 750 | 15 |
| 6 | 190 | 500 | 600 | 800 | 0 |

[a]Design information for the pumping unit is not available at this time.

[b]Horizontal length is lineal feet of pipe per ton. Vertical length is lineal feet of borehole per ton.

[c]Percentage by volume.

*Source:* Command-Aire.

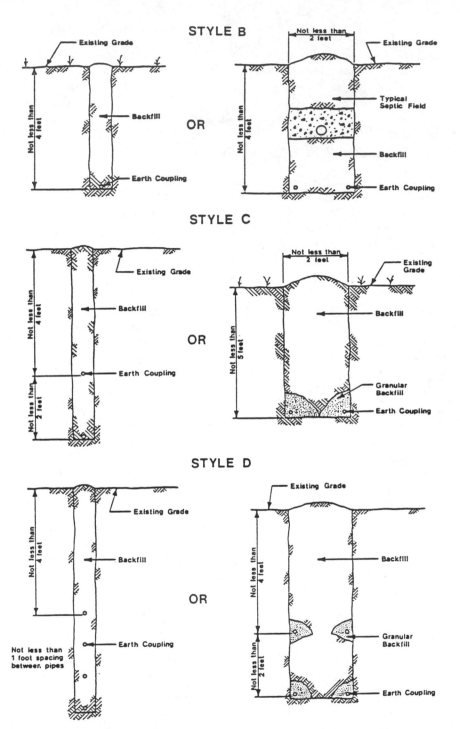

**FIGURE 5-2** Horizontal Piping Applications (Courtesy of Command-Aire Corporation)

# EARTH-COUPLED SYSTEM DESIGN _____

The design of an earth-coupled system is divided up into the following steps: (1) determine the structure design heating load in Btu/h loss and design cooling load in Btu/h gain, (2) select a water-source heat pump, and (3) select the earth-type coil and materials to be used.

1. *Determine the structure design heating load in Btu/h loss and design cooling load in Btu/h gain.* An accurate heat load calculation is a must if the system is to perform satisfactorily. Therefore, it is recommended that some nationally accepted method be used for these calculations.

2. *Select a water-source heat pump.* The water-source heat pump selected for use on an earth-coupled system may be required to operate at entering water temperatures which vary between 30 and 100°F ($-1.67$ and 37.78°C). Because of this, it is necessary that the minimum and maximum temperatures of the water source selected remain within that range through all seasons. There are several models on the market which have a much smaller operating range of entering water temperature. Some of these units are not satisfactory for use on earth-coupled heat pump installations.

The heating and cooling capacities of water-source heat pumps should be determined from the manufacturer's specifications for the local groundwater temperatures and design conditions. Following is a suggested method of sizing water-source heat pump systems.

*Heat Pumps Sized for Cooling:* The sensible heat output capacity of the cooling equipment selected should in no case be less than the calculated total sensible heat load. Also, it should never be more than 25% greater than the calculated sensible heat load. The latent heat capacity should never be less than the calculated total latent heat load. The sensible and latent heat capacities of the equipment should be obtained from the performance data listed in the manufacturer's catalog. The capacities of the equipment should be verified for the local indoor design conditions and the local groundwater temperature.

*Heat Pumps Sized for Heating:* Systems that are designed for heating-only applications should provide a minimum of 75% and a maximum of 115% of the total Btu/h heat loss. In applications where the system is undersized, auxiliary heat strips must be installed to bring the system up to the calculated heat loss of the structure. In some localities emergency heat which will provide either a large part or the total heat loss must be provided depending upon the local requirements.

*Heat Pumps Designed for Both Heating and Cooling:* Heat pump systems which are designed to provide both heating and cooling must be sized to satisfy the cooling requirements of the structure. When designed in this manner the heating thermal balance point will be determined by the cooling requirements. In applications when a lower balance point is desired, the system may be oversized by a maximum of 25% of the system capacity. Auxiliary heat strips are then used to provide the amount of heat that is lacking between the balance point and the output of the heat pump system. Some localities may require that emergency heat be provided that will produce the total or some percentage of the heat loss of the structure.

3. *Select the earth-type coil and materials to be used.* In earth-coupled systems, the earth coupling is a particular method which uses water or another type of solution that is circulated through special pipes buried in the ground. The heat is transferred to and from the water through the walls of the pipe. These types of systems are used in areas where there is an

insufficient amount of available groundwater, or where the drilling of a well would be impractical. The amount of required piping may be buried in either the vertical or the horizontal method.

Of particular importance when designing an earth-coupled heat pump system is the balancing between the heat pump unit and the earth-coupled loop. When this balance is achieved, the earth-coupled loop will remove all the heat that is transferred to the water by the heat pump during the cooling cycle, and will provide all the heat that is required to heat the water during the heating cycle. The final result of a perfectly balanced system is that the change in water temperature through the heat pump system is offset by an equal and opposite change in temperature through the earth-coupled loop. As an example, if the heat pump unit when operating in the cooling cycle causes a temperature rise of the circulating water of 15 °F (8.4 °C), the loop must respond with a corresponding but opposite temperature drop of 15 °F (8.4 °C).

It should be remembered that even though the earth-coupled loop is designed for a balanced rise and fall of the water temperature, which suggests that the net average loop water temperature remains constant, because the ground temperature may vary plus or minus 15 °F (8.4 °C) from season to season, the loop water temperature may vary plus or minus 20 °F (11.2 °C) from the balance-point temperature. This is because the ground is able to overcool the loop water during the heating season, and may undercool it during the cooling season. For this reason, the water temperature entering the water-source heat pump system may drop below 30 °F (−1.11 °C) during the heating season or rise above 100 °F (37.78 °C) during the cooling season. This range in the temperature of the entering water is extremely important because water-source heat pump systems are designed to operate within specific operating temperature ranges. The manufacturer's specifications for the particular unit being considered must be checked to make certain that the unit is designed to operate within the entering water temperature range at the point of application. The temperature ranges are established to protect both the heat pump unit and the water loop piping. Also, these temperature ranges are generally based on water only passing through the system. The low-temperature limit of 40 °F (4.44 °C) in a water-source heat pump unit is established to protect the loop water from freezing. Again, this low limit is assumed for water only circulating through the system. However, if the circulating water is mixed with a nontoxic antifreeze solution, the entering water temperature can be allowed to fall to 30 °F (−1.11 °C).

## HEAT PUMP _____

Use only a water-source heat pump that can be operated on loop water temperatures that fall below 40 °F (4.44 °C) and perhaps down to 25 °F (−3.89 °C). Information on when to use an antifreeze solution in a ground-coupled water-source heat pump system is given in the following discussions on vertical and horizontal installation methods.

*Pipe:* It is generally preferred that polybutylene (PB) or polyethylene (PE) pipe be used for horizontal coils, vertical U-bend wells, and for service lines to the wells and lake heat exchangers. IPS (iron pipe size) PB pipe is used with insert fittings and clamps. CTS (copper tube size) PB pipe is fused together with the appropriate fittings using a fusion tool. PE pipe is heat-butt fused together with the appropriate fittings using a fusion tool.

*Cleanliness:* During the installation procedure it is very important that all trash, soil, and small animals be kept out of the piping system. Leave the tape on the ends of the pipe until they are to be connected to the service lines or the equipment room piping.

*Pressure Testing:* The plastic pipe assemblies should be pressure tested at twice the anticipated system operating pressures before the backfilling procedure is started. Normal static equipment room pressure is 50 psig (344.738 kPa).

*Backfill:* The narrow trenches that are made with a chain trenching machine may be backfilled with the tailings provided that there are no sharp rocks present. The wider trenches made with a backhoe may be backfilled with the dirt that was removed from the trench if it is a loose granular type of material. If there are clumps of clay or rocks in the removed material, the plastic pipe must be covered first with sand before filling the trench with the clumps and rocks.

The drilled boreholes, which are 4 to 6 in. (101.60 to 152.40 mm) in diameter, are commonly used for vertical geothermal wells. The backfill may be of any granular material that does not contain sharp rocks or other sharp objects. This material includes such as drilling tailings, sand, pea gravel, or bentonite mud.

*Location Markers:* The locations of important points such as well heads must be marked for subsequent recovery. A steel rod driven to a point just below the surface of the earth is a good way to identify these features or mark the outline of an entire serpentine earth coil.

*As-Built Plans:* The components of the earth-coupling unit should be drawn on a site plan as it will be installed, if at all possible. This step will aid in the location of the key components.

A simple way to locate key components is to make two measurements (sides of a triangle) from two corners of a building to the specific component being identified. These measurements should be recorded in the table on the plans.

## REASONS FOR USING AN EARTH-COUPLED SYSTEM ____

Some of the major reasons for using earth-coupled systems are as follows:

1. Unlike a standard solar system, the loop operates day or night, rain or shine all year long, delivering heat to and from the heat pump unit.
2. It is cost-effective in either northern or southern climates.
3. Because the water circulates through a sealed closed loop of high-strength plastic pipe, it eliminates scaling, corrosion, water shortage, pollution, waste, and disposal problems which are possible in some open well water systems.

## VERTICAL EARTH—COUPLED SYSTEM ____

A vertical earth-coupled system consists of one or more vertical boreholes through which water flows in a plastic pipe. A distinct advantage of a vertical system over a horizontal system is that the vertical system requires less surface area (acreage). In areas where the ambient ground-water (average well water) temperature is less than 60 °F (15.56 °C), the use of an anitifreeze solution, such as propylene glycol, to avoid freezing the loop is recommended.

Boreholes are drilled 5 to 6 in. (127.00 to 152.40 mm) in diameter for $1\frac{1}{2}$ -in. (38.10-mm)-diameter pipe. For $\frac{3}{4}$ -in. (19.05-mm)-diameter pipe loop systems, the vertical loops are connected in parallel to a $1\frac{1}{2}$ -inch. (38.10-mm)-diameter pipe header. A borehole of 3 to 4 in. (76.20 to 101.60 mm) in diameter is used for $\frac{3}{4}$ -in. (19.05-mm)-diameter loops, this procedure lowers the drilling costs. The $\frac{3}{4}$ -in. (19.05-mm)-diameter pipe also costs less per ton of heat pump capacity. The smaller pipe is easier to handle, yet there is no sacrifice in the pressure rating. Also, two loops in

one hole reduces borehole length. The depth for these systems usually between 80 and 180 ft (24.384 and 54.864 m).

The basic components of a vertical earth-coupled system are detailed in Figure 5-3. Each borehole contains a double length of pipe with a U-bend fitting at the bottom. Multiple boreholes may be joined in series or in parallel. (Figure 5-4). Sand or gravel packing is required around the piping to assure heat transfer. In addition, the bore around the pipes immediately below the service (connecting) lines must be cemented closed to prevent surface-water contamination of an aquifer in accordance with local health department regulations (Figure 5-5).

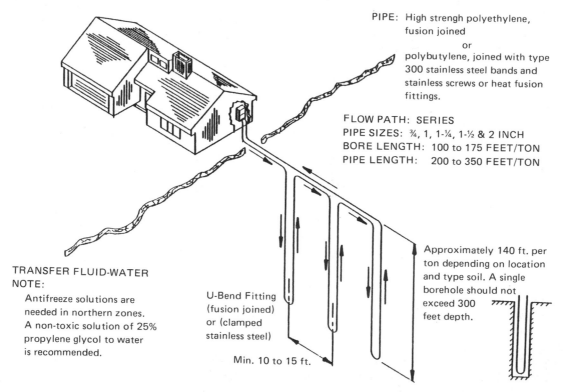

PIPE: High strengh polyethylene, fusion joined

or

polybutylene, joined with type 300 stainless steel bands and stainless screws or heat fusion fittings.

FLOW PATH: SERIES
PIPE SIZES: ¾, 1, 1-¼, 1-½ & 2 INCH
BORE LENGTH: 100 to 175 FEET/TON
PIPE LENGTH: 200 to 350 FEET/TON

Approximately 140 ft. per ton depending on location and type soil. A single borehole should not exceed 300 feet depth.

TRANSFER FLUID-WATER NOTE:

Antifreeze solutions are needed in northern zones. A non-toxic solution of 25% propylene glycol to water is recommended.

U-Bend Fitting (fusion joined) or (clamped stainless steel)

Min. 10 to 15 ft.

FIGURE 5-3  Vertical (Series) System (Courtesy of Bard Manufacturing Company)

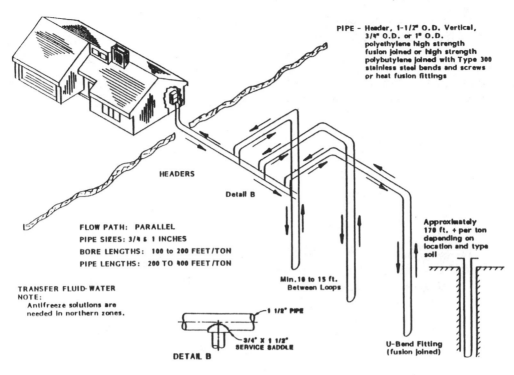

PIPE - Header, 1-1/2" O.D. Vertical,
3/4" O.D. or 1" O.D.
polyethylene high strength
fusion joined or high strength
polybutylene joined with Type 300
stainless steel bends and screws
or heat fusion fittings

HEADERS

Detail B

FLOW PATH:  PARALLEL
PIPE SIZES: 3/4 & 1 INCHES
BORE LENGTHS:  100 to 200 FEET/TON
PIPE LENGTHS:  200 TO 400 FEET/TON

Approximately
170 ft. + per ton
depending on
location and type
soil

TRANSFER FLUID- WATER
NOTE:
    Antifreeze solutions are
    needed in northern zones.

Min.10 to 15 ft.
Between Loops

1 1/2" PIPE

3/4" X 1 1/2"
SERVICE SADDLE

DETAIL B

U-Bend Fitting
(fusion joined)

**FIGURE 5-4**  Vertical (Parallel) System (Courtesy of Bard Manufacturing Company)

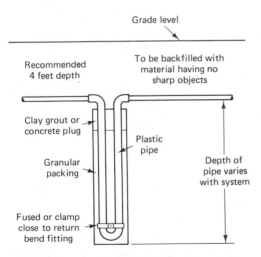

Grade level

Recommended
4 feet depth

To be backfilled with
material having no
sharp objects

Clay grout or
concrete plug

Plastic
pipe

Granular
packing

Depth of
pipe varies
with system

Fused or clamp
close to return
bend fitting

**FIGURE 5-5**  Vertical Borehole and Piping Detail for Earth-Coupled System

## SERIES U-BEND SYSTEM _____

A series U-bend earth coupling is one in which all the water flows through all the pipe, progressively traveling down and then up each well bore. Series wells need not be of equal length.

*Pipe:*  The $1\frac{1}{2}$ -in. (38.10-mm) CTS (copper tube size) or IPS (iron pipe size) polybutylene or polyethylene pipe is commonly used in 5- and 6-in. (127.00- and 152.40-mm) boreholes. IPS PB pipe is used with insert fittings and clamps. Turn the clamps so that they face inward and will not be chaffed by the well bore. Tape the clamped section of the U-bend with duct tape to provide added protection to the clamps while the pipe is being installed into the well. CTS

PB pipe is heat fused together with fittings. PE pipe is heat fused together with butt joints.

*Stiffener:* Tape the last 10 to 15 ft (3.048 to 4.572 m) of pipe above the U-bend together to a rigid piece of pipe or conduit. This will make installing the pipe into the well easier.

*Fill and Pressure Test:* Fill the system with water and pressure test before lowering the U-bend into a well bore. When drilling with air, a bore can be completed that contains no water. If unfilled plastic pipe is lowered into the bore, it will be crushed as the hole slowly fills.

*Multiple Wells:* Multiple 100-ft (30.48-m) wells connected in series are the easiest to drill and install in most areas. It will be difficult to sink water filled plastic U-bends into mud-filled holes over 150 (45.72 m) deep without weights. Wells are generally spaced 10 ft (3.048 m) apart in residential systems.

*Service Lines:* Follow the guidelines for the horizontal earth coil when installing the service lines to and from the U-bend well.

## PARALLEL U-BEND SYSTEM

A parallel U-bend earth coupling is one in which the water flows out through one header, is divided equally, and flows simultaneously down two or more U-bends. It then returns to the other header. Headers are reverse-return plumbed so that equal-length U-bends have equal flow rates. Lengths of individual parallel U-bends must be within 10% of each other to ensure equal flow in each well.

*Pipe:* Either $1\frac{1}{2}$-in. (38.10-mm) polybutylene or polyethylene pipe is used for the headers with 1- or $\frac{3}{4}$-in. (25.40- or 19.05-mm) pipe used for the U-bends. Boreholes of 4 in. (101.60 mm) are sufficient for placement of the 1-in. (25.40-mm) U-bends. Follow the instruc-

**TABLE 5–2**
Minimum Diameters for Boreholes

| Nominal pipe size (in.) | Single U-bend (in.) | Double U-bend (in.) |
|---|---|---|
| $\frac{3}{4}$ | $3\frac{1}{4}$ | $4\frac{1}{2}$ |
| 1 | $3\frac{1}{2}$ | $5\frac{1}{2}$ |
| $1\frac{1}{4}$ | 4 | $5\frac{3}{4}$ |
| $1\frac{1}{2}$ | $4\frac{3}{4}$ | 6 |
| 2 | 6 | 7 |

*Source:* Bard Manufacturing Company.

tions given under "Series U-Bend System" for the stiffener, fill and pressure test, multiple wells, and service lines. For the minimum diameters for boreholes, see Table 5–2.

## HORIZONTAL EARTH-COUPLED SYSTEM

A horiontal earth-coupled system is similar to a vertical system in that water circulates through underground piping. However, the piping in this type of system is buried in a trench (Figures 5–6 to 5–8). The pipe depths in the northern zone of the United States should be 3 to 5 ft (0.9144 to 1.524 m) (Figure 5–9). Burying the pipes at excessive depths will reduce the ability of the sun to recharge the heat used during the heating cycle. Pipe depths in the southern zone of the United States should be 4 to 6 ft (1.2192 to 1.8288 m), so that the high temperature of the soil in late summer time will not seriously affect the performance of the system.

Antifreeze will be necessary in the northern zone to prevent freezing of the circulated water and to allow the system to gain capacity and efficiency, by using the large amount of heat released when the water contained in the soil is frozen. The antifreeze solutions used in these systems are nontoxic propylene glycol or calcium chloride.

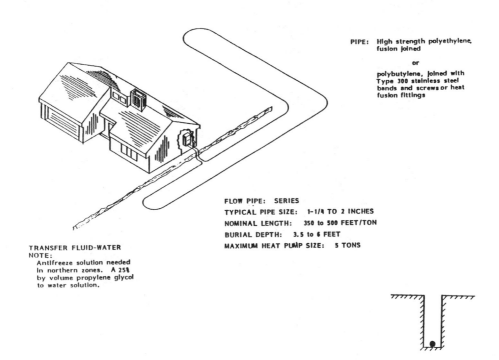

PIPE: High strength polyethylene, fusion joined

or

polybutylene, joined with Type 300 stainless steel bands and screws or heat fusion fittings

FLOW PIPE: SERIES
TYPICAL PIPE SIZE: 1-1/4 TO 2 INCHES
NOMINAL LENGTH: 350 to 500 FEET/TON
BURIAL DEPTH: 3.5 to 6 FEET
MAXIMUM HEAT PUMP SIZE: 5 TONS

TRANSFER FLUID-WATER
NOTE:
Antifreeze solution needed in northern zones. A 25% by volume propylene glycol to water solution.

**FIGURE 5-6** Horizontal (Series) System: One Pipe in Trench (Courtesy of Bard Manufacturing Company)

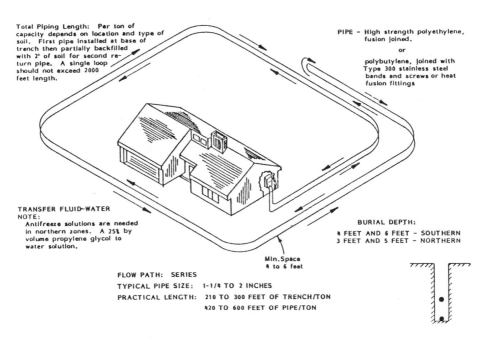

Total Piping Length: Per ton of capacity depends on location and type of soil. First pipe installed at base of trench then partially backfilled with 2' of soil for second return pipe. A single loop should not exceed 2000 feet length.

PIPE - High strength polyethylene, fusion joined.

or

polybutylene, joined with Type 300 stainless steel bands and screws or heat fusion fittings

TRANSFER FLUID-WATER
NOTE:
Antifreeze solutions are needed in northern zones. A 25% by volume propylene glycol to water solution.

BURIAL DEPTH:
4 FEET AND 6 FEET - SOUTHERN
3 FEET AND 5 FEET - NORTHERN

Min. Space 4 to 6 feet

FLOW PATH: SERIES
TYPICAL PIPE SIZE: 1-1/4 TO 2 INCHES
PRACTICAL LENGTH: 210 TO 300 FEET OF TRENCH/TON
420 TO 600 FEET OF PIPE/TON

**FIGURE 5-7** Horizontal (Series) System: Two Pipes in Same Trench (Courtesy of Bard Manufacturing Company)

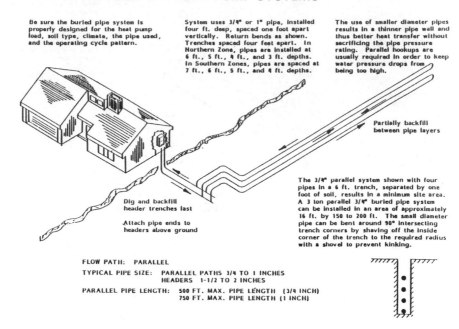

Be sure the buried pipe system is properly designed for the heat pump load, soil type, climate, the pipe used, and the operating cycle pattern.

System uses 3/4" or 1" pipe, installed four ft. deep, spaced one foot apart vertically. Return bends as shown. Trenches spaced four feet apart. In Northern Zone, pipes are installed at 6 ft., 5 ft., 4 ft., and 3 ft. depths. In Southern Zones, pipes are spaced at 7 ft., 6 ft., 5 ft., and 4 ft. depths.

The use of smaller diameter pipes results in a thinner pipe wall and thus better heat transfer without sacrificing the pipe pressure rating. Parallel hookups are usually required in order to keep water pressure drops from being too high.

Partially backfill between pipe layers

Dig and backfill header trenches last

Attach pipe ends to headers above ground

The 3/4" parallel system shown with four pipes in a 6 ft. trench, separated by one foot of soil, results in a minimum site area. A 3 ton parallel 3/4" buried pipe system can be installed in an area of approximately 16 ft. by 150 to 200 ft. The small diameter pipe can be bent around 90° intersecting trench corners by shaving off the inside corner of the trench to the required radius with a shovel to prevent kinking.

FLOW PATH: PARALLEL

TYPICAL PIPE SIZE:   PARALLEL PATHS 3/4 TO 1 INCHES
                     HEADERS  1-1/2 TO 2 INCHES

PARALLEL PIPE LENGTH:  500 FT. MAX. PIPE LENGTH  (3/4 INCH)
                       750 FT. MAX. PIPE LENGTH (1 INCH)

**FIGURE 5-8** Horizontal Multilevel (Parallel) System (Courtesy of Bard Manufacturing Company)

Northern zone
3 ft to 5 ft

Southern zone
4 ft to 6 ft

**FIGURE 5-9** Pipe Depths for Different Regions

The use of multiple pipes in a single trench substantially reduces the total trench length. If a double layer of pipe is laid in the trench, the two layers should be set 2 ft (0.6096 m) apart to minimize thermal interference (Figure 5–10). For example, a $1\frac{1}{2}$ -in. (38.10-mm) series horizontal system with pipes at 5- and 3-ft (1.524- and 0.9144-m) depths. After installing the first pipe at 5 ft, partially backfill to the 3-ft depth using a depth-gauge stick before installing the second pipe in the trench. Arrange with the return line running closest to the surface and the supply line running below it. This arrangement will maximize the overall system efficiency by providing warmer water in the heating mode and cooler water for the cooling mode. Connect the pipe ends to the heat pump after the pipe temperature has stabilized, so that shrinkage will not pull the pipe loose.

Two pipes in the same trench, one above the other, separated by 2 ft (0.6096 m) of earth require a trench 60% as long as a single-pipe system. The total length of pipe would be 120% as long as a single-pipe system due to the heat transfer effect between the pipes.

In addition, when laying a double layer of pipe, be careful to avoid kinks when making the return bend (Figure 5–11). Backfill the trench by hand when changing direction. If it is necessary to join two pipes together in the trench, use the fusion technique for IPS 304 stainless steel or brass fittings for greater

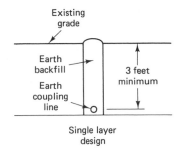

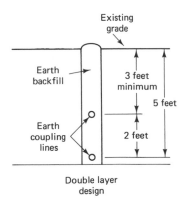

**FIGURE 5-10** Separation of the Horizontal Earth-Coupled Pipes in the Trench

strength and durability, then mark the fitting locations for future reference by inserting a steel rod just below the grade (Figure 5–12). The steel rod will aid in finding the location of the fittings with a metal detector.

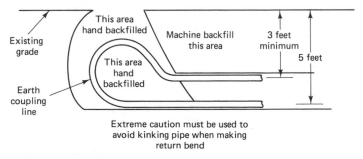

**FIGURE 5-11** Detail of Narrow Trench Return Bend Used on Double-Layer Horizontal Earth-Coupling System

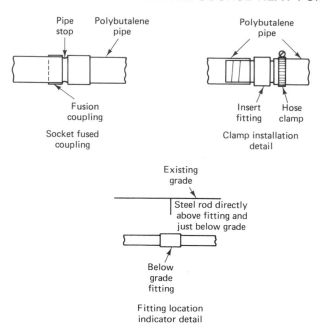

**FIGURE 5-12** Hose Fitting and Location Detail

Trenches can be located closer together if the pipe in the previous trench can be tested and covered before the next trench is started. This also makes backfilling easier. Four- to 5-ft (1.2192- to 1.524-m) spacing between the pipes is good.

In those areas with dry climates and heavy clay soil, any heat that is dissipated into the soil may reduce the thermal conductivity of the soil significantly. In such cases, the designer may specify additional feet of pipe per ton of heat pump capacity. A few inches (millimeters) of sand may also be put in with the pipe, or a drip irrigation pipe buried with the top pipe to add occasional amounts of water to the soil.

Series horizontal earth couplings are the ones in which all the water flows through all the pipe. These may be made of 1-, 1$\frac{1}{2}$ and 2-in. (25.40, 38.10, and 50.80-mm) pipe either insert coupled or fused.

*Narrow Trenches:* Narrow trenches are dug with trenching machines. The trenches are usually 6 in. (152.40 mm) wide. Generally speaking, the trencher will require about 5 ft (1.524 m) between trenches. This is sufficient spacing for horizontal earth coils.

The pipe can be coiled into an adjoining trench. Since the trencher spaces the trenches about 5 ft (1.524 m) apart, looping the coil from one trench to another will give a return of large enough diameter. The end trench would be backhoed to give enough room for the large-diameter bend.

If the pipe is brought back in the same trench, bend the pipe over carefully to avoid kinking the pipe and handfill the area around the return bend (Figure 5–11).

To reduce the bend radius, elbows may be used. However, keeping the number of fittings underground to a minimum may be preferable since the potential for leaks is reduced.

If a double layer of pipe is used, the incoming water to the heat pump unit should be from the deepest pipe (Figure 5–13). This provides the heat pump with the coolest water during the cooling season and the warmest water during the heating season.

*Backhoe Trenches:* If a backhoe is used, the trench will probably be about 2 ft (0.6096 m) wide. In a wide-backhoed trench, two pipes may be placed side by side, one on each side

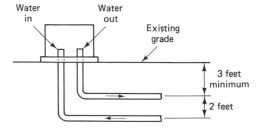

**FIGURE 5-13** Double-Layer Horizontal Earth-Coil Water Flow Connection Schematic

of the trench. The pipes in the trench must be at least 2 ft (0.6096 m) apart.

Backfill carefully around the pipe with fine soil or sand. Do not drop clumps of clay or rock onto the pipe (Figure 5–14). A pit may be dug at the end of the trench to acommodate a 4-in. (101.60-mm)-diameter return bend (Figure 5–15).

*Service Lines:* The recommendations for the horizontal earth coils also apply for the installation of the service lines to and from the U-bend wells and pond or lake heat exchanger.

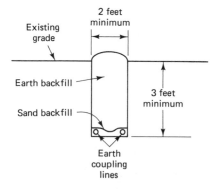

**FIGURE 5-14**  Backhoed Trench

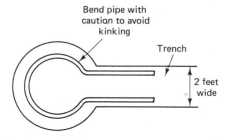

Pit excavated to at least 4 feet deep with a diameter large enough to place a full diameter loop of pipe.

**FIGURE 5-15**  Detail of Wide Trench Return for Either a Single- or Double-Layer Horizontal Earth Coil

Bury the service lines a minimum of 3 ft (0.9144 m) for single-layer pipe and 3 ft and 5 ft (0.9144 and 1.520 m) for double-layer installations. If two pipes are buried in the same trench, keep them 2 ft (0.6096 m) apart.

A parallel horizontal earth coupling is one in which the water flows out through a supply header is then divided equally and then flows simultaneously into two or more earth coils. The water then returns to the other header. Headers are reverse-return plumbed so that equal-length earth coils have equal flow rates. The lengths of individual parallel earth coils must be within 10% of each other to ensure equal flow in each coil. Follow the "Horizontal Earth-Coupled System" instructions on narrow trenches and backhoe trenches.

## FREEZE PROTECTION _____

The antifreeze solutions that are used in earth loop systems must be nontoxic and noncorrosive: nontoxic because should a leak develop in the loop system, the groundwater will not be contaminated, and noncorrosive for the protection of the metal components of the system which come into contact with the circulating solution.

In situations when the local well water temperature is below 60 °F (15.56 °C), the water in the earth loop should be protected from freezing at temperatures down to 18 °F (-7.78 °C). The recommended antifreeze solution is propylene glycol. The amount of antifreeze needed for the earth loop is determined as follows: Calculate the approximate volume of water in the loop system by using Table 5–3. This table gives the gallon of water per 100 ft (30.48 m) of pipe. Add 2 gallons (7.57 liters) for the equipment room devices and heat pump.

**TABLE 5–3**

Volume of Water in the Loop System

| Pipe material | Nominal pipe size (in.) | Gallons (Liters) per 100 ft of pipe[a] |
|---|---|---|
| Polyethylene | | |
| SDR-11 | $\frac{3}{4}$ (19.05 mm) | 3.02 (11.43 l) |
| SDR-11 | 1 (25.40 mm) | 4.73 (17.9 l) |
| SDR-11 | $1\frac{1}{4}$ (31.75 mm) | 7.52 (28.46 l) |
| SDR-11 | $1\frac{1}{2}$ (38.1 mm) | 9.85 (37.29 l) |
| SDR-11 | 2 (50.8 mm) | 15.40 (58.29 l) |
| SCH 40 | $\frac{3}{4}$ (19.05 mm) | 2.77 (10.48 l) |
| SCH 40 | 1 (25.40 mm) | 4.49 (16.99 l) |
| SCH 40 | $1\frac{1}{4}$ (31.75 mm) | 7.77 (29.41 l) |
| SCH 40 | $1\frac{1}{2}$ (38.1 mm) | 10.58 (40.05 l) |
| SCH 40 | 2 (50.8 mm) | 17.43 (65.98 l) |
| Polybutylene | | |
| SDR-17 IPS | $1\frac{1}{2}$ (38.1 mm) | 11.46 (43.38 l) |
| SDR-17 IPS | 2 (50.8 mm) | 17.91 (67.79 l) |
| SDR-13.5 CTS | 1 (25.40 mm) | 3.74 (14.15 l) |
| SDR-13.5 CTS | $1\frac{1}{4}$ (31.75 mm) | 5.59 (21.16 l) |
| SDR-13.5 CTS | $1\frac{1}{2}$ (38.1 mm) | 7.83 (29.64 l) |
| SDR-13.5 CTS | 2 (50.8 mm) | 13.38 (50.65 l) |
| Copper | 1 (25.40 mm) | 4.3 (16.27 l) |

[a]Add 2 gallons (7.5708 l) for the equipment room devices and heat pump.

*Propylene Glycol:* When the groundwater at a depth of 100 ft (30.48 m) is 66 °F (18.89 °C) or less, a 20% solution by volume of propylene glycol is required to provide the necessary protection. The percentage of antifreeze solution depends on the geographical location. A 20% by volume of propylene glycol solution will provide freeze protection down to 18 °F (-7.78 °C). For example, for a system that holds 100 gallons (378.54 liters) of water in the loop, 20 gallons (75.708 literes) of propylene glycol is needed to provide the desired protection.

*Adding the antifreeze.* Two small pieces of hose, a bucket, and a small submersible pump are needed to add the antifreeze to the system. Block the system by closing a ball valve. Blocking the flow prevents the antifreeze from being pumped into one boiler drain and out the other. Attach the hoses to the boiler drains. Run the uppermost hose to a drain. Connect the other hose to the submersible pump in the bucket. Put full-strength propylene glycol in the bucket and pump the desired amount into the system. When the required amount has been pumped in, turn off the pump, close the boiler drains, disconnect the hoses, and open the isolation flange or gate valve.

*Calcium Chloride:* A 20% by weight solution of calcium chloride and water may also be used as an antifreeze in the earth-coupled heat pump system. It is also nontoxic, a better heat conductor, and less expensive than propylene glycol. However, it is mildly corrosive. To determine the amount of calcium chloride needed for the system, multiply the gallons (liters) of water in the loop system by 1.4841 to find the pounds (kilograms) of 94 to 98% pure calcium chloride required to provide the desired 18 °F (− 7.78 °C) freeze protection.

## CLOSED-LOOP LAKE HEAT EXCHANGER _____

The closed-loop groundwater heat pump systems that use a pond or a lake for their heat source are by far the most economical of all closed-loop earth-coupled systems. The use of pond loops for residential applications is restricted to bodies of water having at least 1 acre of surface area and a depth of 10 ft (3.048 m) or more at the location of the loop during low-water-level conditions. The acceptibility of any body of water varies depending on the size, average depth, and water source (springs, river or creek, runoff, etc.) Most manufacturers recommend that they be consulted to discuss any application larger than 5 tons or any commercial installation.

Also, there are special considerations that must be given to pond loop applications. *Never,* under any circumstances, place a loop in a river or a stream, because the drifting objects at flood conditions will in all probability cause damage to the loop. The pond loop gets colder than either a horizontal or vertical earth coupling, so regardless of the location, the sytem must use a 20% by volume antifreeze solution. Evaluate the location of the pond loop in relation to the building. There is no need to install a pond loop when the trench length to the pond and back to the building is equal to or greater than that required for a horizontal earth coil.

*Heat Exchanger:* The length of the loop is determined to a great extent by the temperature of the water in the pond. For example, if the well water temperature is between 50 and 68 °F (10 and 20°C), use 60 ft (18.288 m) of $\frac{3}{4}$ -in. (19.05-mm) copper pipe per nominal ton of refrigeration. This information is available from the local well water association or from the National Water Well Association. In other areas where the well water temperature is indicated to be either above 68 °F (20°C) or below 50 °F (10°C), use 80 ft (24.384 m) of $\frac{3}{4}$ -in. (19.05-mm) copper pipe per nominal ton of refrigeration (Figure 5–16).

**FIGURE 5-16** Typical Lake Heat Exchanger (Courtesy of Command-Aire Corporation)

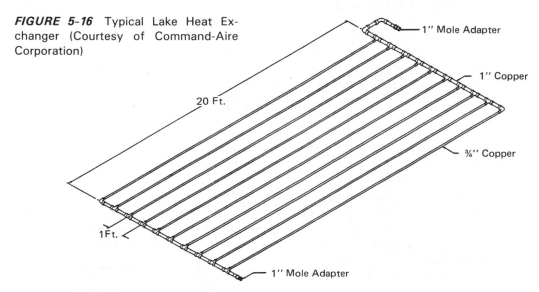

*Location:* The pond loop will not perform correctly if it is allowed to settle down into the mud. It should be supported at least 1 ft (0.3048 m) off the bottom, and at least 9 ft (2.7432 m) below the surface of the water. Suspending the pond loop under a pier is an easy application method. Another is to secure an old tire to each of the corners and allow the pond loop to settle to the top of the tires. The tires act as spacers to keep the loop about 1 ft (0.3048 m) or so off the bottom of the pond.

*Installation:* The service lines for the pond loop should be buried 4 ft (1.2192 m) deep or below the frost line, whichever is the deepest. The lines should be about 2 ft (0.6096 m) apart in a wide trench. If a narrow trench is used, the pipes should be at depths of 4 and 6 ft (1.2192 and 1.8288 m). Do not attempt to use the pond loop if the service lines cannot be buried across the shore of the lake and out into the water. *Never* bring service lines up from the loop and run them underneath the pier—always leave the lines in the water and as deep as possible.

## OPEN-LOOP (GROUNDWATER) APPLICATION _____

In the open-loop type of system the water is taken into the system, passed through the heat pump, and then discharged. It is not recirculated as in the closed-loop groundwater systems.

Since the water is a crucial element in the operation of these types of systems, many manufacturers recommend that the services of an experienced local well driller be obtained. His expertise is valuable in assessing an existing well and equipment or making recommendations for a new installation. If no local driller is available, contact your state well drilling association or the National Water Well Association at 6375 Riverside Drive, Dublin, Ohio

43017 for referral. Their telephone number is (614) 761-1711.

The water used in these systems does not need to be suitable for human consumption; however, it should meet certain standards (Table 5-4). The local or state health department can run water tests to see if it will meet the manufacturer's recommendations.

Most heat pumps require a water flow rate of approximately 2 gpm (7.5708 liters/m) per nominal ton of refrigeration. The required flow rate may need to be somewhat higher in northern climates during winter operation. Under extreme weather conditions the heat pump(s) may run continuously for extended periods of time. Therefore, the water supply must be able to deliver the appropriate continuous water flow for 24 hours or more. A thorough "drawdown" testing of the well, according to the methods described by the National Water

**TABLE 5-4**

Water Standards for Open-Loop (Groundwater) Heat Pump Application

| | |
|---|---|
| Scaling | |
| Calcium and magnesium salts (hardness) | Less than 350 ppm |
| Iron oxide | Low |
| Corrosion | |
| pH | 5–10 |
| Hydrogen | Less than 50 ppm |
| Carbon dioxide | Less than 75 ppm |
| Chloride | Less than 600 ppm |
| Total dissolved solids | Less than 1500 ppm |
| Biological growth: iron bacteria | Low |
| Suspended solids | Low |

*Source:* Command-Aire Corporation.

Well Association, will indicate if it has the capacity to handle the extreme condition demand. If the well is intended to supply both the heat pump and the household water supply, be sure to conduct the drawdown test using the combined water requirements, not just the heat pump water flow. It is the responsibility of the driller to be knowledgeable of all applicable water well codes and regulations and perform the installation accordingly.

A final consideration for the water well is its pump sizing and depth. Careful evaluation of the pumping power requirements may prevent an installation where all the energy saved by the heat pump is consumed by the well pump.

After an adequate water supply is assured, the water must be delivered to the heat pump. When tapping into an existing pressure line, the tap should be made immediately after the pressure tank prior to down line taps. The tank should be sufficiently large to prevent rapid cycling of the well pump. If the tank is not of sufficient size, a new tank must be installed; or it might be desirable to bypass the tank and cycle the pump with the heat pump unit. Remember, the electrical power consumed by the pump is an important consideration. These are decisions that need to be made for each installation based on the existing equipment and water availability.

*Control Valve Installation:* Because of the open-loop nature of all natural well applications, the installer should take precautions against scaling and fouling of the water-to-refrigerant heat exchanger. By keeping the well water under pressure and not introducing air into the system, some of the more severe problems can be avoided. Installation of water controls on the discharge side of the heat pump will keep the water under pressure in the heat exchanger.

Pete's Plugs, boiler drains, and ball or gate

valves should be installed in the "water in" and "water out" heat pump piping runs (Figure 5-17). The Pete's Plugs are self-sealing orifices that allow a probe-type thermometer or pressure gauge to the inserted directly into the system water. The temperature differential and the pressure drop across the heat exchanger are easily measured using one thermometer or pressure gauge. Measuring and recording the pressure drop across the heat exchanger provides a reference point to use when checking for scaling problems. The gate or ball valves and boiler drains will permit isolation and cleaning of the heat exchanger should scaling occur.

The water flow to the heat pump can be controlled by a slow-closing solenoid valve, a motorized slow-closing valve, or a pressure regulating valve assembly. Some manufacturers

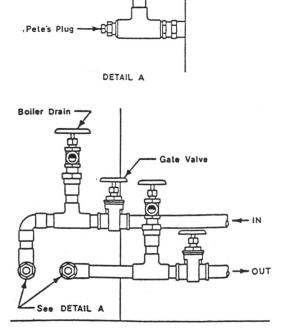

*FIGURE 5-17* Heat Exchanger Valve Location (Courtesy of Command-Aire Corporation)

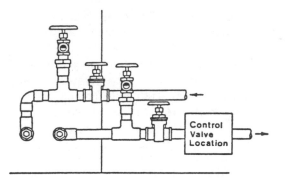

**FIGURE 5-18** Control Valve Location (Courtesy of Command-Aire Corporation)

recommend the size and type of valve to be used on their equipment. Their recommendations should be followed.

The solenoid or motorized valve should be installed in the proper location (Figure 5–18). Based on normally closed valves, the wiring for the solenoid valve is shown in Figure 5–19. The alternate motorized valve should be wired as shown in Figure 5–20. Some units make use of a pressure-regulating valve to control the flow of water through the unit (Figure 5–21).

Once the valve assembly is installed, connect the sensing-line tee to the refrigerant port provided on most residential units. If there is

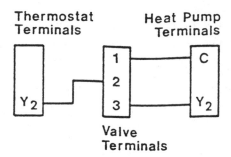

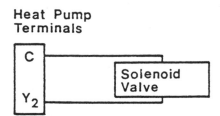

**FIGURE 5-19** Solenoid Valve Wiring Diagram (Courtesy of Command-Aire Corporation)

**FIGURE 5-20** Motorized Valve Wiring Diagram (Courtesy of Command-Aire Corporation)

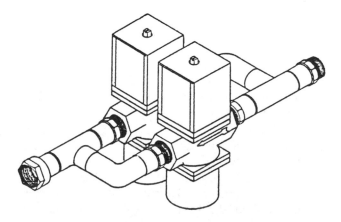

**FIGURE 5-21** Pressure-Regulating Valve Assembly (Courtesy of Command-Aire Corporation)

no port, a port must be added to the refrigerant line between the reversing valve and the water-to-refrigerant heat exchanger. The assembly contains two valves, one that opens in response to an increase in discharge pressure during the cooling cycle and the other which opens in response to a decrease in suction pressure during the heating cycle. The correct flow may be established by setting the opening point to its lowest setting if the ambient temperature is below the set point. If the ambient temperature is too low to allow the valve to open, the system may be manually flushed by lifting the range spring follower with a screwdriver at two sides of the lower spring cap to open the valve. This will not affect the valve adjustment. To optimize performance, with the refrigerant gauges attached, operate the heat pump in the heating mode. Adjust the heating-cycle water-regulating valve for an increase in the suction pressure. Adjust the valve to the point where there is little or no increase in the pressure reading by further opening. Any increase beyond this setting will be wasting water. The object is to gain a higher pressure reading for optimal heating capacity, and this is gained by opening the valve for heating.

Now operate the unit in the cooling mode and adjust the cooling cycle valve according to Table 5–5. Be sure to use the recommended pressure settings for the unit in consideration.

Each manufacturer will have his own settings to follow.

After completing these adjustments, remove your refrigerant gauges and tightly replace the caps on the Schrader service ports to avoid refrigerant leakage. It is the responsibility of the installer to be knowledgeable of all applicable plumbing and electrical codes and regulations and perform the installation accordingly.

*Water Discharge:* Three common methods of discharge disposal are:

1. *Recharge well.* In most cases, the use of a recharge well is recommended for water from the heat pump. There should be at least 50 ft (15.24 m) between the supply and the recharge well to prevent thermal interference between the wells.

2. *Drain fields.* In most areas, the use of a drain field is not recommended. Drain fields require soil conditions which allow a very rapid percolation. Remember—if a 5-ton heat pump runs continuously for 24 hours, it will discharge 14,400 gallons (54,509.9 liters) of water.

3. *Surface discharge.* Trouble free, with no clogged screen, percolation, or ser-

**TABLE 5–5**

Water-Regulating Valve Operating Pressures

| Water (°F) (°C) | Cooling 80°F (26.67°C) return air discharge pressure | Heating 70°F (21.11°C) return air suction pressure |
|---|---|---|
| 45 (7.22°C) | 160 min. | 46–54 (317.159–372.317 kPa) |
| 50 (10.00°C) | 160 min. | 52–60 (358.527–413.685 kPa) |
| 55 (12.78°C) | 160 min. | 56–66 (386.106–455.054 kPa) |
| 60 (15.56°C) | 166–190 | 60–72 (413.685–496.423 kPa) |
| 65 (18.33°C) | 178–204 | 62–74 (427.475–510.212 kPa) |

*Source:* Command-Aire Corporation.

vice problems, the surface discharge application is very simple. Regardless of dry land or surface water discharge, the requirements are as follows:

a. Discharge a minimum of 1 ft (0.3048 m) above flood level at the discharge point. This prevents organic growth or animal entry into the system. Siphoning due to a loss of pressure is also prevented.

b. Protect the discharge pipe end with a screen or flap valve to keep foreign objects out of the pipe.

*Note:* Surface discharge with a secondary discharge to allow use of the water is an excellent idea. Uses such as irrigation or livestock watering allow optimum utilization of the dollars spent for pumping power. Be sure that the secondary circuit is properly designed and does not restrict the required flow of water to the heat pump.

Acceptance of these methods of disposal varies nationwide, so it is the responsibility of the installer to be knowledgeable of all applicable codes and regulations and perform the installation accordingly.

*Special Applications:* The majority of special applications are the result of attempting to use surface water to supply a heat pump or having insufficient water supply from the well. Surface water should not be used! Installations are attempted quite often with what initially seems to be great success, only to fail later. The following problems make long-term success almost impossible:

1. Suspended solids damage or destroy water controls or heat exchangers.

2. Plant life, small shellfish, and fish always appear to be able to bypass pro-

tective devices to reach and restrict or clog a heat exchanger.

3. Temperature! Heated water rises and cooled water sinks. Any body of water, regardless of current or wave motion, is moving continually due to the effects of air and ground temperatures. These currents continually change the water temperature throughout the body of water. For this reason, it is difficult to rely on surface water to continually supply water no colder than 42°F (5.56°C). Entering water less than 42°F (5.56°C) will trip the freezestat and shut down the heat pump.

When an insufficient water supply is the problem, solutions can vary. Some manufacturers would possibly recommend a closed-loop earth-coupled application. If you do not select that type of system, the following are basic descriptions of alternative methods for utilizing low-flow wells.

The most common solution is to supply from and return to the same well, you can utilize one of the following methods to supply the heat pump(s).

1. Use the well as a heat exchanger. If the well is deep enough to have a large volume of water, you may be able to supply from and return to the same well without wasting any water. This method requires that at least 100 ft (30.48 m) of standing water per ton and the owner's willingness to accept fluctuations in household water temperature if it comes from the same well.

2. Waste some—save some. If the well volume is too small to allow its use as a heat exchanger, you can return to the well while discharging some portion elsewhere. Typical examples are:

a. Any time that the heat pump is running, the primary discharge circuit will return most of the water to the well, while a secondary circuit wastes the remainder. This prevents depletion of the well by allowing its limited capacity to replenish what is wasted.

b. By using temperature sensors, the secondary circuit mentioned previously will only waste water if the supply temperatures become colder or hotter than desired. As the well replaces the wasted water, it also moderates the overall well-water temperature.

In all cases, the well return discharge point should be as far from the pump intake as possible. Never discharge above the well's water level since aeration can cause clouding and increase the possibility of scaling and bacteria problems. Other solutions can be used with normal discharge methods or with the "supply from and the return to" methods mentioned previously. Depending on water temperature and system requirements, you can utilize one of the following methods to conserve water.

1. Where the supply water temperature at the first heat pump is 50°F (10.00°C) or warmer, the water can be run through two heat pumps in series prior to disposal. The water flow provided should be the amount required by the largest heat pump if they are different sizes. Figure 5–22 is a plumbing schematic for this type of application. Figure 5–23 is a wiring schematic using a 24-V solenoid valve.

2. When a mixing tank is used, a portion of the water already circulated through the heat pump can be mixed with the well supply water to provide a larger volume than the well can deliver. Figure

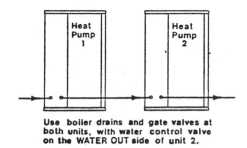

Use boiler drains and gate valves at both units, with water control valve on the WATER OUT side of unit 2.

**FIGURE 5-22** Plumbing Schematic for Multiple Heat Pumps in Series (Courtesy of Command-Aire Corporation)

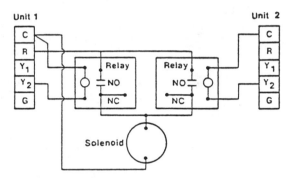

**FIGURE 5-23** Wiring Schematic for Multiple Heat Pumps in Series Using a Solenoid Valve (Courtesy of Command-Aire Corporation)

5–24 is a plumbing schematic for this type of application. Figure 5–25 is a wiring schematic using a Grundfos $\frac{1}{12}$-hp pump.

*Heat Exchanger Cleaning:* If scaling is a problem, the coil can be cleaned with a solution of an inhibited acid to minimize corrosion to copper or cupronickel heat exchanger tubing. Use the gate or ball valves shown in Figure 5–18 to isolate the heat pump from the remainder of the water system. Then, using a bucket and pump, circulate the acid solution through the heat exchanger by utilizing both boiler drains. Allow several hours for cleaning.

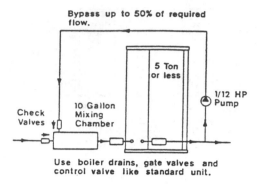

**FIGURE 5-24** Heat Pumps Using a Mixing Tank (Courtesy of Command-Aire Corporation)

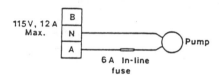

**FIGURE 5-25** Pump Wiring Schematic (Courtesy of Command-Aire Corporation)

——————— **SUMMARY** ———————

Water-source heat pumps are generally referred to as closed-loop earth-coupled and open-loop (water well or groundwater) systems.

The basic operating principle of this system is that the heat is rejected to the water from the air being cooled during the cooling cycle, and heat is gained from the water by the air to be heated during the heating cycle.

This type of system is most popular for heating and cooling residential structures, where an adequate source of water or ground area is available.

In the closed-loop earth-coupled heat pump a solution is circulated through special pipes buried in the ground.

There are three ways used in installing these types of heat pump systems: (1) horizontal, (2) vertical, and (3) lake loops.

The horizontal system is used in applications where the building site is large enough to allow the proper amount of pipe to be buried in a horizontal manner.

The horizontal type of system is especially advantageous when drilling a hole is not desirable.

The choice of vertical, horizontal, or lake loop earth-coupling should be based on the characteristics of each application.

The three earth-coupling methods should be considered for each application, with the most cost-effective method chosen after all have been evaluated.

The design of the earth-coupled system is divided up into the following steps: (1) determine the structure design heating load in Btu/h loss and design cooling load Btu/h gain, (2) select a water source heat pump, and (3) selection of the earth-type coil and materials used.

The sensible heat output capacity should in no case be less than the calculated total sensible heat load. Also, it should never be more than 25% greater than the calculated sensible heat load.

Systems that are designed for heating-only applications should provide a minimum of 75% and a maximum of 115% of the total heat loss.

Heat pump systems which are designed to provide both heating and cooling must be sized to satisfy the cooling requirements of the structure.

Of particular importance when designing an earth-coupled heat pump system is the balancing between the heat pump unit and the earth-coupled loop.

The final result of a perfectly balanced system is that the change in water temperature through the heat pump system is offset by an equal and opposite change in temperature through the earth-coupled loop.

Use only a water-source heat pump that can be operated on loop water temperatures that fall below 40°F (4.44°C) and perhaps down to 25°F (-3.89°C).

It is generally preferred that polybutylene (PB) or polyetheylene (PE) pipe be used for horizontal coils, vertical U-bend wells, and for sevice lines to the wells and lake heat exchangers.

During the installation it is very important that all trash, soil, and small animals be kept out of the piping system.

The plastic pipe assemblies should be pressure tested at twice the anticipated system operating pressures before the backfilling procedure is started.

The locations of important points such as wellheads must be marked for subsequent recovery.

A vertical earth-coupled system consists of one or more vertical boreholes through which water flows in a plastic pipe.

In each borehole sand or gravel backing is required around the piping to assure heat transfer.

A series U-bend earth coupling is one in which all the water flows through all of the pipe, progressively traveling down and then up each well bore.

A parallel U-bend earth coupling is one in which the water flows out through one header, is divided equally, and flows simultaneously down two or more U-bends.

A horizontal earth-coupled system is similar to a vertical system in that the water circulates through underground piping buried in a trench.

Burying the pipe at excessive depths will reduce the ability of the sun to recharge the heat used during the heating cycle.

Series horizontal earth couplings are the ones in which all the water flows through all of the pipe.

If a double layer of pipe is used, the incoming water to the heat pump unit should be from the deepest pipe.

The antifreeze solutions that are used in earth loop systems must be nontoxic and noncorrosive: nontoxic because should a leak develop in the loop system, the groundwater will not be contaminated.

The closed-loop groundwater heat pump systems that use a pond or lake for their heat source are by far the most economical of all closed-loop earth-coupled systems.

Never, under any circumstances, place a loop in a river or stream, because the drifting objects at flood conditions will in all probability cause damage to the loop.

The length of the loop is determined by a great extent by the temperature of the water in the pond.

The pond loop will not perform correctly if it is allowed to settle down into the mud. It should be supported at leat 1 ft (0.3084 m) off the bottom, and at least 9 ft (2.7432 m) below the surface of the water.

The service lines for the pond loop should be buried 4 ft (1.2192 m) deep or below the frost line, whichever is the deepest.

In the open-loop type of system, the water is taken into the system, passed through the equipment, and then discharged from the system.

Since the water is a crucial element in the operation of these types of systems, many manufacturers recommend that the services of an experienced well driller be obtained.

The water used in these systems does

not need to be suitable for human consumption; however, it should meet certain standards.

Most heat pumps require a water flow rate of approximately 2 gpm (7.5709 liters/m) per nominal ton of refrigeration.

A final consideration for the water well is its pump sizing and depth. Careful evaluation of the pumping power requirements may prevent any installation where all the energy saved by the heat pump is consumed by the well pump.

The installation of the water controls on the discharge side of the heat pump will keep the water under pressure in the heat exchanger and help to prevent scaling of the equipment.

The water flow to the heat pump can be controlled by a slow-closing solenoid valve, a motorized slow-closing valve, or a pressure-regulating valve assembly.

To optimize performance, with the refrigerant gauges attached, operate the heat pump in the heating mode. Adjust the heating-cycle water-regulating valve for an increase in the suction pressure. Adjust the valve to the point where there is little or no increase in the pressure reading by further opening. Any increase beyond this setting will be wasting water.

The three common methods of discharge disposal are: (1) recharge well, (2) drain fields, and (3) surface discharge.

If scaling is a problem, the coil can be cleaned with a solution of an inhibited acid to minimize corrosion to copper or cupro-nickel heat exchanger tubing.

## ——— REVIEW QUESTIONS ———

1. What is the basic operating principle of the water-source heat pump system?
2. In what application is water-source heat pump systems most popular?
3. Name the three types of installations for water-source heat pump systems.
4. What types of systems may use either vertical or horizontal tubing buried in the ground?
5. In what types of situations can the horizontal system be used?
6. What should be the deciding factor when choosing which type of system to be used on a particular installation?
7. Name the three steps used in designing an earth-coupled heat pump system.
8. What should be the output capacity of equipment sized for cooling-only applications?
9. What is done to bring a heating-only heat pump system up to the needed capacity?
10. How are heat pumps designed for both heating and cooling sized?
11. How much emergency heat is required by some localities?
12. What factor is of importance when designing an earth-coupled heat pump system?
13. If the circulating water is mixed with a nontoxic antifreeze solution, to what temperature can the entering water temperature be allowed to fall?
14. What method is used to connect PE pipe together?
15. What should be done to the ends of the pipe during the installation procedure?
16. At what pressure should the plastic pipe assemblies be tested before burying them?
17. Of what may backfill consist?
18. What is a good method of marking important points?
19. List three major reasons for using earth-coupled systems.
20. Describe a vertical earth-coupled heat pump system.
21. What is a distinct advantage of a vertical system over a horizontal system?
22. How may multiple boreholes be joined together?
23. How does the water flow through a series U-bend earth-coupling system?
24. In what type of system does the water flow out through one header?
25. What is the difference between the horizontal

and the vertical earth-coupled heat pump systems?

26. When running pipes in the same trench, how far apart must they be?

27. When two pipes are buried in the same trench, which one should be the nearer to the surface?

28. Describe the water flow through a series horizontal earth-coupled system.

29. In a horizontal earth coupling using a double layer of pipe, which pipe should feed the heat pump?

30. What two characteristics should the antifreeze used in earth-coupled heat pump systems have?

31. Of the water-source heat pump systems, which one is the most economical to operate?

32. Is a stream or river a good place to install a loop?

33. To a great extent, what determines the length of pond loop heat exchanger?

34. In what type of system is the water not recirculated through the system?

35. What is the approximate water flow rate required of water-source heat pump systems?

36. What precautions should the installer take to prevent when installing the control valve?

37. What are Pete's Plugs?

38. Name three methods of controlling the flow of water to the equipment?

39. What is the object when optimizing the system performance on the heating cycle?

40. Name three common methods of water disposal from a heat pump?

41. Where should the water discharge be in relation to the water intake?

# 6

# Heat Pump System Design

## INTRODUCTION

Previously, designers of air-conditioning and heating systems made guesstimates of the cooling and heating loads of a building. Because of this method, many systems were undersized while others were grossly oversized. Today, estimators cannot afford to use the guesstimation method of sizing equipment. Most builders and building owners require accurate cost and equipment size estimates before giving their authorization to install a unit. To do this, an accurate calculation of the heat loss and heat gain of the building must be completed to determine the size of the equipment required.

It is no longer safe to guess at the cooling and heating requirements, because the use of standardized building materials, which helped to make guesstimates possible, is a thing of the past. A wide variety of building materials, such as insulation, double-glazed windows, insulating boards, and insulation-type roofing materials, is being used in all types of building construction today.

## THE HEAT LOAD

The only safe and correct method of estimating the heat load of a building is to calculate the British thermal units (Btu) gained or lost through the building materials. Precise calculations of both cooling and heating loads are required before the proper selection of a heat pump can be made. The R and U factors, along with the factors used for insulation types, window types, physical location, the local climate design conditions, inside area, and number of occupants, are very critical in the proper calculation of the heat load.

*Calculating Building Heat Loss:* Many forms are available for calculating the heat load of a building. There is usually space on these forms for calculating both the heat loss and heat gain of a structure. The instructions for each form should be followed. In our examples, we will use the form recommended by Amana Refrigeration, Inc.

In our example, the outside design temperature is based on 97.5% of the average temperature for the area being considered. The inside design temperature will be 70°F (21.11°C); Table 6–1. The difference between the inside and outside design conditions will be used in calculations on the heat load forms.

*Orientation* refers to the position of the building on the site with respect to the north and south direction and the prevailing winds. Heat losses may be reduced by placing windows to the south rather than to the north.

*Exterior colors* can be used to reduce the heat losses of a building. The use of dark colors on exterior walls and roofs will help in reducing heat losses.

*Insulation* is a good investment. Proper insulation of ceilings and walls and the use of storm windows is an investment that will begin to pay dividends upon installation. Using the recommended thickness of attic insulation can provide more savings in heating costs than insulation in other places. Wall insulation and storm windows will also reduce heating costs.

*Method of Calculation:* The calculation of heat losses and gains is not an extremely complicated process. However, it can at times be rather time consuming. Like all processes, the more times calculations are made, the easier they are to complete, especially when the same form is used each time.

*General information* is usually the first entry on the form (Figure 6–1). Fill in the top of the form by providing the information requested. Enter the inside design temperature in the proper space. The outside design temperature in our example in 97.5% of the average low temperature. These two entries are found in Table 6–1. The design temperature difference is the inside design temperature minus the outside design temperature and should be entered in the appropriate space. Enter the job (customer) name, location, and date in the places indicated. The name of the person who makes the computation should be entered in the appropriate space so that anyone requiring more information about the job will know who to contact.

*Items 1, 2,* and *3* on the form are the actual measurements of the building. These measurements should be taken with care so that a great degree of accuracy can be maintained. These are usually taken as inside measurements. In *item 1*, place the room dimensions (length, width, and height). *Item 2* requests the linear feet of the exposed wall (linear feet of wall exposed to the outside temperatures). *Item 3* requests the floor area of the room (length by width). These data must be filled in for all rooms in the structure. On our example form there are two spaces left blank for additional rooms.

**TABLE 6–1**

Winter Weather Conditions

| State and city [a] | Medium of annual extremes | 97.5% | Coincident wind velocity [b] |
|---|---|---|---|
| Alabama | | | |
|   Birmingham AP | 14 | 22 | L |
|   Mobile AP | 21 | 29 | M |
|   Montgomery AP | 18 | 26 | L |
| Arizona | | | |
|   Flagstaff AP | −10 | 5 | UL |
|   Phoenix AP | 25 | 34 | UL |
|   Tucson AP | 23 | 32 | UL |
|   Winslow AP | 2 | 13 | UL |
|   Yuma AP | 32 | 40 | UL |
| Arkansas | | | |
|   Fort Smith AP | 9 | 19 | M |
|   Little Rock AP | 13 | 23 | M |
| California | | | |
|   Bakersfield AP | 26 | 33 | UL |
|   Fresno AP | 25 | 31 | UL |
|   Los Angeles CO | 38 | 44 | UL |
|   Oakland AP | 30 | 37 | UL |
|   Sacramento AP | 24 | 32 | UL |
|   San Diego AP | 38 | 44 | UL |

**TABLE 6-1**

Winter Weather Conditions (continued)

| State and city[a] | Medium of annual extremes | 97.5% | Coincident wind velocity[b] |
|---|---|---|---|
| San Francisco CO | 38 | 44 | UL |
| Colorado | | | |
| Denver AP | − 9 | 3 | L |
| Grand Junction AP | 12 | 11 | UL |
| Pueblo AP | − 14 | − 1 | UL |
| Connecticut | | | |
| Hartford, | | | |
| Brainard Field | − 4 | 5 | M |
| New Haven AP | 0 | 9 | H |
| D.C. Washington | | | |
| National AP | 12 | 19 | M |
| Florida | | | |
| Jacksonville AP | 26 | 32 | L |
| Miami AP | 39 | 47 | M |
| Tampa | 32 | 39 | M |
| Georgia | | | |
| Atlanta AP | 14 | 23 | H |
| Macon AP | 18 | 27 | L |
| Savannah, | | | |
| Travis AP | 21 | 27 | L |
| Idaho | | | |
| Boise AP | 0 | 10 | L |
| Idaho Falls AP | − 17 | − 6 | UL |
| Illinois | | | |
| Chicago, | | | |
| Midway AP | − 7 | 1 | M |
| Moline AP | − 12 | − 3 | M |
| Peoria AP | − 8 | 2 | M |
| Springfield AP | − 7 | 4 | M |
| Indiana | | | |
| Evansville AP | 1 | 10 | M |
| Indianapolis AP | − 5 | 4 | M |
| Iowa | | | |
| Cedar Rapids AP | − 14 | − 4 | M |
| Des Moines AP | − 13 | − 3 | M |
| Ottumwa AP | − 12 | − 2 | M |
| Sioux City AP | − 17 | − 6 | M |
| Marshalltown | − 16 | − 6 | M |
| Kansas | | | |
| Topeka AP | − 4 | 6 | M |
| Wichita AP | − 1 | 9 | H |
| Kentucky | | | |
| Louisville AP | 1 | 12 | L |
| Louisiana | | | |
| Lake Charles, AP | 25 | 33 | M |
| New Orleans AP | 29 | 35 | M |

**TABLE 6-1**

Winter Weather Conditions (continued)

| State and city[a] | Medium of annual extremes | 97.5% | Coincident wind velocity[b] |
|---|---|---|---|
| Shreveport AP | 18 | 26 | M |
| Maine | | | |
| Bangor, Dow AFB | − 14 | − 4 | M |
| Maryland | | | |
| Baltimore, CO | 12 | 20 | M |
| Massachusetts | | | |
| Boston AP | − 1 | 10 | H |
| Michigan | | | |
| Detroit Metcap | 0 | 8 | M |
| Sault Ste Marie | | | |
| AP | − 18 | − 8 | L |
| Minnesota | | | |
| Minneapolis AP | − 19 | − 10 | L |
| Rochester AP | − 23 | − 13 | M |
| Mississippi | | | |
| Biloxi Keesler | | | |
| AFB | 26 | 32 | M |
| Jackson AP | 17 | 24 | L |
| Missouri | | | |
| Kansas City AP | − 2 | 8 | M |
| St. Louis CO | 1 | 11 | M |
| Springfield AP | 0 | 10 | M |
| Montana | | | |
| Billings AP | − 19 | − 6 | L |
| Butte AP | − 34 | − 16 | UL |
| Great Falls AP | − 29 | − 16 | L |
| Nebraska | | | |
| Lincoln CO | − 10 | 0 | M |
| North Platte AP | − 13 | − 2 | M |
| Omaha AP | − 16 | − 4 | M |
| Nevada | | | |
| Las Vegas AP | 18 | 26 | UL |
| Reno | | | UL |
| New Hampshire | | | |
| Concord AP | − 17 | − 7 | M |
| New Jersey | | | |
| Newark AP | 6 | 15 | M |
| Trenton CO | 7 | 16 | M |
| New Mexico | | | |
| Albuquerque AP | 6 | 17 | L |
| Roswell Walker | | | |
| AFB | 5 | 19 | L |
| New York | | | |
| Albany AP | − 14 | 0 | L |
| Buffalo AP | − 3 | 6 | M |
| Elmira AP | − 5 | 5 | L |

**TABLE 6-1**

Winter Weather Conditions (continued)

| State and city[a] | Medium of annual extremes | 97.5% | Coincident wind velocity[b] |
|---|---|---|---|
| New York | | | |
| LaGuardia AP | 7 | 16 | H |
| Rochester AP | − 5 | 5 | M |
| Syracuse AP | − 10 | 2 | M |
| North Carolina | | | |
| Charlotte AP | 13 | 22 | L |
| Greensboro AP | 9 | 17 | L |
| Raleigh/ | | | |
| Durham AP | 13 | 20 | L |
| Wilmington AP | 19 | 27 | L |
| North Dakota | | | |
| Bismark AP | − 31 | − 19 | UL |
| Fargo AP | − 28 | − 17 | L |
| Grand Forks AP | − 30 | 23 | L |
| Ohio | | | |
| Akron/Canton AP | − 5 | 6 | M |
| Cincinnati CO | 2 | 12 | L |
| Cleveland AP | − 2 | 7 | M |
| Columbus AP | − 1 | 7 | M |
| Dayton AP | − 2 | 6 | M |
| Youngstown AP | − 5 | 6 | M |
| Oklahoma | | | |
| Ardmore | 9 | 19 | H |
| Oklahoma City AP | 4 | 15 | H |
| Tulsa | 4 | 16 | H |
| Oregon | | | |
| Baker AP | − 10 | 1 | UL |
| Eugene AP | 16 | 26 | UL |
| Medford AP | 15 | 23 | UL |
| Portland CO | 21 | 29 | L |
| Pennsylvania | | | |
| Erie AP | 1 | 11 | M |
| Harrisburg AP | 4 | 13 | L |
| Philadelphia AP | 7 | 15 | M |
| Pittsburgh AP | − 1 | 9 | M |
| Rhode Island | | | |
| Prividence AP | 0 | 10 | M |
| South Carolina | | | |
| Charleston AFB | 19 | 27 | L |
| Columbia AP | 16 | 23 | L |
| South Dakota | | | |
| Huron AP | − 24 | − 12 | L |
| Rapid City AP | − 17 | − 6 | M |
| Tennessee | | | |
| Chattanooga AP | 11 | 19 | L |
| Knoxville AP | 9 | 17 | L |

**TABLE 6-1**

Winter Weather Conditions (continued)

| State and city[a] | Medium of annual extremes | 97.5% | Coincident wind velocity[b] |
|---|---|---|---|
| Memphis AP | 11 | 21 | L |
| Texas | | | |
| Abilene AP | 12 | 21 | M |
| Amarillo AP | 2 | 12 | M |
| Austin AP | 19 | 29 | M |
| Brownsville AP | 32 | 40 | M |
| Dallas AP | 14 | 24 | H |
| El Paso AP | 16 | 25 | L |
| Ft. Worth AP | 14 | 24 | H |
| Houston CO | 24 | 33 | M |
| Laredo AFB | 29 | 36 | L |
| San Antonio AP | 22 | 30 | L |
| Utah | | | |
| Ogden CO | − 3 | 11 | UL |
| Salt Lake City AP | − 2 | 9 | L |
| Vermont | | | |
| Burlington AP | − 18 | − 7 | M |
| Virginia | | | |
| Norfolk AP | 18 | 23 | M |
| Richmond AP | 10 | 18 | L |
| Roanoke AP | 9 | 18 | L |
| Washington | | | |
| Seattle CO | 14 | 32 | L |
| Spokane AP | − 5 | 4 | UL |
| West Virginia | | | |
| Charleston AP | 1 | 14 | L |
| Wisconsin | | | |
| Green Bay AP | − 16 | − 7 | M |
| La Crosse AP | − 18 | − 8 | M |
| Madison AP | − 13 | − 5 | M |
| Milwaukee AP | − 11 | − 2 | M |
| Wyoming | | | |
| Cheyenne AP | − 15 | − 2 | M |
| Rock Springs AP | − 16 | − 1 | UL |

[a]When airport temperature observations were used to develop design data.

"AP" follows city, "AFB" follows air force bases, and "CO" follows office locations within an urban area.

[b]Coincidental wind velocities derived from approximately coldest 600 hours out of 20,000 hours of December through February data per station. VL, very light, 70% or more cold extreme hours ≤ 7 mph. L, light, 50 to 69% cold extreme hours ≤ 7 mph. M, moderate, 50 to 74% cold extreme hours ≤ mph. H, high. 75% or more cold extreme hours ≤ 7 mph (50% are ≤ 12 mph).

*Source:* Amana Refrigeration, Inc.

Job Name: P. Smith
Location: CINCINNATI, OHIO
Date: 1/1/75
Computed By: C.C.S.

Inside Design Temp. 70° F.
Outside Design Temp. −5° F.
Temp. Difference 75° F.
Temp. Difference +10 _____

| Room | | Living Room | | Dining Room | | | Kitchen | | | Bedroom 1 | | | Bedroom 2 | | | Bath | | | Rooms Btu Totals |
|---|---|---|---|---|---|---|---|---|---|---|---|---|---|---|---|---|---|---|---|
| | Factor | Ar. or Ln. Ft. | BTU | Ar. or Ln. Ft. | BTU | | Ar. or Ln. Ft. | BTU | | Ar. or Ln. Ft. | BTU | | Ar. or Ln. Ft. | BTU | | Ar. or Ln. Ft. | BTU | | |
| 1. Room Size (LxWxH) | | 40x29x8 | | | | | | | | | | | | | | | | | |
| 2. Linear Ft. Exp. Wall | | 136 | | | | | | | | | | | | | | | | | |
| 3. Floor Area | | 1120 | | | | | | | | | | | | | | | | | |
| 4. Windows | 1.13 | 135 | 152.55 | | | | | | | | | | | | | | | | |
| 5. Doors | .52 | 38.5 | 20.02 | | | | | | | | | | | | | | | | |
| 6. Exp. Wall | .11 | 944.5 | 100.6 | | | | | | | | | | | | | | | | |
| 7. Ceiling | .09 | 1120 | 100.8 | | | | | | | | | | | | | | | | |
| 8. Infiltration a | .22 | 205.5 | 46.21 | | | | | | | | | | | | | | | | |
| b | — | — | — | | | | | | | | | | | | | | | | |
| c | .60 | 39 | 23.4 | | | | | | | | | | | | | | | | |
| d | .13 | 30 | 3.9 | | | | | | | | | | | | | | | | |
| 9. SUB TOTALS | | | 446.98 | | | | | | | | | | | | | | | | |
| 10. Sub-Total Btu for T.D. 75° | | | 33986 | | | | | | | | | | | | | | | | |
| 11. Floor Btu for T.D. 3.4 | | 1120 | 3808 | | | | | | | | | | | | | | | | |
| 12. Partition Btu for T.D. | | — | | | | | | | | | | | | | | | | | |
| 13. Room Total Btu Loss | | | 37294 | | | | | | | | | | | | | | | | |

14. +30% Duct Loss 11188.2
15. Grand Total Btu Load 48982

T.D. = TEMPERATURE DIFFERENCE

**Windows**

| | U Factors |
|---|---|
| Single Pane | (1.13) |
| Double Pane, Sealed | .57 |
| Storm Windows, Tight | .45 |
| Storm Windows, Loose | .75 |
| Glass Block | .50 |

**Wood Doors**

| Nom. Thickness | U Factors No Storm | Storm Door |
|---|---|---|
| 1" | .69 | .35 |
| 1½" | (.52) | .30 |
| 2" | .46 | .28 |
| 2½" | .38 | .25 |

**Exp. Walls** U Factors

| Frame Construction With: | Clapboard or Shingle | Brick Venr. | Stone Venr. |
|---|---|---|---|
| No Insulation | .25 | .28 | .30 |
| ¾" Insl. Board | .19 | .21 | .22 |
| 2" Insl. Batts | (.11) | .12 | .12 |
| 3½" Insl. Batts | .09 | .10 | .10 |

**Ceilings** (Vent: Attic Space Above) U Factors

| | |
|---|---|
| No Insulation | .50 |
| 2" Insulation | .10 |
| 4" Insulation | (.09) |
| 6" Insulation | .05 |

**CONCRETE FLOOR ON GROUND**
Btu PER HR. PER LINEAL FT. OF EDGE

| Outside Design Temp. | 25 | 20 | 15 | 10 | 5 | 0 | −5 | −10 | −15 | −20 | −25 | −30 |
|---|---|---|---|---|---|---|---|---|---|---|---|---|
| 1" Vert. Insul. Extending Down 18" Below Surface | 55 | 62 | 68 | 75 | 80 | 85 | 90 | 95 | 100 | 105 | 110 | 115 |
| 1" L Type Insul. Extending 12" Down and 12" Under | 50 | 57 | 63 | 70 | 75 | 80 | 85 | 90 | 95 | 100 | 105 | 110 |
| 2" L Type Insul. Extending 12" Down and 12" Under | 40 | 45 | 50 | 55 | 60 | 65 | 70 | 75 | 80 | 85 | 90 | 95 |
| No edge insulation and no heat in slab | | | | | This construction not recommended. | | | | | | | |

% Duct Loss Schedule

| | 1 Story | 1½ Story | 2 Story |
|---|---|---|---|
| Ducts Insul. | 10 | 8 | 6 |
| Ducts No Insul. | (30) | 25 | 20 |

Infiltration (factor times lineal ft. of crack)

| | | Weather Stripped | Not Weather Stripped |
|---|---|---|---|
| Double Hung Window | (a) | (.22) | .35 |
| Pivoting Windows | (b) | .30 | .50 |
| Doors | (c) | (.60) | 1.00 |
| Fixed Window | (d) | (.13) | .13 |

Double Wood Floors or Partition

| With Air Space | 10 | 20 | 30 | 40 | 50 |
|---|---|---|---|---|---|
| Temp. Diff. | | | | | |
| Factor No Insul. | (3.4) | 6.8 | 10.2 | 13.6 | 17.0 |
| Factor 2" Insul. | 1 | 2 | 3 | 4 | 5 |

U Factors

For additional coefficients or more complete descriptions of these elements of structure, consult the Htg. Ventil. & Air Cond. Guide.

Note (The above lists of factors is comprised of only those most commonly used.)

**FIGURE 6-1** Residential Heat Loss Tabulation (Courtesy of Amana Refrigeration, Inc.)

*Factors* for the building are the design factors, which must be selected from the chart at the bottom of the form (Figure 6–1), in accordance with the building structure and materials. These are the U factors for a given type of building material. A U factor is the amount of heat transfer per square foot per degree of temperature difference between the inside and outside of the building. When a U factor has been determined that closely matches the structure, it should be placed on the chart in the factor column next to the structure area.

*Items 4 through 8* deal with the windows, doors, exposed wall, ceiling, and infiltration of the building. In *item 4*, figure the area of each window by multiplying the height by the width. Add the areas of all the windows in each room and place the total in the column labeled Ar. (area) or Ln. ft. (linear feet) under the proper room heading.

*Item 5* deals with outside doors. Multiply the height by the width of the outside doors. Then add all the areas of all the doors in each room and place the total in the column labeled Ar. (area) or Ln. ft. (linear feet) under the proper room heading.

*Item 6* deals with the outside walls. Multiply the height by the width of the outside wall. The area obtained by this calculation is the net wall area. Therefore, the window and door areas must be subtracted from the gross wall area. The net outside wall area should be placed in the column labeled Ar. or Ln. ft. under the room being considered.

*Item 7* deals with the ceiling. The ceiling area of the room in most cases will be the same as the floor area calculated in item 3. However, if the ceiling is a cathedral type, the area must be calculated by multiplying the width by the length of the ceiling plus the ends if they are exposed to an unconditioned space (Figure 6–2). Enter the total exposed area under the proper room in the Ar. or Ln. ft. column.

*Item 8* deals with infiltration. Four general

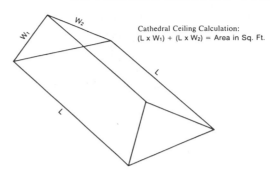

**FIGURE 6-2** Cathedral Ceiling Area Calculation Method (Courtesy of Amana Refrigeration, Inc.)

areas of infiltration must be considered: (1) double-hung windows, (2) pivoting windows, (3) doors, and (4) fixed windows. If a room has one or more of these items in an exterior wall, the infiltration for that area must be calculated. The factors shown in the infiltration chart should be placed in the factor column next to that type of infiltration area. The linear feet of crack area must be added together and entered under the column Ar. or Ln. ft.

*Items 9 through 15* deal with calculating the Btu loss of the building. The areas or linear feet in each item are multiplied by their U factor and the results placed in the column marked Btu.

*Item 9* deals with the Btu subtotals of the room or building being considered. Add up the Btu from items 4 through 8 and enter in item 9.

*Item 10* deals with the Btu subtotal for the design temperature difference. Multiply the Btu in item 9 by the temperature difference shown at the top of the form. Enter the answer in item 10.

*Item 11* deals with the Btu loss through the floor of the room or building. Place the floor area from item 3 into the column Ar. or ln. ft. and multiply the area by the factor for either slab floor construction or floor over a basement or crawl space. Place the answer in the appropriate Btu column.

*Item 12* deals with partition walls. If a room has a wall adjoining a nonconditioned area, the wall is to be treated as a partition wall. Multiply the wall area by the factor for the temperature difference shown at the bottom of the form.

*Item 13* is the room's total Btu loss. Add items 10, 11, and 12 and enter the total here. The various room Btu totals should be added and the total placed in the far right column.

*Item 14* is the duct heat loss. This section is the percentage of heat lost through the duct-work. Multiply this number (from the table) by the total of the room heat loss.

*Item 15* is the total heat loss of the building. It is the subtotals added for all rooms plus the heat required for duct heat loss.

## HEAT LOSS CALCULATION EXAMPLE

Using the preceding instructions and the Amana residential heat loss tabulation form, calculate one room of a structure with the following characteristics (Figure 6–3). Use the factors from Tables 6–2 through 6–9.

Inside design temperature—75 °F

Outside design temperature—0 °F

Temperature difference—75 °F

All windows—single-pane glass—not weather stripped

Door—not weather stripped—no storm door Exposed wall—clapboard or shingle with 2 in. of insulation

Ceiling—4 in. of insulation, vented attic space above

Floor—double wood with air space, no insulation over 60 °F basement

### STEP 1

*Room Size:* List the length, width, and height of room (23 ' × 12 ' × 8 ').

### STEP 2

*Linear Feet of Exposed Wall:* Add the lengths of all walls exposed to the outdoors and enter the sum in the space provided (23 ' + 12 ' = 35 ').

### STEP 3

*Floor Area:* Multiply the room length by the room width (12 ' × 23 ' = 256 square feet). Enter the total in the space provided.

### STEP 4

*Btu Loss for Windows:* Select and enter the proper U factor from Table 6–2. Single-pane glass, not weatherstripped (1.13). Enter this figure in the Factor column. Calculate the total of the window areas, width, width times height $(2\frac{1}{3}' \times 4') + (4' \times 10') + (1' \times 1') = 50$ square feet. Enter this figure in the Area column. Calculate the Btu loss for 1° of temperature difference by multiplying the factor by the area ($1.13 \times 50 = 56.6$). Enter this figure in the Btu column.

### STEP 5

*Btu Loss for Doors:* Select and enter the proper U factor from Table 6–3. Remember, the door is not weatherstripped and has no storm door (0.52). Enter this figure in the Factor column. Calculate the net door area, width times height less the area of any glass in the door $(3' \times 6\frac{2}{3}') - 1' = 19$ square feet. Enter this figure in the Area column. Calculate the Btu loss for 1° of temperature difference by multiplying the factor times the area ($0.52 \times 19 = 9.9$). Enter this figure in the Btu column.

### STEP 6

*Btu Loss for Exposed Walls:* Select and enter the proper U factor from Table 6–4. The exposed walls are made from clapboard or shingle with 2 in. of insulation (0.11). Enter this figure in the Factor column. Calculate the net wall area: linear feet of exposed wall times the height, less window and door area ($35' \times 8'$) −50 square feet − 19 square feet = 211 square feet. Enter this figure in the Area column. Calculate the Btu loss for 1° of temperature difference by multiplying the factor times the area ($0.11 \times 211 = 23.2$). Enter this figure in the Btu column.

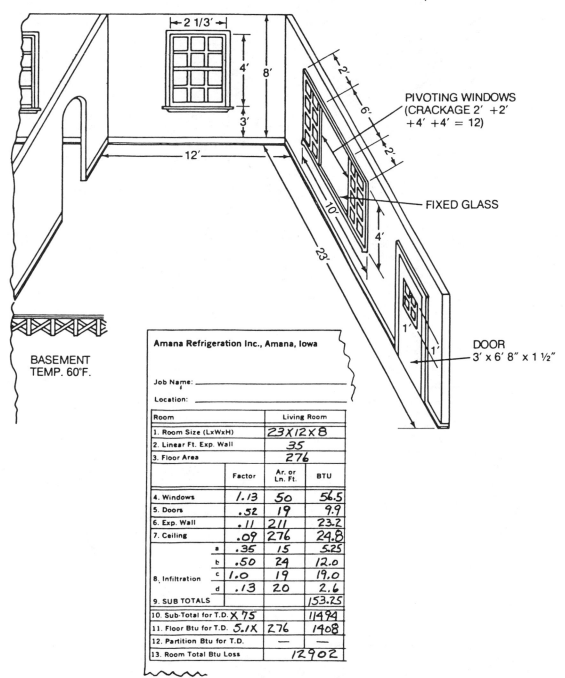

**2 1/3'**

**4'** **8'**

**3'**

**12'**

PIVOTING WINDOWS
(CRACKAGE 2' +2'
+4' +4' = 12)

2'

6'

2'

FIXED GLASS

10'

4'

23'

1'

1'

DOOR
3' x 6' 8" x 1 ½"

BASEMENT
TEMP. 60°F.

**Amana Refrigeration Inc., Amana, Iowa**

Job Name: _____

Location: _____

| Room | | Living Room | |
|---|---|---|---|
| 1. Room Size (LxWxH) | | 23X12X8 | |
| 2. Linear Ft. Exp. Wall | | 35 | |
| 3. Floor Area | | 276 | |
| | Factor | Ar. or Ln. Ft. | BTU |
| 4. Windows | 1.13 | 50 | 56.5 |
| 5. Doors | .52 | 19 | 9.9 |
| 6. Exp. Wall | .11 | 211 | 23.2 |
| 7. Ceiling | .09 | 276 | 24.8 |
| 8. Infiltration    a | .35 | 15 | 5.25 |
| b | .50 | 24 | 12.0 |
| c | 1.0 | 19 | 19.0 |
| d | .13 | 20 | 2.6 |
| 9. SUB TOTALS | | | 153.25 |
| 10. Sub-Total for T.D. X 75 | | | 11494 |
| 11. Floor Btu for T.D. 5.1X | | 276 | 1408 |
| 12. Partition Btu for T.D. | — | | — |
| 13. Room Total Btu Loss | | 12902 | |

**FIGURE 6-3** Room for Example Heat Loss Calculations
(Courtesy of Amana Refrigeration, Inc.)

## STEP 7

*Btu Loss for Ceiling:* Select and enter the proper U factor from Table 6–5. The ceiling has 4 in. of insulation with a vented attic space above (0.09). Enter this figure in the Factor column. Calculate the total ceiling area, length times width, which is usually the same as the floor area (12′ × 23′ = 276 square feet). Enter this figure in the Area column. Calculate the Btu loss for 1° of temperature difference by multiplying the factor times the area (0.09 × 276 = 24.8). Enter this figure in the Btu column.

**TABLE 6–2**

Factors for Windows

| | |
|---|---|
| Single pane | 1.13 |
| Double pane, sealed | 0.55 |
| Storm windows, tight | 0.45 |
| Storm windows, loose | 0.75 |
| Glass block | 0.50 |

*Source:* Amana Refrigeration, Inc.

**TABLE 6–3**

Factors for Wood Doors

| Nominal thickness of door (in.) | Without storm door | With storm door |
|---|---|---|
| 1 | 0.69 | 0.35 |
| $1\frac{1}{2}$ | 0.52 | 0.30 |
| 2 | 0.46 | 0.28 |
| $2\frac{1}{2}$ | 0.38 | 0.25 |

*Source:* Amana Refrigeration, Inc.

**TABLE 6–4**

Factors for Exposed Walls

| Frame construction with: | Clapboard or shingle | Brick veneer | Stone veneer |
|---|---|---|---|
| No insulation | 0.25 | 0.28 | 0.30 |
| $\frac{3}{4}$ in. of insulation board | 0.19 | 0.21 | 0.22 |
| 2 insulation batts | 0.11 | 0.12 | 0.12 |
| $3\frac{5}{8}$ in. of insulation batts | 0.09 | 0.10 | 0.10 |

*Source:* Amana Refrigeration, Inc.

## STEP 8

*Btu Loss in Infiltration:* Select and enter the proper U factors from Table 6–6.

1. Double-hung windows, not weather stripped (0.35)
2. Pivoting windows, not weather stripped (0.50)
3. Doors, not weatherstripped (1.0)
4. Fixed windows, not weatherstripped (0.13)

Enter these figures in the proper Factor columns. Calculate the total linear feet of crackage:

1. Double-hung windows, linear feet of crack: $2\frac{1}{3}′ + 2\frac{1}{3}′ + 2\frac{1}{3}′ + 4′ + 4′ = 15′$.
2. Pivoting windows, linear feet of crack: $(2′ + 2′ + 4′ + 4′) × 2$ windows = 24′
3. Door, linear feet of crack: $3′ + 3′ + 6\frac{2}{3}′ = 19′$
4. Fixed windows, linear feet of crack: $6′ + 6′ + 4′ + 4′ = 20′$

Enter these figures in the proper Area columns.

Calculate the Btu loss for 1° of temperature difference by multiplying the factor times the linear feet of crack.

1. Double-hung windows (0.35 × 15 = 5.25)
2. Pivoting windows (0.50 × 24 = 12)
3. Doors (1.0 × 19 = 19)
4. Fixed windows (0.13 × 20 = 2.6)

Enter these figures in the proper Btu columns.

## STEP 9

*Subtotals:* The subtotal is the sum of the Btu values calculated in steps 4 through 8 (56.5 + 9.9 + 23.2 + 24.8 + 5.25 + 12 + 19 + 2.6 = 153.25). Enter this figure in the proper column.

## STEP 10

*Subtotal Btu for Temperature Difference:* This subtotal is calculated by multiplying Btu from step 9 by the temperature difference taken from the upper right corner of the Heat Loss Tabulation Form (75 °F inside temperature design − 0 °F outdoor temperature design = 75 × 153.25 = 11,494). Enter this figure in the proper column.

## STEP 11

*Floor Btu for Temperature Difference:* Select the proper U factor from Table 6–7, with air space and no insulation with a 15 °F temperature difference (room temperature 75 °F − basement temperature 60 °F = 15 °F). Since there is no factor for 15 °F, we must interpolate. The 10 °F column equals 3.4 and the 20 °F column equals 6.5. Therefore, the 15 °F temperature difference equals

$$6.8 - 3.4 = 3.4 \div 2 = 1.7 + 3.4 = 5.1$$

Calculate the Btu loss by multiplying this factor times the floor area (276 × 5.1 = 1408). If this were a concrete floor on the ground we would use the linear feet, calculated in step 2, in place of the floor area, calculated in step 3, and select the U factor from Table 6–8. Enter this figure in the proper column.

## STEP 12

*Partition Btu for Temperature Difference:* Select the proper U factor from Table 6–7. Calculate the Btu

### TABLE 6–5

Factors for Ceilings[a]

| | |
|---|---|
| No insulation | 0.50 |
| 2 in. of insulation | 0.10 |
| 4 in. of insulation | 0.09 |
| 6 in. of insulation | 0.05 |
| 10 in. of insulation | 0.042 |

[a]With ventilated attic space above.

*Source:* Amana Refrigeration, Inc.

### TABLE 6–6

Infiltration Factors[a]

| | Weatherstripped | Not weatherstripped |
|---|---|---|
| Double-hung windows | 0.22 | 0.35 |
| Pivoting windows | 0.30 | 0.50 |
| Doors | 0.60 | 1.00 |
| Fixed window | 0.13 | 0.13 |

[a]Factor times lineal feet of crack.

*Source:* Amana Refrigeration, Inc.

### TABLE 6–7

Floor Loss Factors for Double Wood Floors or Partitions

| Air space temperature difference | 10 | 20 | 30 | 40 | 50 |
|---|---|---|---|---|---|
| No insulation | 3.4 | 6.8 | 10.2 | 13.6 | 17.0 |
| 2 in. of insulation | 1 | 2 | 3 | 4 | 5 |

*Source:* Amana Refrigeration Inc.

**TABLE 6–8**

Heat Loss Factors for Concrete Floor on Ground (Slab)[a]

| | Outside design temperature (°F) | | | | | | | | | | | |
| | 25 | 20 | 15 | 10 | 5 | 0 | −5 | −10 | −15 | −20 | −25 | −30 |
|---|---|---|---|---|---|---|---|---|---|---|---|---|
| 1 in. of vertical insulation extending down 18 in. below surface | 55 | 62 | 68 | 75 | 80 | 85 | 90 | 95 | 100 | 105 | 110 | 115 |
| 1 in. of L-type insulation extending 12 in. down and 12 in. under | 50 | 57 | 63 | 70 | 75 | 80 | 85 | 90 | 95 | 100 | 105 | 110 |
| 2 in. of L-type insulation extending 12 in. down and 12 in. under | 40 | 45 | 50 | 55 | 60 | 65 | 70 | 75 | 80 | 85 | 90 | 95 |
| No edge insulation and no heat in slab | This construction not recommended[b] | | | | | | | | | | | |

[a]Btu per hour per lineal foot of edge.

[b]Approximately double of factors 1 in. of vertical insulation down 18 in. below surface.

*Source:* Amana Refrigeration, Inc.

**TABLE 6–9**

Percentage of Duct Loss

| | 1-story | 1½-story | 2-story |
|---|---|---|---|
| Ducts insulated | 10 | 8 | 6 |
| Ducts not insulated | 30 | 25 | 20 |

*Source:* Amana Refrigeration, Inc.

by multiplying the U factor times the area of the partition. In our example the partition is adjoining a conditioned space, so no factor is considered.

**STEP 13** ————————

*Total Room Btu Loss:* The total room Btu loss is the sum of the Btu calculated in steps 10 to 12 (11,494 + 1408 = 12,902). Enter this figure in the proper column.

**STEP 14** ————————

*Percentage of Duct Loss:* A certain percentage of heating capacity must be provided because of heat loss through the heat duct system to the rooms. This percentage is determined by whether or not the ducts are insulated. See Table 6–9. In our example, let's say the ducts were not insulated. Therefore, a 30% factor is used and entered on line 14. This factor is multiplied times the Btu on line 13 and entered on line 14 (30% × 12,902 = 3,870.6).

This represents the total heat loss for one room. To calculate the total heat loss for the entire building, each room must be surveyed; then all are added together for the total. The equipment is then selected to provide the desired amount of heat for the entire building.

## HEAT GAIN CALCULATION EXAMPLE ————————

The following is a relatively simple but adequate method of calculating the heat gain for a given residence. This method is strictly for residential buildings and is based on 24-hour/day operation of the air-conditioning equipment. No allowances are included for instantaneous cooling effect. To use this method, you should be familiar with the terms used in air-conditioning work, such as wet and dry bulb temperatures, dew-point temperature, condensation, heat, Btu, specific heat, sensible heat, latent heat, heat transfer, and relative humidity.

When considering air conditioning for a residence, many factors are involved that affect the heat gain load. These are discussed next.

*Orientation:* Building orientation refers to the position in which the house is placed on the site with respect to north and south and to the hot summer sun and prevailing winds. The heat gain load from the sun can be reduced by using solid walls, or at least a minimum of windows. A garage on the west side of the building will also help. Also, advantage should be taken of existing windbreaks and shade trees on the west side of the house.

*Shading:* The use of awnings, large roof overhangs, porches, and sun screens will reduce heat gain loads through windows and doors. Roof overhangs are effective shading devices, especially on southern exposures. However, they are less effective on east and west exposures.

*Exterior Colors:* Building heat gain loads can be effectively reduced by using white or light-colored roofs and exterior walls. White can reflect approximately 55% more of the sun's heat than black, blue, and other dark colors. Pastel colors can reduce the sun's heat load by approximately 35 to 40%.

*Attic Ventilation:* In an unventilated attic under a dark roof, temperatures can reach 160°F and add significantly to the heat gain of a building. Adequate ventilation between the roof and finished ceiling can reduce the heat that enters the building through the roof by 20% or more.

Attic vents for gable roofs should never be smaller than 1 square foot of free inlet area. The best attic vent for natural ventilation should be 1 square foot of free area for each 300 square feet of attic floor area. Homes with flat roofs or very low pitched roofs should have ventilators twice as large as those for average or high-pitched roofs. Continuous soffit slats are recommended for attic ventilation for buildings using hip roofs with a roof overhang. Power ventilation of attic spaces can reduce the heat gain load to that of transmission gain only, thereby completely eliminating solar heat loads.

*Insulation:* Proper insulation of the ceiling, walls, and windows is just as good an investment for cooling as it is for heating. Dividends can be realized as soon as the insulation is installed.

*Outside Wet Bulb Temperatures:* The local outside wet bulb temperatures must be taken into consideration when calculating heat gain loads. Dry areas such as those along the California coast, Oregon, and Washington, along with the dry areas of the southwestern states, require more cfm per ton of cooling than the wet areas along the Gulf Coast and the Atlantic Coast.

*Daily Outside Temperature Range:* The daily outside temperature range is another very important factor to be considered when computing building heat gains. High daily temperature ranges occur quite frequently in the far western states and low daily temperature ranges occur along the coastal areas. High temperature ranges result in larger heat gains; low temperature ranges result in lower heat gains.

*Occupancy:* People produce heat, which adds to the total heat gain load of the building. This load is made up of both sensible and latent heat. A person at rest gives off about 400-Btu total heat; a child playing gives off about 700 Btu. Therefore, the activity and number of occupants must be considered when calculating heat gain into a building.

*Ventilation:* The air in a residence must be kept fresh and free of odors, and the carbon dioxide content must be kept to a minimum. To control odors (mostly tobacco, food, body, and waste material), fresh outside air should be introduced into the home. Most older homes will allow sufficient quantities of outside air for ventilation due to air leakage through cracks around the windows and doors. However, newer homes are usually more tightly constructed, and some means of providing fresh outside air is desirable. The introduction of fresh outside air into the home adds both sensible and latent heat gains for the equipment to remove.

*Temperature Differential:* The temperature differential between the outside temperature and the inside temperature is one of the most important factors to be considered when calculating the total heat gain load. Reliable comfort-zone charts indicate that the average person is more comfortable and has better health when living in an air temperature 20 to 30 °F below 98.6 °F, or normal body temperature.

## PROCEDURE FOR CALCULATING RESIDENTIAL HEAT GAIN ————————

Use the Btu tabulation cooling (residential) form (Figure 6–4).

*Item 1, outside design temperature.* From Table 6–10, select the proper outside design temperature and the desired inside design temperature for the specific area in which the building is being constructed. Determine the temperature difference and record these data in the upper-right corner of the form.

*Item 2, humidity factor.* From Table 6–10, select the percent of outside relative humidity and convert it to the proper humidity factor, as shown in the lower-right corner of Figure 6–4. Record this factor on the form.

*Item 3, daily temperature range.* From Table 6–10, determine the daily temperature range and record it in the upper-right corner of Figure 6–4.

*Item 4, people load.* Determine the people load. Allow one person for each bedroom and three persons in the living room or den areas. Record this figure in the proper place on the form.

*Item 5, room dimensions.* List the dimensions of each room (length, width, and height). Determine the linear feet of exposed wall for each room and record these dimensions in the proper column.

*Item 6, ceiling and floor areas.* Determine the ceiling and floor area of each room (length by width). Record these dimensions in the proper column on the form.

*Item 7, building orientation and color.* Record the color of the roof and walls and the building orientation in the proper columns on the form.

*Item 8, attic ventilation.* Determine if the attic space ventilation is adequate. Make a note on the form stating whether or not the ventilation is adequate.

(BASED ON 24 HOUR PER DAY OPERATION OF EQUIPMENT)

Job Name __P. Smith__   Date __1/1/75__

Location __CINCINNATI  OHIO__   Computed By __C.L.S.__   Outside Humidity Factor __1.35__

Outside Design Temp __97°F.__
Inside Design Temp __75°F.__
Temp. Diff. __22°F.__
Daily Temp Range __20°F.__

| ROOM | LIVING ROOM FACTOR | AREA OR QUAN | BTUH | DINING ROOM AREA OR QUAN | BTUH | KITCHEN AREA OR QUAN | BTUH | BED ROOM 1 AREA OR QUAN | BTUH | BED ROOM 2 AREA OR QUAN | BTUH | BATH AREA OR QUAN | BTUH | BTUH TOTALS |
|---|---|---|---|---|---|---|---|---|---|---|---|---|---|---|
| ENTIRE HOUSE | | | | | | | | | | | | | | |
| Room Size (LxWxH) | | 40X28X8 | | | | | | | | | | | | |
| Linear Ft. Exposed Wall | | 136 | | | | | | | | | | | | |
| Floor-Ceiling Area | | 1120 | | | | | | | | | | | | |
| Windows DIRECTION N Shade | 21.5 | 12.5 | 268.75 | | | | | | | | | | | |
| Windows DIRECTION E | 41 | 36.25 | 1486.25 | | | | | | | | | | | |
| Windows DIRECTION W | 41 | 94.75 | 3884.75 | | | | | | | | | | | |
| Windows DIRECTION S | 35.5 | 30 | 1065 | | | | | | | | | | | |
| Walls Shade | 1.55 | 211.5 | 327.83 | | | | | | | | | | | |
| Walls Sunlit | 3.1 | 703 | 2179.3 | | | | | | | | | | | |
| Ceiling | 4.2 | 1120 | 4709 | | | | | | | | | | | |
| Cooking | | | 1200 | | | | 1200 | | | | | | | |
| People | 380 | 3+3 | 2280 | | | | | 1 | | 1 | | | | |
| Room Sensible Heat | | | 17396 | | | | | | | | | | | 17396 |
| Room CFM | | | | | | | | | | | | | | |

HOUSE TOTAL SENS. BTUH _____ 17396

DUCT GAIN __18__ % _____ 3131

VENTILATION LOAD BTUH _____ 3589

TOTAL SENSIBLE BTUH _____ 24111

GRAND TOT. = TOT. SENS. x HUM. FACTOR __1.35__ _____ 32550

**% DUCT GAIN**

| | 1 Story | 1½ Story | 2 Story |
|---|---|---|---|
| Ducts Insulated | 8 | 6 | 4 |
| Ducts Not Insulated | (18) | 15 | 12 |

NOTE: All ducts in attic spaces must be insulated with 3" minimum thickness insulation with vapor barrier. Ducts in crawl spaces and damp basement must be covered with 2" minimum thickness insulation with vapor barrier.
Do not figure duct gain for concrete slab under floor ducts.

**HUMIDITY FACTORS**

| % Outside Hum. | Below 40 | 40 to 45 | Above 45 |
|---|---|---|---|
| Factor | 1.25 | 1.30 | (1.35) |

**VENTILATION LOAD**

House Volume Cu. Ft. (LxWxH) x .40 BTUH/CU. Ft.

40X28X8 X .40 = 3584

**WINDOW & DOOR FACTORS**

| Temp. Diff. F | 15 | 20 | 25 | 30 |
|---|---|---|---|---|
| North | 12 | 18 | 25 | 31 |
| NE & NW | 23 | 29 | 36 | 42 |
| East & West | 32 | 38 | 44 | 51 |
| SE & SW | 35 | 41 | 48 | 55 |
| South | 26 | 32 | 39 | 45 |

Above Factors Assume Shades Or Venetian Blinds. If No Shades Double Factors.

For Outside Shading Or Awnings Multiply Factor x .60.

For Double Glazed Or Storm Windows Multiply Factor x .80.

NOTES:

PEOPLE LOAD — One Person in each bedroom and three persons in living room.

ROOF OVERHANG — South walls only. 36" overhang provides complete shade to wall and windows. Use shade values for windows and walls.

DOORS — Treat all outside doors as windows.

**WALL FACTORS**

Light Color

| Temp Diff. F | 15 | 20 | 25 | 30 |
|---|---|---|---|---|
| 15 Daily Temp. Range No Insul. | 3.6 | 5.4 | 7.2 | 9.0 |
| 1½" Insul. | 2.1 | 3.0 | 4.0 | 5.1 |
| 3" Insul. | 1.5 | 2.1 | 2.9 | 3.6 |
| 20 Daily Temp. Range No Insul. | 2.7 | 4.5 | 6.5 | 8.2 |
| 1½" Insul. | 1.5 | 2.6 | 3.6 | 4.7 |
| 3" Insul. | 0.9 | 1.8 | 2.6 | 3.3 |
| 25 Daily Temp. Range No Insul. | 1.4 | 3.6 | 5.4 | 7.4 |
| 1½" Insul. | 0.9 | 2.1 | 3.0 | 4.1 |
| 3" Insul. | 0.6 | 1.5 | 2.1 | 2.9 |

Dark Color

| Temp Diff. F | 15 | 20 | 25 | 30 |
|---|---|---|---|---|
| 15 Daily Temp. Range No Insul. | 5.1 | 6.9 | 8.9 | 10.7 |
| 1½" Insul. | 2.7 | 3.6 | 4.7 | 5.1 |
| 3" Insul. | 2.0 | 2.6 | 3.3 | 4.1 |
| 20 Daily Temp. Range No Insul. | 4.2 | 6.0 | 7.8 | 9.8 |
| 1½" Insul. | 2.1 | 3.2 | 4.2 | 5.1 |
| 3" Insul. | 1.5 | 2.3 | 3.0 | 3.6 |
| 25 Daily Temp. Range No Insul. | 3.3 | 5.1 | 6.9 | 8.9 |
| 1½" Insul. | 1.7 | 2.7 | 3.6 | 4.7 |
| 3" Insul. | 1.2 | 2.2 | 2.6 | 3.3 |

For Masonry Walls Multiply Factors x 1.2. For North or Shaded Wall Figure ½ Value in Table.

**CEILING FACTORS**

| Temp. Diff. F | 15 | 20 | 25 | 30 |
|---|---|---|---|---|
| 15 Daily Temp. Range 2" Insul. | 5.7 | 6.9 | 8.7 | 9.6 |
| 4" Insul. | 3.6 | 4.4 | 5.1 | 5.7 |
| 6" Insul. | 3.2 | 4.0 | 4.6 | 5.1 |
| 20 Daily Temp. Range 2" Insul. | 5.0 | 6.2 | 7.5 | 8.6 |
| 4" Insul. | 3.2 | 3.7 | 4.7 | 5.4 |
| 6" Insul. | 2.9 | 3.5 | 4.2 | 4.9 |
| 25 Daily Temp. Range 2" Insul. | 4.4 | 5.5 | 6.9 | 7.6 |
| 4" Insul. | 2.9 | 3.6 | 4.4 | 5.0 |
| 6" Insul. | 2.6 | 3.2 | 4.0 | 4.5 |

Factors assume dark color roof with attic having vents. If light color roof — take 75% of values in tables. If no vents in attic space double values.

*FIGURE 6-4* Btu Tabulation Cooling (Residential) (Courtesy of Amana Refrigeration, Inc.)

Note (The above list of factors are comprised of only those most commonly used.) For additional coefficients or more complete descriptions of these elements of structure, consult the Htg. Ventil. & Air Cond. Guide.

**TABLE 6–10**

Summer Weather Conditions

| State and city[a] | Outside % R.H. | Summer design dry bulb (°F) | Daily temp. range (°F) | Summer design wet bulb (°F) |
|---|---|---|---|---|
| Alabama | | | | |
| Birmingham | 49 | 94 | 20 | 78 |
| Mobile | 55 | 93 | 20 | 79 |
| Montgomery | 48 | 95 | 20 | 79 |
| Arizona | | | | |
| Flagstaff | 28 | 82 | 30 | 60 |
| Phoenix | 22 | 106 | 30 | 76 |
| Tucson | 21 | 102 | 25 | 73 |
| Winslow | 17 | 95 | 30 | 65 |
| Yuma | 22 | 109 | 25 | 78 |
| Arkansas | | | | |
| Fort Smith | 47 | 99 | 25 | 78 |
| Little Rock | 43 | 96 | 20 | 79 |
| California | | | | |
| Bakersfield | 19 | 101 | 30 | 71 |
| Fresno | 20 | 99 | 35 | 72 |
| Los Angeles, CO | 41 | 90 | 20 | 70 |
| Oakland | 47 | 81 | 20 | 63 |
| Sacramento | 24 | 97 | 35 | 70 |
| San Diego | 53 | 83 | 15 | 70 |
| San Francisco, CO | 34 | 77 | 20 | 62 |
| Colorado | | | | |
| Denver | 18 | 90 | 30 | 64 |
| Grand Junction | 15 | 94 | 30 | 63 |
| Pueblo | 24 | 94 | 30 | 67 |
| Connecticut | | | | |
| Hartford, Brainard Field | 62 | 88 | 20 | 76 |
| New Haven | 73 | 86 | 20 | 76 |
| D.C., Washington, National AP | 47 | 92 | 15 | 77 |
| Florida | | | | |
| Jacksonville | 50 | 94 | 20 | 79 |
| Miami | 63 | 90 | 15 | 79 |
| Tampa | 53 | 91 | 15 | 80 |
| Georgia | | | | |
| Atlanta | 47 | 92 | 20 | 77 |
| Macon | 50 | 96 | 20 | 79 |
| Savannah, Travis AP | 55 | 94 | 20 | 80 |
| Idaho | | | | |
| Boise | 20 | 93 | 30 | 66 |
| Idaho Falls | 28 | 88 | 40 | 64 |
| Illinois | | | | |
| Chicago, Midway | 43 | 92 | 20 | 76 |
| Moline | 46 | 91 | 25 | 77 |
| Peoria | 48 | 92 | 20 | 77 |
| Springfield | 47 | 92 | 20 | 78 |

**TABLE 6-10** (cont.)

Summer Weather Conditions

| State and city[a] | Outside % R.H. | Summer design dry bulb (°F) | Daily temp. range (°F) | Summer design wet bulb (°F) |
|---|---|---|---|---|
| Indiana | | | | |
| Evansville | 44 | 94 | 20 | 78 |
| Indianapolis | 57 | 91 | 20 | 77 |
| Iowa | | | | |
| Cedar Rapids | 48 | 90 | 20 | 76 |
| Des Moines | 49 | 92 | 20 | 77 |
| Marshalltown | 48 | 91 | 25 | 77 |
| Ottumwa | 48 | 93 | 20 | 78 |
| Sioux City | 48 | 93 | 20 | 77 |
| Kansas | | | | |
| Topeka | 40 | 96 | 25 | 78 |
| Wichita | 33 | 99 | 20 | 76 |
| Kentucky | | | | |
| Louisville | 55 | 93 | 20 | 78 |
| Louisiana | | | | |
| Lake Charles | 51 | 93 | 15 | 79 |
| New Orleans | 57 | 91 | 15 | 80 |
| Shreveport | 48 | 96 | 20 | 80 |
| Maine | | | | |
| Bangor, Dow AFB | 55 | 85 | 20 | 73 |
| Maryland | | | | |
| Baltimore, CO | 58 | 92 | 20 | 78 |
| Massachusetts | | | | |
| Boston | 53 | 88 | 15 | 74 |
| Michigan | | | | |
| Detroit | 44 | 88 | 20 | 75 |
| Sault Ste. Marie | 57 | 81 | 25 | 71 |
| Minnesota | | | | |
| Minneapolis | 52 | 89 | 20 | 75 |
| Rochester | 45 | 88 | 20 | 75 |
| Mississippi | | | | |
| Biloxi, Keesler AFB | 63 | 92 | 15 | 81 |
| Jackson | 50 | 96 | 20 | 78 |
| Missouri | | | | |
| Kansas City | 37 | 97 | 20 | 77 |
| St. Louis, CO | 40 | 94 | 15 | 78 |
| Springfield | 37 | 94 | 20 | 77 |
| Montana | | | | |
| Billings | 28 | 91 | 30 | 66 |
| Butte | 21 | 83 | 30 | 59 |
| Great Falls | 20 | 88 | 30 | 63 |
| Nebraska | | | | |
| Lincoln, CO | 37 | 96 | 25 | 77 |
| North Platte | 29 | 94 | 25 | 73 |
| Omaha | 40 | 94 | 20 | 78 |

**TABLE 6–10** (cont.)

Summer Weather Conditions

| State and city[a] | Outside % R.H. | Summer design dry bulb (°F) | Daily temp. range (°F) | Summer design wet bulb (°F) |
|---|---|---|---|---|
| Nevada | | | | |
| Las Vegas | 16 | 106 | 30 | 71 |
| Reno | 13 | 92 | 30 | 62 |
| New Hampshire | | | | |
| Concord | 55 | 88 | 25 | 73 |
| New Jersey | | | | |
| Newark | 51 | 91 | 20 | 76 |
| Trenton | 53 | 90 | 20 | 77 |
| New Mexico | | | | |
| Albuquerque | 18 | 94 | 25 | 65 |
| Roswell, Walker AFB | 22 | 99 | 25 | 70 |
| New York | | | | |
| Albany | 50 | 88 | 20 | 74 |
| Buffalo | 53 | 86 | 20 | 73 |
| Elmira | 56 | 90 | 20 | 73 |
| New York, La Guardia AP | 47 | 90 | 15 | 76 |
| Rochester | 53 | 88 | 20 | 74 |
| Syracuse | 57 | 87 | 20 | 74 |
| North Carolina | | | | |
| Charlotte | 52 | 94 | 20 | 77 |
| Greensboro | 54 | 91 | 20 | 76 |
| Raleigh/Durham | 49 | 92 | 20 | 78 |
| Wilmington | 66 | 91 | 15 | 81 |
| North Dakota | | | | |
| Bismarck | 38 | 91 | 30 | 72 |
| Fargo | 37 | 88 | 25 | 74 |
| Grand Forks | 41 | 87 | 25 | 72 |
| Ohio | | | | |
| Akron/Canton | 57 | 87 | 20 | 73 |
| Cincinnati, CO | 46 | 92 | 20 | 77 |
| Cleveland | 43 | 89 | 20 | 75 |
| Columbus | 52 | 88 | 20 | 76 |
| Dayton | 48 | 90 | 20 | 75 |
| Youngstown | 60 | 86 | 20 | 74 |
| Oklahoma | | | | |
| Ardmore | 43 | 101 | 20 | 78 |
| Oklahoma City | 35 | 97 | 20 | 77 |
| Tulsa | 37 | 99 | 20 | 78 |
| Oregon | | | | |
| Baker | 27 | 92 | 30 | 65 |
| Eugene | 38 | 88 | 30 | 67 |
| Medford | 28 | 94 | 35 | 68 |
| Portland | 33 | 88 | 20 | 68 |
| Pennsylvania | | | | |
| Erie | 66 | 85 | 20 | 74 |
| Harrisburg | 50 | 89 | 20 | 75 |
| Philadelphia | 56 | 90 | 20 | 77 |
| Pittsburg | 46 | 87 | 20 | 74 |

**TABLE 6-10** (cont.)
Summer Weather Conditions

| State and city[a] | Outside % R.H. | Summer design dry bulb (°F) | Daily temp. range (°F) | Summer design wet bulb (°F) |
|---|---|---|---|---|
| Rhode Island | | | | |
| Providence | 47 | 86 | 15 | 75 |
| South Carolina | | | | |
| Charleston, AFB | 57 | 92 | 15 | 80 |
| Columbia | 57 | 96 | 20 | 79 |
| South Dakota | | | | |
| Huron | 33 | 93 | 30 | 75 |
| Rapid City | 28 | 94 | 25 | 71 |
| Tennessee | | | | |
| Chattanooga | 48 | 94 | 20 | 78 |
| Knoxville | 43 | 92 | 20 | 76 |
| Memphis | 44 | 96 | 20 | 79 |
| Texas | | | | |
| Abilene | 34 | 99 | 20 | 75 |
| Amarillo | 29 | 96 | 25 | 71 |
| Austin | 40 | 98 | 20 | 78 |
| Brownsville | 54 | 92 | 15 | 80 |
| Dallas | 38 | 99 | 20 | 78 |
| El Paso | 22 | 98 | 25 | 69 |
| Ft. Worth | 32 | 100 | 20 | 78 |
| Houston | 47 | 94 | 15 | 80 |
| Laredo, AFB | 40 | 101 | 20 | 78 |
| San Antonio | 37 | 97 | 20 | 77 |
| Utah | | | | |
| Ogden, CO | 25 | 92 | 30 | 64 |
| Salt Lake City | 18 | 94 | 30 | 66 |
| Vermont | | | | |
| Burlington | 50 | 85 | 25 | 73 |
| Virginia | | | | |
| Norfolk | 49 | 91 | 15 | 78 |
| Richmond | 57 | 93 | 20 | 78 |
| Roanoke | 53 | 91 | 20 | 75 |
| Washington | | | | |
| Seattle, CO | 42 | 76 | 20 | 65 |
| Spokane | 20 | 90 | 25 | 64 |
| West Virginia | | | | |
| Charleston | 45 | 90 | 20 | 75 |
| Wisconsin | | | | |
| Green Bay | 42 | 85 | 20 | 73 |
| La Crosse | 46 | 88 | 20 | 76 |
| Madison | 58 | 88 | 20 | 75 |
| Milwaukee | 54 | 87 | 20 | 75 |
| Wyoming | | | | |
| Cheyenne | 22 | 86 | 25 | 62 |
| Rock Springs | 13 | 84 | 25 | 57 |

[a]All dry bulb temperatures taken at airport unless otherwise noted: CO, urban area; AFB, air force base.
*Source:* Amana Refrigeration, Inc.

**TABLE 6–11**

Window and Door Factors

| | Temperature difference (°F) | | | |
|---|---|---|---|---|
| | 15 | 20 | 25 | 30 |
| North | 12 | 18 | 25 | 31 |
| NE and NW | 23 | 29 | 36 | 42 |
| East and West | 32 | 38 | 44 | 51 |
| SE and SW | 35 | 41 | 48 | 55 |
| South | 26 | 32 | 39 | 45 |

*Source:* Amana Refrigeration, Inc.

*Item 9, heat transfer (Btu) factors.* Select the proper Btu/h factors for the windows, doors, exposed walls, and ceiling (Tables 6–11 through 6–13). Record these factors in the proper columns on the form. If the building has a 36″ overhang on the southern exposure, consider the south wall and windows as shaded and use the north wall and north window factors. All window factors are based on inside shaded or venetian blinds fully drawn. If the windows have no inside shades, double the window factor values. If the windows have outside full shading, such as awnings or closed white blinds,

**TABLE 6–12**

Wall Factors[a]

| | | Temperature difference (°F) for: | | | | | | | |
|---|---|---|---|---|---|---|---|---|---|
| | | Light color | | | | Dark color | | | |
| | | 15 | 20 | 25 | 30 | 15 | 20 | 25 | 30 |
| 15° daily | No insulation | 3.6 | 5.4 | 7.2 | 9.0 | 5.1 | 6.9 | 8.9 | 10.7 |
| range | $3\frac{5}{8}$ in. of insulation | 1.5 | 2.1 | 2.9 | 3.6 | 2.0 | 2.6 | 3.3 | 4.1 |
| 20° daily | No insulation | 2.7 | 4.5 | 6.5 | 8.2 | 4.2 | 6.0 | 7.8 | 9.8 |
| range | $3\frac{5}{8}$ in. of insulation | 1.1 | 1.8 | 2.6 | 3.3 | 1.5 | 2.3 | 3.0 | 3.6 |
| 25° daily | No insulation | 1.4 | 3.6 | 5.4 | 7.4 | 3.3 | 5.1 | 6.9 | 8.9 |
| range | $3\frac{5}{8}$ in. of insulation | 0.6 | 1.5 | 2.1 | 2.9 | 1.2 | 2.0 | 2.6 | 3.3 |

[a]For masonry walls, multiply factors by 1.2; for north or shaded walls, use $\frac{1}{2}$ value in table.
*Source:* Amana Refrigeration, Inc.

**TABLE 6–13**

Ceiling Factors[a]

| | | Temperature difference (°F) | | | |
|---|---|---|---|---|
| | | 15 | 20 | 25 | 30 |
| 15° daily | 4 in. of insulation | 3.6 | 4.4 | 5.1 | 5.7 |
| range | 6 in. of insulation | 3.2 | 4.0 | 4.6 | 5.1 |
| 20° daily | 4 in. of insulation | 3.2 | 3.7 | 4.7 | 5.4 |
| range | 6 in. of insulation | 2.9 | 3.5 | 4.2 | 4.9 |
| 25° daily | 4 in. of insulation | 2.9 | 3.6 | 4.4 | 5.0 |
| range | 6 in. of insulation | 2.6 | 3.2 | 4.0 | 4.5 |

[a]Factors assume dark-color roof with attic having adequate vents. If light-color roof, take 75% of values in tables. If no vents in attic space, double values.
*Source:* Amana Refrigeration, Inc.

**TABLE 6–14**

Percentage of Duct Gain[a]

|  | 1-story | 1½-story | 2-story |
|---|---|---|---|
| Ducts insulated | 8 | 6 | 4 |
| Ducts uninsulated | 18 | 15 | 12 |

[a]All ducts in attic spaces must be insulated with 3 in. minimum thickness insulation with vapor barrier. Ducts in crawl spaces and damp basements must be covered with 2 in. minimum thickness insulation with vapor barrier. Do not figure duct gain for concrete slab under floor ducts.

*Source:* Amana Refrigeration, Inc.

use 60% of the window factor values. If the windows are double glazed or are covered with storm windows, use 80% of the window factor value.

*Item 10, areas for windows, doors, exposed walls, and ceilings.* List the areas for these elements in the Area or Quantity column. The area of the exposed walls to be listed is the total area of the wall less the area of windows and doors in the wall. The factor is in Btu/h and is to be recorded in the Btuh column.

*Item 11, heat gains.* The Btu/h heat gains for each room are totaled in the Btuh Totals column, and the result is recorded in the House Total Sensible Btuh space.

*Item 12, percent of duct gain.* Select the proper factor from Table 6–14 and multiply this factor by the house total sensible Btu/h.

*Item 13, fresh-air requirements.* Determine the fresh-air requirements or ventilation load by multiplying the total building volume (length × width × height = cubic feet) by 0.4. This gives you the total Btu/h load for the ventilation requirements. The results should be listed in the Ventilation Load Btuh space on the form.

*Item 14, total house sensible Btu/h.* The sum of the house total sensible Btu/h the percent of duct gain Btu/h and the ventilating load

Btu/h equals the total Btu/h sensible load. Add these items together and enter the result in the proper column.

*Item 15, humidity factor.* Select the proper humidity factor from Table 6–15. Multiply this factor by the total sensible Btu/h load. These results are the grand total load.

*Btu/h Factors for Calculating Residental Heat Gain:* All factors contained on the Btu tabulation cooling (residential) form for estimating are expressed in Btu/h.

**STEP 1** ————————

The fresh-air requirements (ventilation) are determined from the volume of the building (length × width × height = cubic feet) multiplied by 0.40.

**STEP 2** ————————

The people load is determined by allowing one per-

**TABLE 6–15**

Humidity Factors[a]

| Percent outside humidity | Factor |
|---|---|
| Below 40 | 1.25 |
| 40 to 45 | 1.30 |
| Above 45 | 1.35 |

[a]Apply to total sensible Btu/h.

*Source:* Amana Refrigeration, Inc.

son per bedroom and three persons in the living-room areas. Multiply the number of people by 380.

## STEP 3
The cooking load in the kitchen area is a constant factor of 1200 Btu/h.

## STEP 4
The shading (roof overhang) is considered on the south wall only if the overhang is 36 in. or more. A 36 in. overhang provides a complete shade to the wall, windows, and doors. Use as factors for shaded areas the north glass and north wall area factors. Shaded wall area factors are one-half the value of the factors shown in Table 6–12.

## STEP 5
Treat all outside door areas as windows.

*Room CFM Factor:* Before calculating the cubic feet per minute (cfm) required for each room, it is necessary to select the proper size of cooling equipment to satisfy the grand total heat gain load. It is important that cooling-equipment size closely match the grand total heat gain load in order to obtain the best results. If this figure is between the capacity ratings of the cooling equipment, choose the higher-capacity equipment. For example, if the total heat gain load equals 26,500 Btu/h, select the 2.5-ton (29,000 Btu/h) equipment.

The amount of cfm to be delivered by the cooling equipment depends on the range of the moisture load encountered in the area being considered. Large volumes of air are usually recommended for dry and arid areas, and smaller volumes are usually required for wet and humid areas. Table 6–10 gives the outside percent of relative humidity for most areas in the country. The humidity factor table (Table 6–15) is based upon the outside percent of relative humidity. It is suggested that the following cfm be used, based on the humidity conditions and the tonnage capacity of the cooling equipment:

| Humidity Conditions | CFM per Ton |
|---|---|
| Dry to arid: below 40% RH | 450 |
| Medium: 40 to 50% Rh | 400 |
| Wet to humid: above 45% RH | 350 |

The cfm required per room is determined by proportioning the room sensible heat gain Btu/h to the total cfm to be delivered by the cooling equipment.

## EXAMPLE 1
A 2.5-ton cooling unit is selected for a wet area. The total cfm required by the unit is approximately 900 (2.5 × 350 = 875 rounded off to 900).

Total house sensible heat gain Btu/h = 20,700
Room sensible heat gain Btu/h = 4180

$$\text{Room cfm} = \frac{4180}{20,700} \times 900 = 182 \text{ cfm}$$

## EXAMPLE 2
A 2.5-ton cooling unit is selected for a dry area. The total cfm required by the unit is 1125 (2.5 × 450 = 1125).

Total house sensible heat gain Btu/h = 20,700
Room sensible heat gain Btu/h = 4180

$$\text{Room cfm} = \frac{4180}{20,700} \times 1125 = 227 \text{ cfm}$$

*Heat Gain Calculation Example.* In our example of calculating heat loss we used a single room. In this example we will use the house-as-a-cube method. We will use the Btu tabulation cooling (residential) form (Figure 6–5) and the house floor plan in Figure 6–6 to complete our example. This method is used to select cooling equipment by quick cost estimating. A complete estimation would involve the building on a room by room basis. *Note:* The house is in Cincinnati, Ohio, and the building construction data are listed on the floor plan of Figure 6–6.

Job Name: P. SMITH  Date: 1/1/75  Computed By: C.L.S.

Location: CINCINNATI OHIO  Outside Humidity Factor: 1.35

Outside Design Temp: 97°F  
Inside Design Temp: 75°F  
Temp. Diff.: 22°F  
Daily Temp Range: 20°F

**ENTIRE HOUSE** (LIVING ROOM column)

| ROOM | FACTOR | AREA OR QUAN | BTUH |
|---|---|---|---|
| Room Size (LxWxH) | | 40X28X8 | |
| Linear Ft. Exposed Wall | | 136 | |
| Floor Ceiling Area | | 1120 | |
| Windows — N | 21.5 | 12.5 | 268.75 |
| Windows — E | 41 | 36.25 | 1486.25 |
| Windows — W | 41 | 94.75 | 3884.75 |
| Windows — S | 36.5 | 30 | 1065 |
| Walls — Shade N | 1.55 | 211.5 | 327.83 |
| Walls — Sunit N | 3.1 | 703 | 2179.3 |
| Ceiling | 4.2 | 1120 | 4704 |
| Cooking | | | 1200 |
| People | 380 | 3+3 | 2280 |
| Room Sensible Heat | | | 17396 |
| Room CFM | | | |

(Other room columns — DINING ROOM, KITCHEN, BED ROOM 1, BED ROOM 2, BATH — each with AREA OR QUAN and BTUH: "1200" noted under People area; "1" noted under People for BED ROOM 1 and BED ROOM 2.)

| | | BTUH TOTALS |
|---|---|---|
| HOUSE TOTAL SENS. BTUH | | 17396 |
| DUCT GAIN  18 % | | 3131 |
| VENTILATION LOAD BTUH | | 3584 |
| TOTAL SENSIBLE BTUH | | 24111 |
| GRAND TOT. = TOT. SENS. x HUM. FACTOR  1.35 | | 32550 |

**% DUCT GAIN**

| | 1 Story | 1½ Story | 2 Story |
|---|---|---|---|
| Ducts Insulated | 8 | 6 | 4 |
| Ducts Not Insulated | (18) | 15 | 12 |

NOTE: All ducts in attic spaces must be insulated with 3" minimum thickness insulation with vapor barrier. Ducts in crawl spaces and damp basements must be covered with 2" minimum thickness insulation with vapor barrier.
Do not figure duct gain for concrete slab under floor ducts.

**HUMIDITY FACTORS**

| % Outside Hum. | Below 40 | 40 to 45 | Above 45 |
|---|---|---|---|
| Factor | 1.25 | 1.30 | (1.35) |

**VENTILATION LOAD**

House Volume Cu. Ft. (LxWxH) x 40 BTUH/CU. Ft.

40X28X8 X .40 = 3584

**WINDOW & DOOR FACTORS**

| Temp. Diff. F | 15 | 20() | 25 | 30 |
|---|---|---|---|---|
| North | 12 | 18() | 25 | 31 |
| NE & NW | 23 | 29 | 36 | 42 |
| East & West | 32 | 38() | 44 | 51 |
| SE & SW | 35 | 41 | 48 | 55 |
| South | 26 | 32() | 39 | 45 |

Above Factors Assume Shades Or Venetian Blinds. If No Shades Double Factors.

For Outside Shading Or Awnings Multiply Factor x .60.

NOTES:
PEOPLE LOAD — One Person in each bedroom and three persons in living room.

ROOF OVERHANG — South walls only. 36" overhang provides complete shade to wall and windows. Use shade values for windows and walls.

DOORS — Treat all outside doors as windows

**WALL FACTORS**

| Temp. Diff. F | 15 | 20() | 25 | 30 |
|---|---|---|---|---|
| **15 Daily Temp. Range** | | | | |
| No Insul. | 3.6 | 5.4 | 7.2 | 9.0 |
| 1½" Insul. | 2.1 | 3.0 | 4.0 | 5.1 |
| 3" Insul. | 1.5 | 2.1 | 2.9 | 3.6 |
| **20 Daily Temp. Range** | | | | |
| No Insul. | 2.7 | 4.5 | 6.5 | 8.2 |
| 1½" Insul. | 1.5 | 2.6() | 3.6 | 4.7 |
| 3" Insul. | 1.1 | 1.8 | 2.6 | 3.3 |
| **25 Daily Temp. Range** | | | | |
| No Insul. | 1.4 | 3.6 | 5.4 | 7.4 |
| 1½" Insul. | 0.9 | 2.1 | 3.0 | 4.1 |
| 3" Insul. | 0.6 | 1.5 | 2.1 | 2.9 |

For Masonry Walls Multiply Factors x 1.2. For North or Shaded Wall Figure ½ Value in Table.

**CEILING FACTORS**

| | Light Color | | | | Dark Color | | | |
|---|---|---|---|---|---|---|---|---|
| Temp. Diff. F | 15 | 20() | 25 | 30 | 15 | 20() | 25 | 30 |
| **15 Daily Temp. Range** | | | | | | | | |
| 2" Insul. | 5.1 | 6.9 | 8.9 | 10.7 | 5.7 | 6.9 | 8.7 | 9.6 |
| 4" Insul. | 2.7 | 3.6 | 4.7 | 5.1 | 3.6 | 4.4 | 5.1 | 5.7 |
| 6" Insul. | 2.0 | 2.6 | 3.3 | 4.1 | 3.2 | 4.0 | 4.6 | 5.1 |
| **20 Daily Temp. Range** | | | | | | | | |
| 2" Insul. | 4.2 | 6.0 | 7.8 | 9.8 | 5.0 | 6.2 | 7.5 | 8.6 |
| 4" Insul. | 2.1 | 3.0 | 4.2 | 5.1 | 3.2 | 3.7() | 4.7 | 5.4 |
| 6" Insul. | 1.5 | 2.3 | 3.0 | 3.6 | 2.9 | 3.5 | 4.2 | 4.9 |
| **25 Daily Temp. Range** | | | | | | | | |
| 2" Insul. | 3.3 | 5.1 | 6.9 | 8.9 | 4.4 | 5.5 | 6.9 | 7.6 |
| 4" Insul. | 1.7 | 2.7 | 3.6 | 4.7 | 2.9 | 3.6 | 4.4 | 5.0 |
| 6" Insul. | 1.2 | 2.0 | 2.6 | 3.3 | 2.6 | 3.2 | 4.0 | 4.5 |

Factors assume dark color roof with attic having vents. If no vents in attic
If light color roof — take 75% of values in tables.

For Double Glazed Or Storm Windows Multiply Factor x .80.

FIGURE 6-5 Btu Tabulation Cooling (Residential) (Courtesy of Amana Refrigeration, Inc.)

Note (The above list of factors are comprised of only those most commonly used.) For additional coefficients or more complete descriptions of these elements of structure, consult the Htg. Ventil. & Air Cond. Guide.

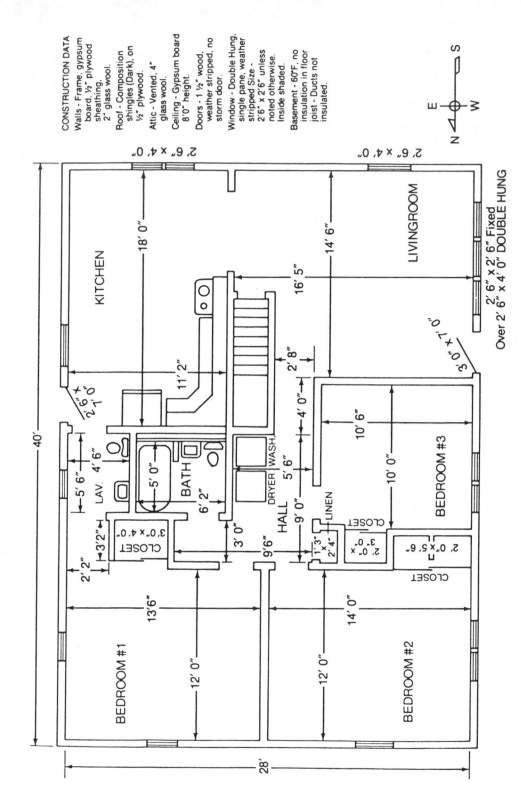

CONSTRUCTION DATA

Walls - Frame, gypsum board, ½" plywood sheathing, 2" glass wool.

Roof - Composition shingles (Dark), on ½" plywood.

Attic - Vented, 4" glass wool.

Ceiling - Gypsum board 8'0" height.

Doors - 1 ½" wood, weather stripped, no storm door.

Window - Double Hung, single pane, weather stripped Size - 2'6" x 2'6" unless noted otherwise. Inside shaded.

Basement - 60°F. no insulation in floor joist - Ducts not insulated.

*FIGURE 6-6* Basic Floor Plan (Courtesy of Amana Refrigeration, Inc.)

## STEP 1 ———————————————

*Job Name and Location:* Fill in the job name and location spaces provided on the form. The date and the estimator's initials should also be recorded.

## STEP 2 ———————————————

*Design conditions:* In Table 6-10 the established design conditions for Cincinnati, Ohio, are 97 °F dry bulb temperature, 79 °F wet bulb temperature, 46% relative humidity, 22 °F design temperature difference suggested, and a 20 °F daily temperature range. These conditions should be entered in the space provided in the upper-right corner of the form. A 22 °F temperature reduction is advised for this locality; therefore, the inside design temperature is equal to 97 °F − 22 °F = 75 °F. The design relative humidity is given as 46%; thus a factor of 1.35 should be selected from the Table 6-15 and entered in the proper space.

## STEP 3 ———————————————

*Building Volume:* The total length, width, and height of the house is next entered on the line labeled Room Size under any column labeled (your choice) Entire House. In this example the building is 40′ long and 28′ wide with an 8′ ceiling (40′ × 28′ × 8′ = 8960). Next the linear feet of exposed wall is recorded. In our example, figuring the entire building, the linear feet of exposed wall is equal to the perimeter of the house or 136 linear feet (40′ + 40′ + 28′ = 136′). Next calculate the floor area, which would be the product of the length and width of the building (40′ × 28′ = 1120 square feet).

## STEP 4 ———————————————

*Determining the Heat Gain Factors:* The factors for calculating the heat gain can now be selected by taking the construction data that are noted on the floor plan (Figure 6-6) and the house orientation, and referring to the proper table for this type of construction. These factors are determined by locating the temperature difference factor of 22 °F on the window factor table (Table 6-11). For example, for north windows, we would first note that 22 °F is not listed. It is approximately halfway between 20 and 25 in the table. The factor for 20° is 18 and the factor for 25° is 25. Therefore, to establish our factor, we must interpolate by subtracting 18 from 25, which is 7; divide 7 by 2 (= 3.5); and add 18 (= 21.5). The same procedure will be used on all factor calculations throughout this example.

$$
\begin{aligned}
\text{Windows, north} &= 21.5 \\
\text{Windows, east} &= 41 \\
\text{Windows, west} &= 41 \\
\text{Windows, south} &= 35.5 \\
\text{Wall, north (shade)} &= 1.55 \\
\text{Walls, sunlit} &= 3.1 \\
\text{Ceiling} &= 4.2
\end{aligned}
$$

Note that these factors for windows and doors assume a shaded condition or venetian blinds, and, in accordance with our basic floor plan, these have inside shades. If no shades were used, these factors would be doubled. Enter these figures in the proper column on the tabulation chart.

The north wall factor is one-half the value we found in Table 6-12, since it is a shaded wall. This is in accordance with the footnote below Table 6-1

## STEP 5 ———————————————

*Determining the Areas or Quantity:* Now that we have determined the factors, we can determine and enter the area or quantity for each required part of the building.

The height of all windows in feet is multiplied by their width in feet and added together for the total area for a given exposure. Note that doors are handled as windows.

*North windows:* 2.5′ × 2.5′ = 6.25 square feet each. There are two north windows; therefore, 2 × 6.25 = 12.5 square feet.

*East windows:* 2.5′ × 2.5′ = 6.25 square feet each. There are three east windows; therefore, 3 × 6.25 = 18.75 square feet.

*East door:* 2.5′ × 7′ = 17.5 square feet.

The total of the east doors and the windows is 17.5 + 18.75 = 36.25 square feet.

*West windows:* 2.5′ × 2.5′ = 6.25 square feet each. There are seven west windows of this size; therefore, 7′ × 6.25 = 43.75 square feet. Also, 2.5′ × 4′ = 10 square feet each. There are a total of 3 west windows of this size; therefore, 10 × 3 = 30 square feet.

*West door:* 3′ × 7′ = 21 square feet.

The total of the west doors and windows is $43.75 + 30 + 21 = 94.75$ square feet.

*South windows:* $2.5' \times 4' = 10$ square feet each. There are three south windows; therefore, $10 \times 3 = 30$ square feet.

The height of the walls in feet is multiplied by their length in feet. The next wall area is then found by subtracting any doors and windows in the wall under consideration.

*Wall, shaded (north):* $8' \times 28' = 224$ square feet $- 12.5$ square feet $= 211.5$ square feet.

*Walls, sunlit (east):* $8' \times 40' = 320$ square feet $- 36.25 = 283.75$ square feet.

*Walls, sunlit (west):* $8' \times 40' = 320$ square feet $- 94.75$ square feet $= 225.25$ square feet.

*Walls, sunlit (south):* $8' \times 28' = 224$ square feet $- 30$ square feet $= 194$ square feet.

The total of sunlit walls is $283.75 + 225.25 + 194 = 703$ square feet.

The ceiling width in feet is multiplied by the length in feet for the total area.

*Ceiling:* $28' \times 40' = 1120$ square feet.

Enter these figures in the proper column in the tabulation chart.

## STEP 6

*Determining Heat Gains for Various Items:* Having determined both the factors and the areas, multiply them to determine the Btu/h heat gains for the various items.

Windows, north $= 21.5 \times 12.5 = 268.75$ Btu/h
Windows, east $= 41 \times 36.5 = 1486.25$ Btu/h
Windows, west $= 41 \times 94.75 = 3804.75$ Btu/h
Windows, south $= 35.5 \times 30 = 1065.00$ Btu/h
Walls, shaded $= 1.55 \times 211.5 = 327.83$ Btu/h
Walls, sunlit $= 3.1 \times 703 = 2179.30$ Btu/h
Ceiling $= 4.2 \times 1120 = 4704.00$ Btu/h
Cooking $= 1200.00$ Btu/h

This 1200.00 Btu/h is taken as the kitchen cooking heat gain, and because the kitchen is part of the home, this quantity will be shown in the appropriate column under the entire house.

People $= 380$ (factor) $\times 6 = 2280.00$ Btu/h

A total of six people has been determined as the occupant load for this home with a factor of 380 per person. Three people in the living room and one for each bedroom.

Total heat gain load $= 17,395.88$ Btu/h

The Btu/h column should be added to determine the total Btu/h sensible heat gain for the entire building and entered on the Room Sensible Heat column.

## STEP 7

*Determining the Sensible Heat Gain:* The sensible heat gain should be entered in the House Total Sensible Btu/h column on the right side of the form.

The duct heat gain percentage should be obtained from Table 6–14. (Ducts not insulated in a single story building $= 18\%$.) The figure obtained from the table should be entered in the Duct Gain Percent column and multiplied by the building total sensible Btu/h ($18\% \times 17,396 = 3131$).

The ventilation heat load Btu/h is obtained by multiplying the cubic feet of the building by 0.40 ($40' \times 28' \times 8' \times 0.40 = 3584$ Btu/h). Enter this figure in the Ventilation Load column.

## STEP 8

*Determining the Total Sensible Btu/h:* The total sensible Btu/h is the sum of all the other calculations:

Total building sensible Btu/h $= 7,396$
Duct percent gain of 18% $= 3,131$
Ventilation load Btu/h $= 3,504$
Total sensible heat Btu/h $= 34,111$

## STEP 9

*Determining the Grand Total Btu/h:* The grand total is equal to the total sensible heat times the humidity factor (figured in step 2): $24,111 \times 1.35 = 32,550$ Btu/h. Enter this figure in the space provided on the form.

The preceding steps cover the entire house as a cube. The same basic instruction would also hold true for calculations done on a room-by-room basis, which is the preferred method because it gives the heat gain of each room.

Thus, the required amount of cooling can be directed to the proper room by the air-distribution system.

## HEAT PUMP SELECTION ———

The following are some of the more pertinent points that will be covered in detail in the following pages to help you to select a good heat pump system.

*Equipment selection.* The equipment selected must match the building heat loss and heat gain. To acomplish this, consideration must be given to the following points:

1. Load calculations must be completed and analyzed.
2. The building must be applied to an equipment performance graph.
3. All the controls must be set at the proper design points.

*Equipment application.* The proper application of equipment is very important. To be certain that the equipment is properly applied, give consideration to the following points:

1. Use only one manufacturer's equipment because both the indoor and outdoor sections have been designed to operate together. Mismatching equipment will prevent efficient operation and cause reduced economy.
2. If at all possible, use quick-connect tubing. This procedure prevents foreign material from entering the system during installation and provides a proper charge of refrigerant.
3. Provide proper air volume. It is best for optimum efficiency to have the proper amount of air flowing through the system.
4. Use emergency heat thermostats. This provides the customer with greater flexibility in system operation.

*Informing the homeowner.* The homeowner should be informed of how the equipment performs as compared to standard equipment. The following are some of the major points to cover:

1. The lower temperature of the air leaving a heat pump in the heating mode as compared to a gas-fired furnace.
2. How to obtain proper control of the indoor and outdoor thermostats.
3. The continuous operation of the compressor at temperatures below the system balance point.

*Selecting the Equipment:* To select the equipment, use the following steps:

1. Both the heat loss and heat gain should be calculated as was done earlier in this chapter.
2. The load form should be analyzed to evaluate if there are areas where a change in the home construction would reduce the equipment size and, therefore, the operating cost.
3. Select the equipment by using the total heat gain (cooling) to match the equipment capacity.
4. When the equipment has been selected, the building heat loss should be plotted onto heat graphs for the particular equipment selected. The system balance point should be below 35 °F (1.67 °C) outside ambient temperature.
5. In using the load calculations that were

figured earlier, select the heat pump equipment from the manufacturer's specification. For example, if we had a house total heat gain of 31,983 Btu and chose to use Amana equipment, there would be three different units that would satisfy our needs:

a. CRH3 and AFCH 3015 with a capacity at 97 °F (36.11 °C) = 33,660 Btu/h.

b. CRH3 and VBCH 35X with a capacity at 97 °F (36.11 °C) = 34,700 Btu/h.

c. PKH3 with a capacity at 97 °F (36.11 °C) = 34,375 Btu/h.

6. To determine the balance point of our system, which should be below 35 °F (1.67 °C) outside ambient temperature for efficient operation, we plot out all these units on the graphs (Figures 6–7 through 6–9). To plot each graph, we assume a building heat loss calculation of 47,312 Btu/h. (The balance point is the temperature below which the heat pump alone will not provide enough heat to warm the building.) To draw in our structure heat loss line, we use two conditions: (1) the total heat loss (47,312) at the outside design temperature of − 5 °F (− 20.55 °C) and (2) the inside design temperature of 70 °F (21.1 °C). We place a dot on the 0 Btu/h line where it intersects the 70 °F (21.1 °C) temperature line and another dot at the 47,312 Btu/h line where it intersects the − 5 °F (− 20.55 °C) temperature line. These two dots are then connected with a straight line, which we will call the *structure heat loss line*. This line represents the amount of heat required to heat the structure to 70 °F (21.1 °C) at the various outdoor ambient temperatures.

The point where the structure heat loss line intersects the compressor line is the *system balance point;* this is the point where the heat pump capacity is equal to the structure heat loss. If the outdoor ambient temperature continues to fall, we will not have enough capacity to heat the structure without including some supplementary heat in the form of electric resistance heaters. The systems we found that could meet our needs had balance points as follows:

a. CRH3 and AFCH 3015 system balance point = 27 °F (− 2.8 °C).

b. CRH3 and VBCH 35X system balance point = 25.5 °F (− 3.6 °C).

c. PKH3 system balance point = 28.5 °F (− 1.95 °C).

Be sure to locate these points on the appropriate graph for each system.

7. We can continue to plot the system operation on each of the graphs to determine at what temperature additional heat is required, what device (indoor or outdoor thermostat) should control the additional heat, the total number of auxiliary heaters required, the total number of controls, and at what temperature to set these controls. You will note in all instances that 15 kW of electric resistance heat is needed. In some instances, outdoor temperature controls (ODT) would be used, but their setting will vary from unit to unit.

8. At this point, if we analyzed the heat load calculation, as suggested in paragraph 2, and made only two structure changes, storm windows and doors and additional insulation in the attic (up to an R-10 factor), we could reduce the equipment size by $\frac{1}{2}$ ton. This in turn would reduce the duct size, wire size, and so on, which would lower the cost for the equipment and installation, plus allow for a lower operating cost for the entire season.

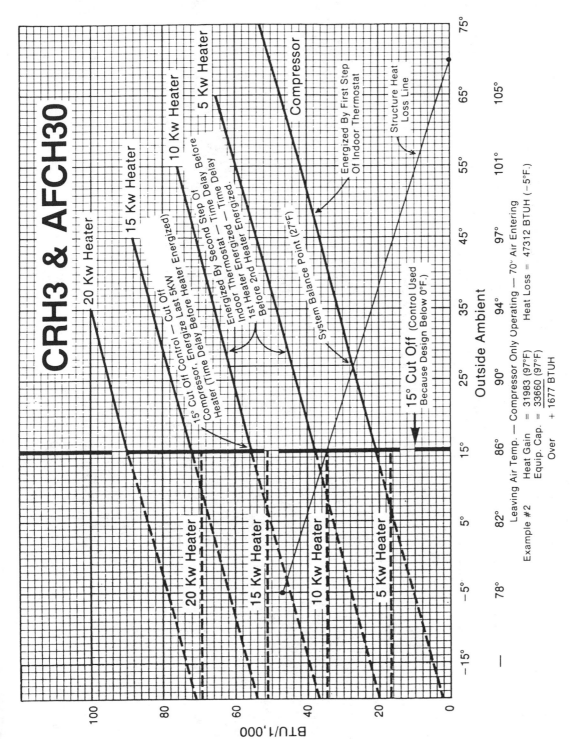

**FIGURE 6-7** Calculating System Balance Point (Courtesy of Amana Refrigeration, Inc.)

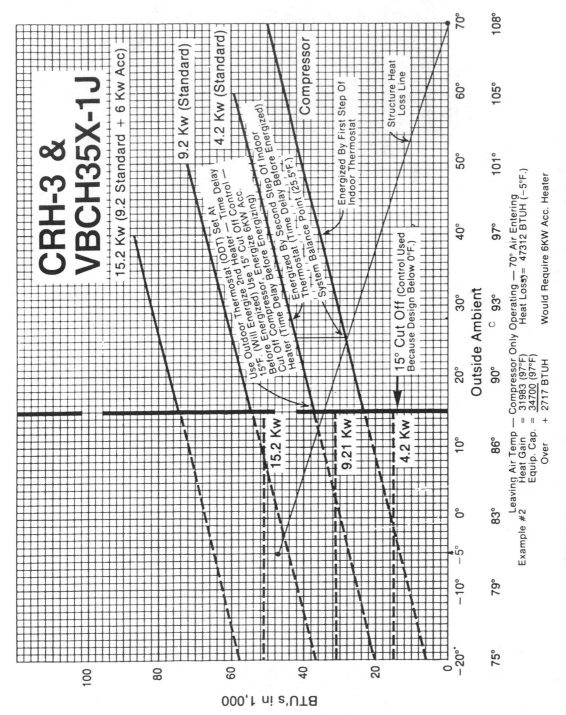

**FIGURE 6-8** Calculating System Balance Point (Courtesy of Amana Refrigeration, Inc.)

136

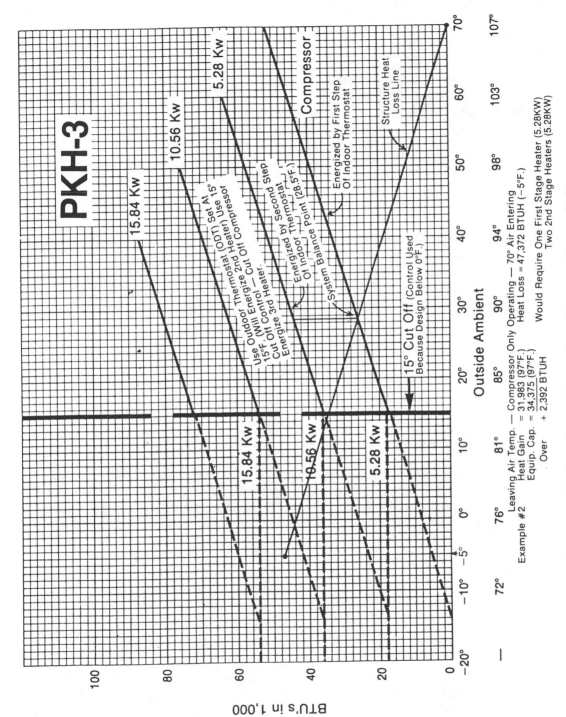

*FIGURE 6-9* Calculating System Balance Point (Courtesy of Amana Refrigeration, Inc.)

137

## SEASONAL PERFORMANCE FACTOR

Now let's calculate the seasonal performance factor (SPF), which shows the equipment efficiency when installed and operating. To make this calculation, we need the average number of heating hours per year at various outdoor ambient temperatures. This information can be secured from the U.S. Weather Bureau or from local sources. See Table 6–16 for the average number of heating hours per year in Cincinnati, Ohio.

To complete the Calculation Form for Seasonal Performance Factor: Heating (Figure 6–10), we need to consider only columns B through M. To fill out these columns, we proceed as follows:

*Column B.* Divide the building total heat loss at the design condition by the temperature difference. In our example the total house heat loss is 47,312 Btu/h. The temperature difference is 70°F (21.1°C) inside minus − 5°F (− 20.55°C), or 75°F (41.25°C). Therefore, 47,312 divided by 75 is 630.82, or 631 − Btu/h loss per degree temperature difference.

*Column C.* In this column place the difference between column A and 65. This gives the number of degrees below 65°F (18.3°C), or 65 − 62 = 3. Calculate this difference for the remainder of the column and enter it on the form.

*Column D.* Multiply column B by column C. This gives us the heat loss in Btu/h for each degree listed in column A. Complete the calculations and enter them in the remainder of the column.

*Column E.* This column is for the heat pump heating capacity in Btu/h and is taken from the manufacturer's specification sheets.

**TABLE 6–16**

Average Number of Heating Hours per year (Cincinnati, Ohio)

| Outdoor temperature (°F) | Heating hours |
|---|---|
| − 15 to − 10 | 1 |
| − 10 to − 5 | 2 |
| − 5 to 0 | 6 |
| 0 to 5 | 10 |
| 5 to 10 | 36 |
| 10 to 15 | 60 |
| 15 to 20 | 101 |
| 20 to 25 | 218 |
| 25 to 30 | 475 |
| 30 to 35 | 709 |
| 35 to 40 | 692 |
| 40 to 45 | 642 |
| 45 to 50 | 619 |
| 50 to 55 | 661 |
| 55 to 60 | 692 |
| 60 to 65 | 738 |

*Source:* Amana Refrigeration, Inc.

Enter these figures for the degree indicated in column A.

*Column F.* This indicates the heat pump running time. Divide column D by column E. Enter these figures fo each degree until a figure of 1 is calculated. Note that at no time can the percentage of running time be greater than 1.

*Column G.* Heat pump electrical input in kilowatts. These figures are taken from the manufacturer's specification sheets. They are normally listed in watts on the specification sheet. To convert to kilowatts, divide watts by 1000.

*Column H.* The seasonal heating hours as obtained from the U.S. Weather Bureau. Enter these figures on the form for each degree indicated in column A.

| | | | | HEAT PUMP ALONE | | |
|---|---|---|---|---|---|---|
| Outdoor Temp. (5° Increments) | BTU/Hr Loss/° (Heat loss ÷ Temp. Diff.) 75° | Outdoor Temp. Below 65° (65 − Column A) | Heat Loss BTU/Hr. (B × C) | Heat Pump Heating Cap. BTU/Hr. Mfg. Data | Heat Pump Running Time (D ÷ E) | Heat Pump Input (KW) Mfg. Data |
| A | B | C | D | E | F | G |
| 62 | 631 | 3 | 1893 | 47000 | .040 | 5.213 |
| 57 | | 8 | 5048 | 44500 | .113 | 5.081 |
| 52 | | 13 | 8203 | 42000 | .195 | 4.947 |
| 47 | | 18 | 11358 | 39200 | .290 | 4.778 |
| 42 | | 23 | 14513 | 36800 | .394 | 4.643 |
| 37 | | 28 | 17668 | 34380 | .514 | 4.521 |
| 32 | | 33 | 20823 | 31920 | .652 | 4.369 |
| 27 | | 38 | 23978 | 29440 | .814 | 4.209 |
| 22 | | 43 | 27133 | 26960 | 1.0 | 4.058 |
| 17 | | 48 | 30288 | 24500 | 1.0 | 3.889 |
| 12 | | 53 | 33443 | 22000 | 1.0 | 3.710 |
| 7 | | 58 | 36598 | | | |
| 2 | | 63 | 39753 | | | |
| −3 | | 68 | 42908 | | | |
| −8 | | 73 | 46063 | | | |
| −13 | | 78 | 49218 | | | |
| −18 | ↓ | 83 | | | | |

HEAT LOSS 47312 BTUH CRH3 & VBCH35X-1J

Annual Requirement Electric
Resistance Heat

$$= \frac{\overset{B}{631} \times \overset{M\,(Total)}{123826}}{3412} = \overset{P}{22899.8}\ \text{KWH}$$

FIGURE 6-10 Calculation Form for Seasonal Performance Factor: Heating (Courtesy of Amana Refrigeration, Inc.)

| HEAT PUMP ALONE | | SUPPLEMENTARY HEAT | | | | |
|---|---|---|---|---|---|---|
| Seasonal Heat Hrs. U.S. Weath. Bur. | Seasonal Heat Pump Input ($F \times G \times H$) | Resistance Heat Input (BTU / Hr) (D minus E) | Resistance Heat Input (KW) ($J \div 3412$) | Seasonal Resist. Heat Input (KW) ($H \times K$) | Degree Hours ($C \times H$) | |
| H | I | J | K | L | M | |
| 738 | 153.888 | | | | 2214 | |
| 692 | 397.314 | | | | 5536 | |
| 661 | 637.644 | | | | 8593 | |
| 619 | 857.699 | | | | 11142 | |
| 642 | 1174.438 | | | | 14766 | |
| 692 | 1608.065 | | | | 19376 | |
| 709 | 2019.649 | | | | 23397 | |
| 475 | 1627.410 | | | | 18050 | |
| 218 | 884.644 | 173 | .051 | 11.118 | 9374 | |
| 101 | 392.789 | 5788 | 1.696 | 171.296 | 4848 | |
| 60 | 222.600 | 11443 | 3.354 | 201.240 | 3180 | |
| 36 | | 36598 | 10.726 | 386.136 | 2088 | |
| 10 | | 39753 | 11.651 | 116.510 | 630 | |
| 6 | | 42908 | 12.576 | 75.456 | 408 | |
| 2 | | 46063 | 13.500 | 27.000 | 146 | |
| 1 | | 49218 | 14.425 | 14.425 | 78 | |
| | 9976.140 TOTAL | | | 1003.181 TOTAL | 123826 TOTAL | |

Annual Requirement Heat Pump System

$$\text{I (Total)} \quad \text{L (Total)} \quad \text{N}$$
$$= \underline{9976.140} + \underline{1003.181} = \underline{10979.3} \text{ KWH}$$

Seasonal Performance Factor

$$\text{S.P.F.} = \frac{P}{N} = \frac{22899.8}{10979.3} = 2.09$$

**FIGURE 6-10** *(cont.)*

*Column I.* The seasonal heat pump input in kilowatt hours. Multiply columns F, G, and H. Enter these figures for each degree as indicated in column A.

*Column J.* The resistance heat input in Btu/h. Subtract column E from column D. Enter these figures in the column for each degree as indicated on the form.

*Column K.* The resistance heat input in kilowatts. Divide column J by 3412 (the number of Btu produced per kilowatt of electricity used). Enter these figures in the column for each degree as indicated on the form.

*Column L.* The seasonal resistance heat input in kilowatts. Multiply column H by column K. Enter these figures in the column for each degree as indicated on the form.

*Column M.* The number of degree hours. Multiply column C by column H. Enter these figures in the column for each degree as indicated on the form.

After all columns are filled in for each outdoor temperature, columns I, L, and M should be totaled. Note that the compressor was cut off at approximately 10°F (− 12.2°C).

To find the annual requirement of electric resistance heat, the Btu/h loss per degree from column B is multiplied by the total of column M. This product is then divided by 3412.

The annual electrical consumption of a heat pump system (heat pump and resistance heating) is the total of column I added to the total of column L.

The seasonal performance factor (SPF) is the result of the annual electrical consumption of the system. Thus, the electric resistance heat P is divided by the annual electric consumption of the heat pump system N. Note that the heat pump and the supplementary electric resistance heat are more than twice as efficient as straight electric resistance heat.

Because the heat pump is more efficient than straight electric resistance heat, it would be logical to use a heat pump when both heating and cooling are desired.

## COST ANALYSIS ————

A cost analysis of the various fuels may be made to calculate the output cost per therm (100,000 Btu). Use Figure 6–11 to make these

**Natural Gas Furnace**
(1) Nat. Gas ———————— (Cost/Therm) ÷ 66% (Furnace Eff.) = ———————— Output Cost/Therm

**Natural Gas Electric-Gas**
(2) Nat. Gas ———————— (Cost/Therm) ÷ 81.2% (EG Eff.) = ———————— Output Cost/Therm

**Oil Furnace**
(3) Oil ———— (Cost/Gal. ÷ 1.4) = ———— ÷ 66% (Furnace Eff.) = ———— Output Cost/Therm

**Electric Furnace**
(4) Electric ———— (Cost/KW × 29.3) = ———— ÷ 1 (Elec. Heat Eff.) = ———— Output Cost/Therm

**Heat Pump**
(5) Electric ———— /Cost(KW × 29.3) = ———— ÷ S.P.F. ———— = ———— Output Cost/Therm

**L.P. Gas Furnace**
(6) L.P. Gas ———— (Cost/Gal. × 1.1) = ———— ÷ 66% (Furnace Eff.) = ———— Output Cost/Therm

**L.P. Gas Electric-Gas**
(7) L.P. Gas ———— (Cost/Gal. × 1.1) = ———— ÷ 81.2% (EG Eff.) = — ———— Output Cost/Therm

**FIGURE 6-11** Cost for 100,000-Btu Output (Courtesy of Amana Refrigeration, Inc.)

analyses. The cost of various fuels and their availability will be the deciding factors on the unit to be installed. To use the table, insert the cost per therm in the proper space and make the calculations as indicated.

## AIR-DISTRIBUTION AND DUCTWORK DESIGN _____

Air distribution is as critical to heat pump operation as any other area because, without the proper amount of air flow on the heating cycle, a high compression ratio (high discharge pressures) situation can develop, which will tremendously shorten the life of the system, especially the compressor. The ductwork rules of thumb that were satisfactory for conventional heating and cooling systems are not adequate for heat pump systems. The demands of the supply and return ducts must not exceed the capabilities of the blower that is built into the system. The manufacturers, distributors, and dealers must stress to the builder and the homeowner that an adequate return air system must be provided to the blower or the air volume will be restricted. We must remember that blowers do not manufacture air; they only move the air that is made available through the return air system. In fact, the air-conditioning system is no better than the air-distribution system.

*Supply Air Ducts:* Because there are many different types of home construction, there are many types of duct systems. In this section, we will discuss four basic air-distribution systems that are designed to fit the variety of construction available, and with proper sizing they will perform satisfactorily. All air-distribution systems should be airtight and insulated in areas that are subject to temperatures higher or lower than the room conditions. The four types of air distribution systems are trunk duct, extended plenum, loop perimeter, and radial.

*Trunk duct system.* This system has a reduction in the main duct for every one or two branch duct take-offs made, thereby maintaining the essential constant velocity. Trunk ducts are commonly used in commercial-type systems. They are also used in residential systems where the savings resulting from the reduction in duct size can justify the cost of making the reduction (Figure 6-12).

*Extended plenum system.* This system is an extension of the equipment discharge air plenum. The extension is maintained at one size throughout its length (Figure 6-13). This type of system is most commonly used in residential applications, particularly to supply air in homes with basements or crawl spaces. The duct is usually run perpendicular to the ceiling or floor joist and is centrally located. Round take-offs can be run from the top, bottom, or sides of the plenum. *Note:* Do not run take-offs from the end of the extended plenum. The take-

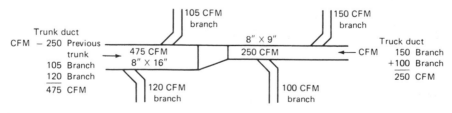

*FIGURE 6-12* Trunk Duct Air-Distribution System (Courtesy of Amana Refrigeration, Inc.)

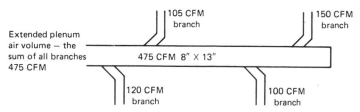

FIGURE 6-13 Extended Plenum Air-Distribution System
(Courtesy of Amana Refrigeration, Inc.)

off should be run between the joist spaces for appearance.

*Loop perimeter supply system.* This system forms a continuous loop around the outside edge of the building and is connected to the unit by several feeder ducts (Figure 6–14). This is an ideal system for heating buildings built on slabs because the heat leakage from the ducts provides more even floor temperatures. It is highly important that the edge of the slab be adequately insulated. Transite and fiber ducts are frequently used in installations of this type.

*Radial supply system.* This system is frequently used in attic installations; it uses high wall registers or ceiling diffusers to add summer cooling to homes with hydronic heat or for year-round use in areas with mild winter seasons (Figure 6–15). This type of system may also be used in a crawl space or slab rather than a loop perimeter system to supply air to the perimeter of the building. It is usually more

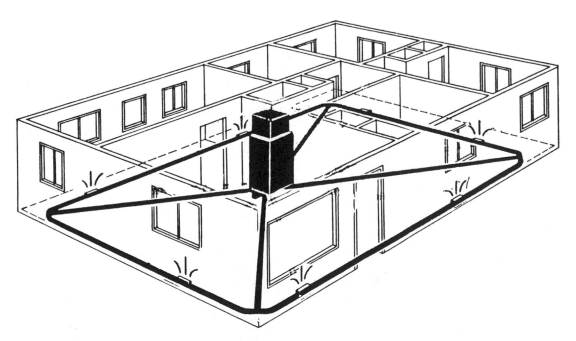

FIGURE 6-14 Loop Perimeter Air-Distribution System

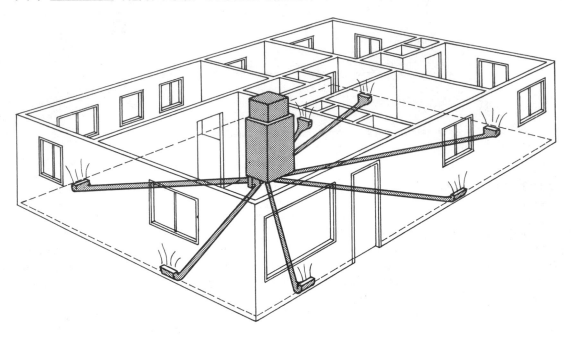

**FIGURE 6-15** Radial Perimeter Air-Distribution System

economical to install than the loop perimeter system, but the floor temperatures are not as even.

*Return Duct System:* Return systems may take the form of any of the types of supply systems. It is possible to reduce the cost of the return air system by causing the air to flow in the joist and stud spaces as part of the return system, which eliminates the cost of metal and duct fabrication. The joist space can be closed on the bottom with sheet metal. Also, crawl spaces can be sealed and used as a return air plenum to eliminate return air ducts completely. The return air systems must be airtight to avoid causing any additional load on the system from areas not to be conditioned. Be sure that all local regulations and building codes are adhered to.

*Existing Duct Systems:* When adding cooling or a heat pump system to an existing

duct system, the existing ducts should be checked completely. The following is a list of the more important points:

1. Check the air static pressure of the longest run.
2. Check to see that there is adequate insulation to accommodate the new system.
3. Make sure that the air registers are correct for the new required throw and spread.
4. Seal all joints to prevent air leakage.
5. The same design criteria used for new duct systems should be used to analyze the existing system.

*Air Quantity:* The air quantity requirements of the equipment dictate the design of the duct system. The duct system must be able to handle an air quantity equal to that required by the equipment. The proper air quantity is one of the most important parts of heating and cooling with a heat pump system.

A low air volume will cause indoor coil freeze-ups during the cooling season and will cause high discharge air temperatures indoors, accompanied by a high compressor discharge pressure and a high compression ratio. A high air volume can cause cool air to be distributed during the heating season and a lack of himidity removal during the summer.

*Resistance to Air Flow:* The ductwork and all the fittings used to change the direction of air flow cause a resistance to air flow that must be overcome by the blower. The total resistance of the duct system and equipment cannot exceed the capacity of the blower used in the system. This air friction can be kept to a minimum with proper duct system design. The friction loss of the system must be calculated before installation. If the static friction loss is higher than that for which the equipment is designed, the duct system must be increased in size.

To determine the total static resistance, a system static chart, as shown in Figure 6–16, or one similar to it, should be used in conjunction with the duct system layout. By doing this, any necessary changes can be determined before the installation is started.

*Duct Sizing and Total Static:* The proper design of the ductwork used in a residence or a commercial installation is a critical part of the overall system design. Several methods are used to design duct systems. They all will assist you in making the proper decisions. In our calculations here we will use the Amana slide rule duct calculator (Figure 6–17).

*Air Volume:* As has been stated before, air volume is very important in heat pump systems. When the equipment size has been calculated, refer to the application data for the cubic feet

per minute (cfm) available from the unit blower. Select a cfm for a static pressure that is 0.1 to 0.15 in. (2.54 to 3.81 mm) water column (WC) below the maximum static pressure shown for that particular unit. By selecting a static point below the maximum blower capacity, the unit will still operate properly when the static pressure increases owing to dirty filters, dirty coil, or dirty blower or if some of the grills are closed.

The cfm calculated for each room from the heat loss form for heat pump systems should be sized in accordance to the cfm available from the heating cfm chart. A cfm at a static pressure of 0.1 to 0.15 in. WC below the maximum capacity of the blower should be selected.

## EXAMPLE _____

*Given:* A residence with a total Btu loss of 25,474 Btu/h (6419.45 kcal). The living room has a heat loss of 7712 Btu/h (1943.42 kcal). The total cfm available from the equipment at 0.25 in (6.35 mm) WC (dry coil) is 1205 cfm. What is the cfm requirement of the living room?

The actual cfm that the ducts to the living room must carry can be calculated as follows:

$$\frac{\text{Living room Btu loss: 7,712 (1943.42 kcal)}}{\text{Total residence Btu loss: 25,474 (6419.45 kcal)}}$$

$$= 0.3027 \times 1205 = 365 \text{ cfm}$$

*Duct Sizing:* The duct size is important to ensure that the proper amount of air flow is attained for satisfactory equipment operation. If one branch duct is to carry 365 cfm, as in our example, the size of the duct would be calculated as follows: Refer to the Amana slide rule (Figure 6–17) at the cfm part B section. (The cfm is printed on the slide portion.) There are two scales; refer to the upper scale. Locate 365 cfm; place the cfm number under the 0.05 in. (1.27 mm) friction per 100 ft (30.48 m) of ductwork (scale A). The duct systems for heat pumps should be sized at 0.05 in. WC friction per 100 ft.

| (A) | (B) | (C) | (D) | (E) | (F) | (G) | Friction | Total |
| | Linear | Duct | Fittings | | Total | Total | Loss | Friction |
| Run | Length | Size | | | Equivalent | (B+F) | Per 100' | Loss |
| | | | | | Length | Length | | |
|-----|-----|-----|-----|-----|-----|-----|-----|-----|
| | | | | | | | | |
| | | | | | | | | |
| | | | | | | | | |
| | | | | | | | | |
| | | | | | | | | |

**SUPPLY**

**RETURN**

*FIGURE 6-16*  System Static Chart (Courtesy of Amana Refrigeration, Inc.)

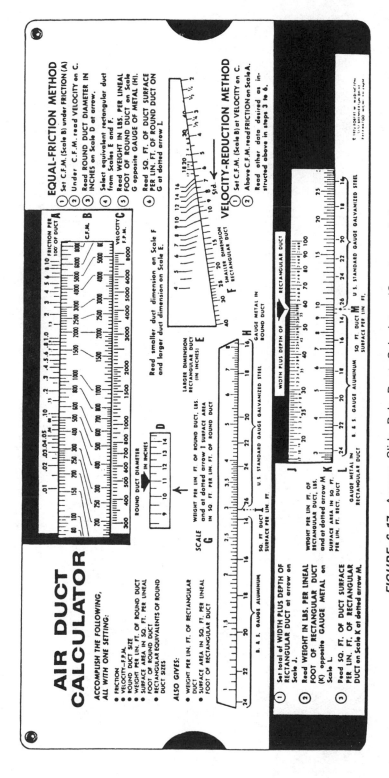

*FIGURE 6-17* Amana Slide Rule Duct Calculator (Courtesy of Amana Refrigeration, Inc.)

When these two figures have been lined up, refer to window D for the size of round duct to be used. Use an 11-in. (279.40 mm) round duct. If a rectangular duct is to be used, refer to window E. The 8-in. (203.20 mm) high standard duct is shown on the botton fixed scale, so the proper size of rectangular duct is 8 in. × 13 in. (203.20 mm × 330.20 mm).

When more than one branch duct is to be used in a single room, divide the air volume (cfm) by the number of branches and calculate the size of each of the branches as shown in our example. Remember, use the 0.05-in. WC friction loss per 100 ft (30.48 m) at all times to size the duct system.

The main duct should be sized by starting at the end of the duct away from the equipment. The air volumes of the branch ducts should be calculated by starting at the end and moving toward the equipment. When a trunk duct system is used, the duct size should be increased in size for each two branches.

## EXAMPLE

The trunk duct is sized the same way as the branch duct. Use the Amana slide rule duct calculator; place the cfm to be carried by the trunk under 0.05 in. WC; read the size of round duct or the size of the rectangular duct. A trunk duct size for 250 cfm would be 8 in. × 9 in. (203.20 mm × 228.60 mm).

Place 250 cfm on scale B under 0.05 on scale A; read in window E for 8-in.-high duct. We find the width to be 9 in. (228.60 mm). Repeat this process for the trunk duct with 475 cfm; we find the duct size to be 8 in. × 16 in. (203.20 mm × 406.40 mm; Figure 6–18).

When an extended plenum system is desired, add up all the branch ducts to be fed by the extended plenum and use the total cfm for sizing the extended plenum. Use 0.08 in. WC friction loss per 100 ft (30.48 m) for the extended plenum, but use 0.05 in. WC friction loss per 100 ft for the branch ducts (Figure 6–19).

The extended plenum size would be determined from the Amana slide rule duct calculator by placing 475 cfm at point B under 0.08 on the A scale.

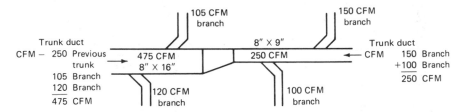

**FIGURE 6-18** Trunk Duct Example (Courtesy of Amana Refrigeration, Inc.)

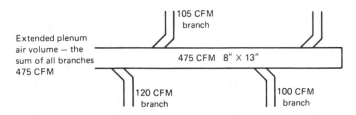

**FIGURE 6-19** Extended Plenum Example (Courtesy of Amana Refrigeration, Inc.)

The rectangular duct size would be found in window E to be 8 in. × 13 in. (203.20 mm × 330.20 mm). Each branch duct for this system would be calculated by the same method used for the trunk duct system using 0.05 on the A scale.

*Duct Static Drop:* This is a measure of resistance to air flow through the air-distribution system. In calculating the static drop of a duct system, use only the longest or the most complicated duct run. The total static drop must include the straight run of ductwork, all fittings, and grills.

When we were sizing the duct system in our previous examples, we based the size of the duct on a friction loss of 0.05 in. WC friction per 100 ft (30.48 m) for the trunk duct and branches. On the extended plenum system we used 0.08 in. WC per 100 ft (30.48 m) of ductwork. To determine the total static drop of the air-distribution system, we must calculate the total length of the duct, including the actual linear feet (meters) plus the equivalent feet (meters) for all the fittings. The static drop of fittings is given in the number of feet (meters) of straight ductwork which the fittings' resistance is equal to.

## EXAMPLE

A 90° elbow is equal to 10 ft (3.0480 m) of straight duct (Figure 6–20). On the extended plenum system, base the static pressure on three-quarters of the total air volume, because the air volume reduces as it travels through the duct, but the duct stays the same size.

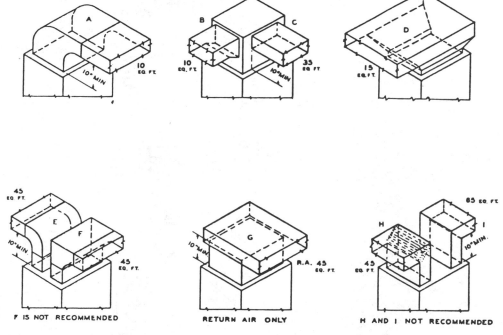

GROUP 1    WARM AIR AND RETURN AIR BONNET OR PLENUM.

*FIGURE 6-20* Equivalent Length of Fittings and Intakes
(Courtesy of Amana Refrigeration, Inc.)

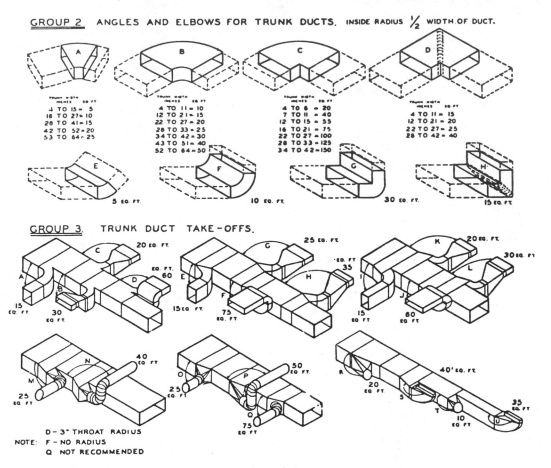

GROUP 2  ANGLES AND ELBOWS FOR TRUNK DUCTS. INSIDE RADIUS ½ WIDTH OF DUCT.

| TRUNK WIDTH INCHES | SQ FT |
|---|---|
| 4 TO 15 = | 5 |
| 16 TO 27 = | 10 |
| 28 TO 41 = | 15 |
| 42 TO 52 = | 20 |
| 53 TO 64 = | 25 |

| TRUNK WIDTH INCHES | SQ FT |
|---|---|
| 4 TO 11 = | 10 |
| 12 TO 21 = | 15 |
| 22 TO 27 = | 20 |
| 28 TO 33 = | 25 |
| 34 TO 42 = | 30 |
| 43 TO 51 = | 40 |
| 52 TO 64 = | 50 |

| TRUNK WIDTH INCHES | SQ FT |
|---|---|
| 4 TO 6 = | 20 |
| 7 TO 11 = | 40 |
| 12 TO 15 = | 55 |
| 16 TO 21 = | 75 |
| 22 TO 27 = | 100 |
| 28 TO 33 = | 125 |
| 34 TO 42 = | 150 |

| TRUNK WIDTH INCHES | SQ FT |
|---|---|
| 4 TO 11 = | 15 |
| 12 TO 21 = | 20 |
| 22 TO 27 = | 25 |
| 28 TO 42 = | 40 |

E  5 SQ. FT.

F  10 SQ. FT.

G  30 SQ. FT.

H  15 SQ. FT.

GROUP 3.  TRUNK DUCT TAKE-OFFS.

D - 3" THROAT RADIUS
NOTE:  F - NO RADIUS
       Q  NOT RECOMMENDED

GROUP 4.  BOOT FITTINGS - FROM BRANCH TO STACK.

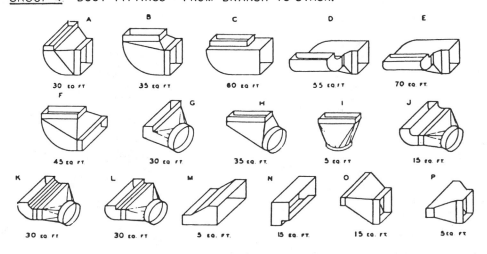

| A | B | C | D | E |
|---|---|---|---|---|
| 30 SQ FT | 35 SQ FT | 60 SQ FT | 55 SQ FT | 70 SQ. FT. |

| F | G | H | I | J |
|---|---|---|---|---|
| 45 SQ FT | 30 SQ FT | 35 SQ. FT. | 5 SQ FT | 15 SQ FT |

| K | L | M | N | O | P |
|---|---|---|---|---|---|
| 30 SQ FT | 30 SQ. FT. | 5 SQ. FT. | 15 SQ. FT. | 15 SQ. FT. | 5 SQ FT |

*FIGURE 6-20  (cont.)*

150

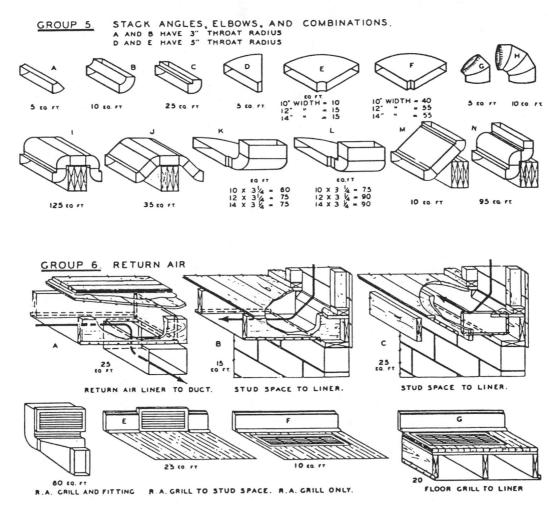

GROUP 5. STACK ANGLES, ELBOWS, AND COMBINATIONS.
A AND B HAVE 3" THROAT RADIUS
D AND E HAVE 5" THROAT RADIUS

A  5 SQ FT
B  10 SQ FT
C  25 SQ FT
D  5 SQ FT
E  10" WIDTH = 10
   12"   "  = 15
   14"   "  = 15
F  10" WIDTH = 40
   12"   "  = 55
   14"   "  = 55
G  5 SQ FT
H  10 SQ FT

I  125 SQ FT
J  35 SQ FT
K  10 X 3¼ = 60
   12 X 3¼ = 75
   14 X 3¼ = 73
L  10 X 3¼ = 75
   12 X 3¼ = 90
   14 X 3¼ = 90
M  10 SQ FT
N  95 SQ FT

GROUP 6. RETURN AIR

A  25 SQ FT   RETURN AIR LINER TO DUCT.
B  15 SQ FT   STUD SPACE TO LINER.
C  25 SQ FT   STUD SPACE TO LINER.

D  60 SQ FT   R.A. GRILL AND FITTING
E  25 SQ FT   R.A. GRILL TO STUD SPACE.
F  10 SQ FT   R.A. GRILL ONLY.
G  20   FLOOR GRILL TO LINER

FIGURE 6-20  (cont.)

## EXAMPLE

To determine how static pressure is calculated, find the static readings in Figure 6-21, and then find the point in the duct system in Figure 6-22.

## EXAMPLE

To determine the calculated static pressure of a return system, find the static readings in Figure 6-23, and then find the point in the duct system in Figure 6-24. Note that the total system static pressure is the sum of the supply static and the return static. In these examples the total static equals 0.219. This would be excellent for heat pump systems.

## EXAMPLE

Using the same return system as in the preceding example, the example of Figures 6-25 and 6-26 would give the total static drop on a trunk duct system.

*Duct and Equipment Layout:* Using the floor plan shown in Figure 6-27, let's examine one method of making a duct and equipment layout. This step is important to the estimator for calculating the cost of the ductwork and to the installer for knowing what the estimator has planned.

151

## SYSTEM STATIC CHART — Extended Plenum

| (A)<br><br>Run | (B)<br>Linear<br>Length | (C)<br>Duct<br>Size | (D)<br><br>Fittings | (E) | (F)<br>Total<br>Equivalent<br>Length | (G)<br>Total<br>(B+F)<br>Length | Friction<br>Loss<br>Per 100' | Total<br>Friction<br>Loss |
|---|---|---|---|---|---|---|---|---|
| **SUPPLY** | | | | | | | | |
| A to B | 20' | 8" × 13" | Unit Take Off<br>Equivalent<br>10' | | 10' | 30' | 356 CFM*<br><br>.045/100' | .014 |
| B to C | 20' | 7"∅ | Duct Take Off<br><br>Equivalent<br>35' | Boot<br><br>Equivalent<br>30' | 65' | 85' | 100 CFM<br><br>.04/100' | .034 |
| | | | | #411 Supply Register 2¼" × 14" (From Catalog) | | | | .04 |
| | | | | | | | Total Supply Static | .088 |

\* 475 CFM @ .08" W.C. = 8" × 13" Main Duct
To calculate static use ¾ (.75) of total air
475 × .75 = 356 CFM
356 CFM in 8" × 13" Duct = .045" W.C./100'
Refer to Pages 3-11, 3-12 and Reverse Side of Duct Calculator

**FIGURE 6-21** Extended Plenum System Supply Static Chart (Courtesy of Amana Refrigeration, Inc.)

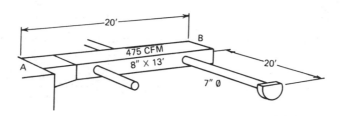

**FIGURE 6-22** Extended Plenum System Supply Example (Courtesy of Amana Refrigeration, Inc.)

## RETURN

| Run | Linear Length | Duct Size | Fittings | | Total Equivalent Length | Total (B+F) Length | Friction Loss Per 100' | Total Friction Loss |
|---|---|---|---|---|---|---|---|---|
| A to B | 20' | 8" × 14" | R.A. Grill To<br>Stud Space<br>25' ▶ | Stud Space<br>To Liner<br>15' ▶ | 40' | 60' | 250 CFM<br>.019/100' | .011 |
| B to C | 20' | 8" × 8" | R.A. Liner<br>To Duct<br>25' ▶ | | 25' | 45' | 250 CFM<br><br>.07/100' | .032 |
| C to D | 7' | 8" × 24" | Return Take<br>Off<br>45' ▶ | Return<br>Connection<br>65' ▶ | 110' | 117' | 1000 CFM<br>.075/100' | .088 |
| | | | | | | | Total Return Static | .131 |

Total System Static = .219

**FIGURE 6-23** Extended Plenum System Return Static Chart (Courtesy of Amana Refrigeration, Inc.)

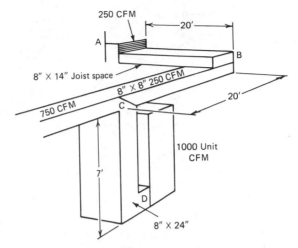

**FIGURE 6-24** Extended Plenum System Return Example (Courtesy of Amana Refrigeration, Inc.)

## SYSTEM STATIC CHART—Trunk Duct

| (A)<br><br>Run | (B)<br>Linear<br>Length | (C)<br>Duct<br>Size | (D)<br><br>Fittings | (E) | (F)<br>Total<br>Equivalent<br>Length | (G)<br>Total<br>(B+F)<br>Length | Friction<br>Loss<br>Per 100' | Total<br>Friction<br>Loss |
|---|---|---|---|---|---|---|---|---|
| | | | **SUPPLY** | | | | | |
| A to B | 10' | 8" × 16" | Unit Take Off Equivalent 10' | | 10' | 20' | 475 CFM .05/100' | .01 |
| B to C | 10' | 8" × 9" | Reduction Equivalent 5' | Turbulence Equivalent 25' | 30' | 40' | 250 CFM .055/100' | .022 |
| C to D | 20' | 7"∅ | Duct Take Off Equivalent 35' | Boot Equivalent 30' | 65' | 85' | 100 CFM .04/100' | .034 |
| | | | #411 Supply Register 2¼" × 14" (From Catalog) | | | | | .04 |
| | | | | | | | Total Supply Static | .106 |
| | | | **RETURN** | | | | | |
| A to B | 20' | 8" × 14" | R.A. Grill To Stud Space 25' | Stud Space To Liner 15' | 40' | 60' | 250 CFM .019/100' | .011 |
| B to C | 20' | 8" × 8" | R.A. Liner To Duct 25' | | 25' | 45' | 250 CFM .07/100' | .032 |
| C to D | 7' | 8" × 24" | Return Take Off 45' | Return Connection 65' | 110' | 117' | 1000 CFM .075/100' | .088 |
| | | | | | | | | |
| | | | | | | | Total Return Static = | .131 |
| | | | | | | | Total System Static = | .237 |

**FIGURE 6-25** Trunk Duct System Static Chart (Courtesy of Amana Refrigeration, Inc.)

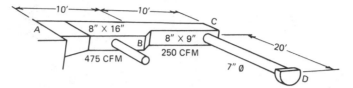

**FIGURE 6-26** Trunk Duct System Example (Courtesy of Amana Refrigeration, Inc.)

After making the heat loss and heat gain calculations and then some building improvements, as shown in Figure 6–28, we have selected a heat pump. From the manufacturer's specification sheet, we note that at a 0.25 in. (6.35 mm) WC external static pressure the air delivery will be approximately 1205 cfm. Using the 0.25 in. (6.35 mm) WC as our duct design criterion, we calculate the required cfm for each room in the heating mode as follows:

**STEP 1** _____

Referring to the heat loss tabulation form, line 13 (Figure 6–26), we find the total Btu loss for each room and the total Btu loss of all the rooms.

**STEP 2** _____

Calculating as follows, we determine the cfm for each room:

$$\text{Living room} =$$
$$\frac{\text{Btu loss}}{\text{total heat loss}} \times \text{total cfm} \times \text{required cfm}$$

$$\text{Living room cfm} =$$
$$\frac{7712}{25,474} \times 1205 = 365 \text{ cfm}$$

**STEP 3** _____

The cfm required for the balance of the rooms would be calculated as in step 2.

We can use this same method to calculate the required cfm for each room in the cooling mode to determine the amount of balancing that will be required. The total cfm will be less during the cooling mode because of the moisture collecting on the indoor coil. The additional resistance caused by the moisture will lower the external static pressure to 0.20 in. (5.08 mm) WC. Again referring to the manufacturer's specification sheet, we see that the total cfm will be approximately 1100 cfm.

Using this as our criterion, we can calculate the required room cfm as follows:

**STEP 1** _____

Referring to the heat gain tabulation form (cooling) (Figure 6–29), we determine the sensible heat gain of each room and the total Btu gain for all rooms.

**STEP 2** _____

Calculating as follows, we determine the cfm for each room:

$$\text{Living room} =$$
$$\frac{\text{sensible heat gain}}{\substack{\text{total building} \\ \text{sensible heat gain}}} \times \text{cfm} = \text{required cfm}$$

$$\text{Living room cfm} =$$
$$\frac{4578}{12,790} \times 1100 = 394 \text{ cfm}$$

**STEP 3** _____

The cfm required for the balance of the rooms would be calculated as in step 2.

Note that some of the rooms will require balancing of the air delivery from winter to summer operation. This is the purpose of installing balancing dampers wherever possible.

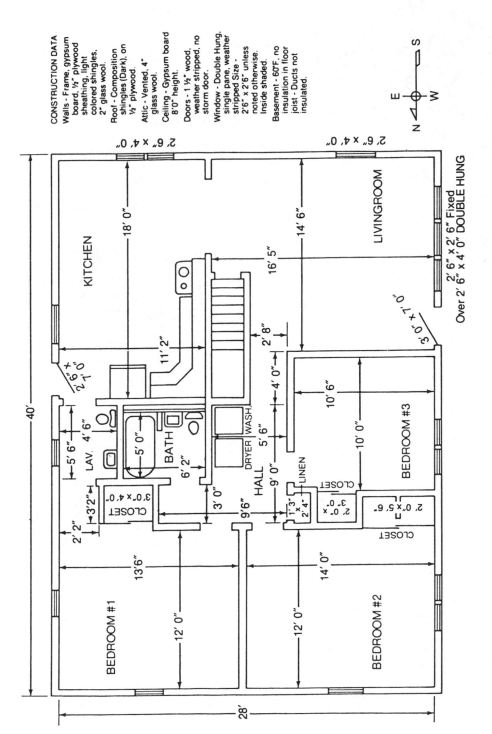

CONSTRUCTION DATA

Walls - Frame, gypsum board, ½" plywood sheathing, light colored shingles, 2" glass wool.

Roof - Composition shingles (Dark), on ½" plywood.

Attic - Vented, 4" glass wool.

Ceiling - Gypsum board 8'0" height.

Doors - 1½" wood, weather stripped, no storm door.

Window - Double Hung, single pane, weather stripped Size - 2'6" x 2'6" unless noted otherwise. Inside shaded.

Basement - 60F, no insulation in floor joist - Ducts not insulated.

NOTE: Made following changes to structure:
1. Added R10 insulation in attic;
2. Installed storm windows and storm doors.

N E S W

KITCHEN

LIVINGROOM

BEDROOM #3

BEDROOM #1

BEDROOM #2

BATH

LAV.

HALL

CLOSET

LINEN

CLOSET

CLOSET

DRYER | WASH.

2' 6" x 4' 0"

2' 6" x 4' 0"

2' 6" x 2' 6" Fixed
Over 2' 6" x 4' 0" DOUBLE HUNG

3' 0" x 7' 0"

3' 0" x 4' 0"

2' 0" x 5' 6"

2' 0" x 3' 0"

1' 3" x 2' 4"

2' 6" x 7' 0"

18' 0"

16' 5"

14' 6"

11' 2"

2' 8"

4' 0"

10' 6"

10' 0"

5' 6"

9' 0"

3' 0"

9' 6"

5' 0"

6' 2"

4' 6"

5' 6"

3' 2"

2' 2"

13' 6"

12' 0"

14' 0"

12' 0"

40'

28'

*FIGURE 6-27* Basic Floor Plan (Courtesy of Amana Refrigeration, Inc.)

155

Job Name: P. Smith

Inside Design Temp. 70°F
Outside Design Temp. -5°F
Temp. Difference 75°F
Temp. Difference +10 ___

Location: CINCINNATI, OHIO   Date 1/1/75   Computed By C.C.S.

| Room | | Living Room | | Dining Room / HALL | | Kitchen | | Bedroom 1 | | Bedroom 2 | | Bedroom 3 / LAV. | | Bath / Bathroom | | Rooms Btu Totals |
|---|---|---|---|---|---|---|---|---|---|---|---|---|---|---|---|---|
| 1. Room Size (LxWxH) | | 16.5x14.5x8 | | 13x5.5x8 | | 18x11.2x8 | | 13.5x12x8 | | 14x12x8 | | 10.5x10x8 | | 6.2x5.5x8 | | |
| 2. Linear Ft. Exp. Wall | | 14.5+14.5=31 | | 71.5+12+3=87 | | 11.2+18=29.2 | | 15.2+13.5=28.7 | | 14+14=28 | | 10 | | 5.5 | | |
| 3. Floor Area | | 2.39 | | | | 2.02 | | 162+12+7=181 | | 168+11=179 | | 105+6=111 | | 31 | | |
| | Factor | Ar. or Ln. Ft. | BTU | Ar. or Ln. Ft. | BTU | Ar. or Ln. Ft. | BTU | Ar. or Ln. Ft. | BTU | Ar. or Ln. Ft. | BTU | Ar. or Ln. Ft. | BTU | Ar. or Ln. Ft. | BTU | |
| 4. Windows | .45 | 58.75 | 26.44 | | | 26.25 | 11.81 | 12.5 | 5.63 | 18.75 | 8.44 | 12.5 | 5.63 | 6.25 | 2.81 | 60.76 |
| 5. Doors | .30 | 21 | 6.3 | | | 17.5 | 5.25 | | | | | | | | | 11.55 |
| 6. Exp. Wall | .11 | 168 | 18.48 | | | 190 | 20.9 | 217 | 23.87 | 205 | 22.55 | 67.5 | 7.43 | 37.8 | 4.16 | 97.39 |
| 7. Ceiling | .047 | 239 | 11.23 | 87 | 4.09 | 202 | 9.49 | 181 | 8.51 | 179 | 8.41 | 111 | 5.22 | 25 | 1.18 | 49.59 |
| a | .22 | 62 | 13.64 | | | 43.5 | 9.57 | 25 | 5.5 | 37.5 | 8.25 | 25 | 5.5 | 12.5 | 2.75 | 43.21 |
| b | — | | | | | | | | | | | | | | | |
| c | .60 | 20 | 12 | | | 19 | 11.4 | | | | | | | | | 23.4 |
| d | .13 | 30 | 3.9 | | | | | | | | | | | | | 3.9 |
| 8. Infiltration | | | | | | | | | | | | | | | | |
| 9. SUB TOTALS | | | 91.99 | | 4.09 | | 68.42 | | 68.92 | | 97.65 | | 23.78 | | 10.9 | 291.8 |
| 10. Sub-Total for T.D. 75° | | | 68.99 | | 3.07 | | 51.32 | | 43.51 | | 73.24 | | 17.84 | | 8.18 | 21,887 |
| 11. Floor Btu for T.D. 3.4 | | 239 | 812.6 | 87 | 295.8 | 202 | 68.8 | 181 | 65.4 | 179 | 68.6 | 111 | 377.4 | 25 | 85 | 3587 |
| 12. Partition Btu for T.D. | | | | | | | | | | | | | | | | |
| 13. Room Total Btu Loss | | | 7112 | | 603 | | 5819 | | 3878 | | 4183 | | 2161 | | 903 | 25979 |
| | | 365 CFM | | 29 CFM | | 275 CFM | | 183 CFM | | 198 CFM | | 102 CFM | | 43 CFM | | 7442 |

14. +30% Duct Loss 33116

All Rm in Ceiling

15. Grand Total Btu Load 33116

**Windows** | **U Factors**
Single Pane ......... 1.13
Double Pane, Sealed. .57
Storm Windows, Tight (.45)
Storm Windows, Loose .75
Glass Block .......... .50
T.D. = TEMPERATURE DIFFERENCE

**Wood Doors** | **U Factors**
| Nom. Thickness | No Storm | Storm Door |
|---|---|---|
| 1" | .69 | .35 |
| 1¼" | .52 | (.30) |
| 2" | .46 | .28 |
| 2¼" | .38 | .25 |

**Exp. Walls** | **U Factors**
| Frame Construction With: | Clapboard or Shingle | Brick Venr. | Stone Venr. |
|---|---|---|---|
| No Insulation | .25 | .28 | .30 |
| ¾" Insl. Board | .19 | .21 | .22 |
| 2" Insl. Batts | .11 | .12 | .12 |
| 3¼" Insl. Batts | .09 | .10 | .10 |

**Ceilings** (Vent: Attic Space Above)
| No Insulation | .50 |
| 2" Insulation | .10 |
| 4" Insulation | (.09) |
| 6" Insulation | .05 |

R = ¾ x 1.69 = 11.11 + 10 = 21.11   U = ½ = ½1.11 = .047

VBCH 30X-1J = 1205 CFM @ .25" W.C.

VBCH 30X-1J = 1205 CFM @ .25 W.C.

**CONCRETE FLOOR ON GROUND**
Btu per hr. per lineal ft. of edge

| Outside Design Temp. | 25 | 20 | 15 | 10 | 5 | 0 | −5 | −10 | −15 | −20 | −25 | −30 |
|---|---|---|---|---|---|---|---|---|---|---|---|---|
| 1" Vert. Insul. Extending Down 18" Below Surface | 55 | 62 | 68 | 75 | 80 | 85 | 90 | 95 | 100 | 105 | 110 | 115 |
| 1" L Type Insul. Extending 12" Down and 12" Under | 50 | 57 | 63 | 70 | 75 | 80 | 85 | 90 | 95 | 100 | 105 | 110 |
| 2" L Type Insul. Extending 12" Down and 12" Under | 40 | 45 | 50 | 55 | 60 | 70 | 75 | 80 | 85 | 90 | 95 |
| No edge insulation and no heat in slab | This construction not recommended. |

Note (The above lists of factors is comprised of only those most commonly used.) For additional coefficients or more complete descriptions of these elements of structure, consult the Htg. Ventil. & Air Cond. Guide.

**Infiltration** (factor times lineal ft. of crack)
| | Weather Stripped | | Not Weather Stripped |
|---|---|---|---|
| Double Hung Window (a) | .30 | | .35 |
| Pivoting Windows (b) | | | .50 |
| Doors (c) | .60 | | 1.00 |
| Fixed Window (d) | .13 | | .13 |

.50
.10
.05

**% Duct Loss Schedule**
| | 1 Story | 1½ Story | 2 Story |
|---|---|---|---|
| Ducts Insul. | 10 | 8 | 6 |
| Ducts No Insul. | (30) | 25 | 20 |

**Double Wood Floors or Partition**
| With Air Space | 10 | 20 | 30 | 40 | 50 |
|---|---|---|---|---|---|
| Temp. Diff. | | | | | |
| Factor No Insul. | (3.4) | 6.8 | 10.2 | 13.6 | 17.0 |
| Factor 2" Insul. | 1 | 2 | 3 | 4 | 5 |

*FIGURE 6-28* Residential Heat Loss Tabulation (Courtesy of Amana Refrigeration, Inc.)

(BASED ON 24 HOUR PER DAY OPERATION OF EQUIPMENT)

Job Name: P. Smith  
Location: Cincinnati, Ohio  
Date: 1/1/75  
Computed By: C.L.S.  
VBCH30X-1J = 1100 CFM @ .20" W.C.  
Outside Humidity Factor: 1.35  

Outside Design Temp: 97°F  
Inside Design Temp: 75°F  
Temp Diff: 22°F  
Daily Temp Range: 20°F  

| ROOM | FACTOR | LIVING ROOM | | HALL DINING ROOM | | KITCHEN | | BED ROOM 1 | | BED ROOM 2 | | Bedroom 3 | | LAV. | | BATH | | BTUH TOTALS |
|---|---|---|---|---|---|---|---|---|---|---|---|---|---|---|---|---|---|---|
| | | AREA OR QUAN | BTUH | AREA OR QUAN | BTUH | AREA OR QUAN | BTUH | AREA OR QUAN | BTUH | AREA OR QUAN | BTUH | AREA OR QUAN | BTUH | AREA OR QUAN | BTUH | AREA OR QUAN | BTUH | |
| Room Size (LxWxH) | | 16.5x14.5x8 | | 13.5x5.5x8 | | 18x11.2x8 | | 13.5x12x8 | | 14x12x8 | | 10.5x10x8 | | 5.5x4.5x8 | | 6.2x5x8 | | |
| Linear Ft. Exposed Wall | | 14.5+16.5=31 | | | | 11.2+18=29.2=28 | | 15.2+13.5=28.7 | | 14+14=28 | | 10.5 | | 5.5 | | — | | |
| Floor Ceiling Area | | 239 | | 71.5+12+3=87 | | 202 | | 162+12+7=181 | | 168+11=179 | | 105+6=111 | | 25 | | 31 | | |
| Windows N | 17.2 | | | | | | | | | | | | | | | | | 216 |
| Windows E | 32.8 | 69.75 | 2288 | | | 23.75 | 779 | 6.25 | 108 | 6.25 | 108 | | | | | | | 1189 |
| Windows W | 32.8 | | | | | | | 6.25 | 205 | 6.25 | 205 | | | 6.25 | 205 | | | 3108 |
| Windows S | 28.9 | 10 | 284 | | | 20 | 568 | | | | | | | | | | | 852 |
| Walls ShadeN | 1.55 | | | | | | | | | | | | | | | | | 322 |
| Sunlit | 3.1 | 168.25 | 522 | | | 189.25 | 589 | 101.75 | 158 | 105.75 | 164 | | | | | | | 2103 |
| Ceiling | 1.44 | 239 | 344 | 87 | 125 | 202 | 291 | 181 | 261 | 179 | 258 | 111 | 410 | 25 | 36 | 31 | 45 | 1520 |
| Cooking | | | 1140 | | | | 1200 | 12.5 | | 12.5 | | 67.5 | 209 | 37.25 | 117 | | | 1200 |
| People | 380 | 3 | 1140 | | | | | 1 | 380 | 1 | 380 | 1 | 380 | | | | | 2280 |
| Room Sensible Heat | | | 4578 | | 125 | | 3427 | | 1970 | | 1628 | | 1159 | | 358 | | 45 | 12790 |
| Room CFM | | | 394 | | 11 | | 295 | | 126 | | 140 | | 100 | | 31 | | 4 | |

HOUSE TOTAL SENS BTUH: 12790  
DUCT GAIN 18 %: 2302  
VENTILATION LOAD BTUH: 3589  
TOTAL SENSIBLE BTUH: 18676  
× 1.35 HUM FACTOR  
GRAND TOT. = TOT SENS × HUM FACTOR: 25213  

**% DUCT GAIN**

| | 1 Story | 1½ Story | 2 Story |
|---|---|---|---|
| Ducts Insulated | 8 | 6 | 4 |
| Ducts Not Insulated | (18) | 15 | 12 |

NOTE: All ducts in attic spaces must be insulated with 3" minimum thickness insulation with vapor barrier. Ducts in crawl spaces and damp basements must be covered with 2" minimum thickness insulation with vapor barrier. Do not figure duct gain for concrete slab under floor ducts.

**HUMIDITY FACTORS**

| % Outside Hum | Below 40 | 40 to 45 | Above 45 |
|---|---|---|---|
| Factor | 1.25 | 1.30 | (1.35) |

**VENTILATION LOAD**

House Volume Cu. Ft. (LxWxH) × 40 BTUH/CU. Ft.  
40x28x8x.40=3584  

**WINDOW & DOOR FACTORS**

| Temp. Diff. °F | 15 | 20 | 25 | 30 |
|---|---|---|---|---|
| North | 12 | 18 | 25 | 31 |
| NE & NW | 23 | 29 | 36 | 42 |
| East & West | 32 | 38 | 44 | 51 |
| SE & SW | 35 | 41 | 48 | 55 |
| South | 26 | 32 | 39 | 45 |

Above Factors Assume Shades Or Venetian Blinds. If No Shades Double Factors.

For Outside Shading Or Awnings Multiply Factor x 60.

For Double Glazed Or Storm Windows Multiply Factor x 80.

NOTES:  
PEOPLE LOAD — One Person in each bedroom and three persons in living room.

ROOF OVERHANG — South walls only. 36" overhang provides complete shade to wall and windows. Use shade values for windows and walls.

DOORS — Treat all outside doors as windows.

Note: (The above list of factors are comprised of only those most commonly used.) For additional coefficients or more complete descriptions of these elements of structure, consult the Htg., Ventil. & Air Cond. Guide.

**WALL FACTORS**

| Temp. Diff. °F | Light Color | | | | Dark Color | | | |
|---|---|---|---|---|---|---|---|---|
| | 15 | 20 | 25 | 30 | 15 | 20 | 25 | 30 |
| 15 Daily Temp. Range No Insul | 3.6 | 5.4 | 7.2 | 9.0 | 5.1 | 6.9 | 8.9 | 10.7 |
| 1½" Insul | 1.5 | 2.1 | 3.0 | 3.6 | 2.7 | 3.6 | 4.7 | 5.1 |
| 3" Insul | 2.7 | 4.5 | 6.5 | 8.2 | 2.0 | 2.6 | 3.3 | 4.1 |
| 20 Daily Temp. Range No Insul | 2.6 | 4.7 | | | 3.7 | 4.7 | 5.4 | |
| 1½" Insul | 1.1 | 1.8 | 2.6 | 3.3 | 3.2 | 4.2 | 5.1 | |
| 3" Insul | 1.4 | 3.6 | 5.4 | 7.4 | 1.5 | 2.3 | 3.0 | 3.6 |
| 25 Daily Temp. Range No Insul | 0.9 | 1.7 | 2.6 | 4.1 | 2.9 | 3.6 | 4.4 | 5.0 |
| 1½" Insul | 0.6 | 1.5 | 2.9 | | 2.6 | 3.2 | 4.0 | 4.5 |

For Masonry Walls Multiply Factors x 12. For North or Shaded Wall Figure ½ Value in Table.

All R10 Insulation

**CEILING FACTORS**

| Temp. Diff. °F | 15 | 20 | 25 | 30 |
|---|---|---|---|---|
| 15 Daily Temp. Range 2" Insul | 5.7 | 6.9 | 8.7 | 9.6 |
| 4" Insul | 3.6 | 4.4 | 5.1 | 5.7 |
| 6" Insul | 3.2 | 4.0 | 4.6 | 5.1 |
| 20 Daily Temp. Range 2" Insul | 5.0 | 6.2 | 7.5 | 8.6 |
| 4" Insul | 3.2 | 3.7 | 4.7 | 5.4 |
| 6" Insul | 2.9 | 3.5 | 4.2 | 4.9 |
| 25 Daily Temp. Range 2" Insul | 4.4 | 5.5 | 6.9 | 7.6 |
| 4" Insul | 2.9 | 3.6 | 4.4 | 5.0 |
| 6" Insul | 2.6 | 3.2 | 4.0 | 4.5 |

Factors assume dark color roof with attic having vents. If light color roof — take 75% of values in tables. If no vents in attic space double values.

4.2+.22=.19; R=V/8; V/8x5.24+10=15.24; 4=V/8x15.24=.064x22=1.44

FIGURE 6-29  Btu Tabulation Cooling (Residential) (Courtesy of Amana Refrigeration, Inc.)

Now that we know the total cfm required and each individual room cfm, we will base our duct design on the heating mode. The heating mode is used in this example because it has the larger air requirement (1205 cfm compared to 1100 cfm).

Under normal circumstances, the air handler (indoor section) would be located at a midpoint in the structure. The required amount of air could then be moved in each direction, which would result in smaller duct trunk systems. Normally, we would try to place the outlet grills in each room so as to blanket the exterior walls to gain better comfort conditions within the building. In doing this, more branch runs will be required.

Starting from each end of the system, we would select a trunk duct size based on the required cfm at 0.05 in. WC static pressure and work back to the indoor section. After each two or three branch runs, the trunk duct size should be increased to keep the air velocity more nearly constant.

After the main duct is sized, each branch duct to each outlet grill at its required cfm could be calculated using a 0.05 in. (1.59 mm) WC static pressure.

To determine the total supply static pressure that the blower must overcome, we must determine which run, including the outlet grill, has the maximum static pressure. This will be either the longest duct run or the most complicated duct run.

## PAYBACK PERIOD

The payback period is the amount of time required to repay the customer for the extra cost of the equipment through the savings realized from using a heat pump system as compared to a conventional system. The payback period is calculated by the following formula:

$$\text{Number of years to payback} = \frac{\text{installed cost of heat} - \text{installed cost of conventional system}}{\text{operation cost of conventional system per year} - \text{operation cost of heat pump per year}}$$

or

$$\text{Pb} = \frac{A - B}{C - D}$$

where
Pb = payback period in years
$A$ = installed cost of heat pump
$B$ = installed cost of conventional system
$C$ = operation cost of conventional system per year
$D$ = operation cost of heat pump per year

## _____ SUMMARY _____

Precise calculations of both cooling and heating loads are required before the proper selection of a heat pump can be made.

The R and U factors, along with the factors used for insulation types, window types, physical location, the local climate design conditions, inside area, and number of occupants, are very critical in the proper calculation of the heat load.

Orientation refers to the position of the building on the site.

Dark exterior colors can be used to reduce the heat loss of a building.

The proper insulation of ceiling and walls and the use of storm windows is an investment that pays dividends upon installation.

General information is usually the first entry on the heat load calculation form.

Items 1, 2, and 3 on the heat load calculation form are the actual measurements of the building.

Factors for the building are the design factors that must be selected from the proper tables in accordance with the building structure and materials.

A U factor is the amount of heat transferred per square foot per degree of temperature difference between the inside and outside of the building.

Items 4 through 8 on the heat load calculation form deal with the windows, doors, exposed walls, ceiling, and infiltration.

Items 9 through 15 on the heat load calculation form deal with calculating the Btu loss of the building. The areas or linear feet in each step are multiplied by their U factor and the results placed in the column marked Btu.

A certain percentage of the total building heating capacity must be provided because of the heat loss through the duct system to the rooms.

A heat loss load calculation is an indication of the amount of heat lost through the structure during the heating season.

A heat gain load calculation is an indication of the amount of heat gained through the structure during the cooling season.

Awnings and sun screens aid in reducing heat gain loads through windows and doors.

Building heat gain loads can be effectively reduced by using white or light-colored roofs and exterior walls.

In an unventilated attic under a dark roof, temperatures can go as high as 160°F (71.11°C) and add significantly to the heat gain of a building.

The proper insulation of ceilings, walls, and windows is just as good an investment for cooling as it is for heating.

The local outside wet bulb temperatures must be taken into consideration when calculating heat gain loads.

The daily outside temperature range is another very important factor to be considered when computing building heat gains.

People produce heat, which adds to the total heat gain load of the building.

To control odors, mostly from tobacco, food, body, and waste material, fresh outside air should be introduced into the building.

The temperature differential between the outside temperature and the inside temperature is one of the most important factors to be considered when calculating the total heat gain load.

The fresh-air requirements (ventilation) are determined from the volume of the building.

Large volumes of air are usually recommended for dry and arid areas; smaller volumes of air are usually required for wet and humid areas.

The heat pump selected must match the building heat gain as closely as possible.

Use only one manufacturer's equipment, because both the indoor and outdoor sections have been designed to operate together. Mismatching equipment will prevent efficient operation and cause reduced economy.

If at all possible, use quick-connect tubing. This procedure prevents foreign material from entering the system during installation and provides a proper charge of refrigerant.

Provide proper air volume. It is best for optimum efficiency to have the proper amount of air flowing through the system.

Use emergency heat thermostats. This provides the customer with greater flexibility in system operation.

The customer should be informed about

the lower temperature of the air leaving a heat pump during the heating mode, how to properly use the indoor and outdoor thermostats, and the continuous compressor operation at temperatures below the system balance point.

The system balance point is the outdoor ambient temperature below which the heat pump compressor alone will not provide sufficient heat to warm the building to the indoor design temperature. It should be below 34 °F (1.67 °C) outside ambient temperature for efficient operation.

The seasonal performance factor (SPF) is an indication of the equipment efficiency when installed and operating. To make this calculation, we need the average number of heating hours per year at various outdoor ambient temperatures.

Because the heat pump is more efficient than straight electric resistance heat, it would be logical to use a heat pump when both heating and cooling are desired.

A cost analysis of the various fuels may be made to calculate the output cost per therm (100,000 Btu). The cost of various fuels and their availability will be the deciding factors on the unit to be installed.

Air distribution is as critical to heat pump operation as any other area because, without the proper amount of air flow on the heating cycle, a high compression ratio (high discharge pressures) situation can develop, which will tremendously shorten the life of the compressor.

The four basic types of air distribution systems are trunk duct, extended plenum, loop perimeter, and radial.

When adding cooling or a heat pump system to an existing duct system, the existing ducts should be checked completely to be sure that an adequate air flow can be accomplished.

The air quantity requirements of the equipment dictate the design of the duct system. The duct system must be able to handle an air quantity equal to that required by the equipment.

The ductwork and all the fittings used to change the direction of air flow cause a resistance to air flow that must be overcome by the blower.

The total resistance of the duct system and equipment cannot exceed the capacity of the blower used in the system.

To determine the total static resistance, a system static chart should be used in conjunction with the duct system layout.

Duct static drop is a measure of resistance to air flow through the air-distribution system. In calculating the static drop of a duct system, use only the longest or the most complicated duct run. The total static drop must include the straight run of duct work, all fittings, and grills.

The duct and equipment layout is important to the estimator for calculating the cost of the ductwork and to the installer for knowing what the estimator has planned.

The payback period is the amount of time required to repay the customer the extra cost of the equipment through the savings realized from using a heat pump system as compared to a conventional system.

## ____ REVIEW QUESTIONS ____

1. What is the correct method used to estimate the heat load of a building?
2. What is the position of the building on the site with respect to the north and south direction and the prevailing winds known as?
3. What type of color can be used on exterior walls and roofs to reduce the heat losses of the building?

4. Why are insulation and storm windows good investments?

5. Define the U factor used in calculating the heat losses and gains of a building.

6. Define net wall area.

7. What measurements are needed to calculate the heat losses and gains for a concrete slab on the ground?

8. What is the term used to indicate the amount of heat lost because of the air flowing through the ducts to the rooms?

9. For what season is heat gain calculated?

10. What effect will shading have on the heat gain load of a building?

11. Will attic ventilation reduce the heat gain of a building?

12. Occupancy refers to what type of heat gain load on a building?

13. To what does the term temperature differential refer?

14. Why do we need to calculate the heat gains and losses of individual rooms? ·

15. How is the people load determined on a heat gain calculation?

16. When calculating the heat gain of a building, do the north walls and windows have a different U factor than the south walls and windows?

17. What number of Btu are taken as the kitchen cooking heat gain in a building?

18. Below what point will the heat pump compressor begin to operate continuously?

19. What is the purpose of the seasonal performance factor (SPF)?

20. What components are used to help the heat pump compressor combat temperatures below the balance point?

21. What is the result of the annual electrical consumption of a heat pump system?

22. Which is more efficient, a heat pump or straight resistance heating?

23. In air-conditioning work, what constitutes a therm?

24. What would be the result of too small a duct system on a heat pump unit?

25. Name the four basic types of air-distribution systems.

26. What dictates the design of the duct system?

27. What component must overcome the resistance of the duct system?

28. What is duct static drop?

29. Why is the duct and equipment layout important to the estimator?

30. What is the term used to indicate the amount of time required to repay the customer the extra cost of equipment through the savings realized from system operation?

# 7

# Heat Pump System Installation

## INTRODUCTION

No system is better than its installation. Therefore, one's best efforts should be given to the installation job. The time has passed for slipshod workmanship in installation work. The proper installation of air-distribution systems will provide air circulation in the desired place and quantity. Also, the installation of the refrigerant lines is critical because of oil return problems and refrigerant flooding the compressor. Equipment manufacturers include installation information with each piece of equipment. These instructions should be followed so that satisfactory operation of the equipment can be achieved. The following pages provide installation data for the different types of equipment. However, before installing a unit, check all codes and ordinances that may affect the installation.

## LOCATING AND INSTALLING THE OUTDOOR UNIT

Several important factors must be given consideration before determining the final location of the outdoor unit; these include wind direction, sound, closeness to the building, electrical power location, drainage of defrost water, ease of installation of the refrigerant lines, and air restrictions.

The recommended method is to select a location as close to the building as possible. However, it should be away from the bedroom and living areas where the sound might be objectionable. Locate the unit so that the noise will not annoy the neighbors.

The prevailing wind direction should not cause the unit to malfunction during periods of high wind velocity. The electrical supply panel should be as close to the outdoor unit as pos-

162

sible to reduce the cost of running the electric wire.

The refrigerant lines to the indoor unit should be as close as possible. This will allow shorter refrigerant lines and fewer fittings, which will reduce the installation cost and the possibility of refrigerant leaks developing at connections. Also, a smaller refrigerant charge is used when this distance is short.

The outdoor unit should be located so that the discharge air will not be recirculated through the coil. Thus, the unit must not be located under the roof overhang or in the corner with a wall on two sides (Figure 7-1). The recirculation of air will reduce unit efficiency.

When heat pump units are operating, they occasionally go into a defrost cycle. The melted water from the defrost cycle must be allowed to drain away from the unit. Otherwise, the water will freeze again on the coil bottom and cause a greater accumulation with each defrost cycle, thus reducing the amount of effective coil surface and the unit efficiency (Figure 7-2).

The user should be cautioned about planting shrubs and flowers where they will restrict air flow, either to the air intake or from the discharge. Leaves from such plants will often

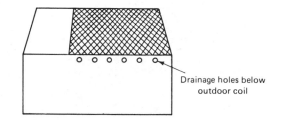

**FIGURE 7-2** Outdoor Unit Drainage Holes

collect on the air inlet side of the coil and must be removed before proper operation can be attained.

In climates where snowfall is likely, certain precautions should be considered to prevent snow from hindering heat pump operation. In milder climates where there is very little or no snowfall, a concrete slab or blocks may be used to support the unit (Figure 7-3). The slab should be installed on a bed of gravel and have an opening through which the defrost water can drain into a gravel bed extending out 12 in. (304.80 mm) from the unit in all directions. The gravel bed will allow the water to drain away from the unit faster and not leave a puddle. The slab should provide about 6 in. (152.40 mm) between the unit base and the ground.

In climates where the annual snowfall is heavy, consideration should be given to the

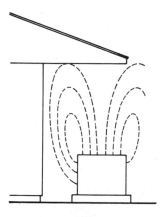

**FIGURE 7-1** Recirculating Air to Outdoor Unit

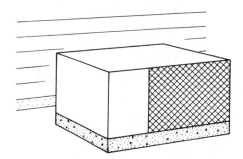

**FIGURE 7-3** Concrete Slab or Blocks Supporting Outdoor Unit

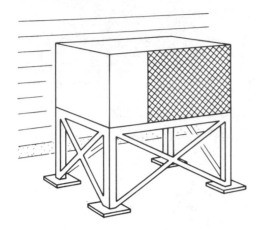

**FIGURE 7-4** Frame to Support Unit in Heavy Snowfall Areas

height of the unit from the ground. A frame support should be built to raise the unit above anticipated snow depth. The legs are usually mounted in concrete to prevent movement (Figure 7-4).

Remember that the outdoor unit is heavy and should be supported above ground and be level. The mounting should be rigid enough to maintain its position. The unit should be installed with the clearances specified in the manufacturer's specifications for air flow and service access. The drainage holes in the unit base pan should be clear for free water drainage.

## LOCATING AND INSTALLING THE INDOOR UNIT

The location of the indoor unit is just as important as the location of the outdoor unit. It may be placed in almost any available place that will allow proper access and clearance, such as a utility room, basement, attached garage, attic, or closet. The installation instructions accompanying the unit will indicate the proper clear-

ances and access and whether the unit can be mounted horizontally or vertically.

The indoor location should be such that a minimum amount of refrigerant tubing between the indoor and outdoor unit is required. The type of duct system will also have a bearing on the location. A radial system would require a central location; other systems are more flexible in unit location. The distance from the electrical service panel should be as short as possible.

When connecting the supply and return ducts, use canvas connectors. Install duct liner insulation in the return air plenum to reduce the fan noise being transmitted to the living area. The unit supports should include vibration isolators or some type of sound-dampening pads to prevent vibration to the building structure.

The flexibility of heat pump application will vary depending on the type of building construction and the location. Therefore, several typical types of installation will be discussed in the material below.

*Crawl Space and Attic:* When installing the indoor unit in crawl space, it is recommended that the unit be suspended from floor joists with hangers (Figure 7-5). Be sure that there is sufficient clearance around the unit to allow servicing.

Units installed in an attic require certain precautions. To comply with FHA requirements, an auxiliary drain pan must be installed under the unit (Figure 7-6). This precaution is to prevent condensate drain water from dripping onto and ruining the ceiling should the regular drain become clogged. Attic-installed units may be suspended from the rafters with hanger straps. An alternative method is to lay the unit on the auxiliary drain pan and not use the hanger straps. Be sure to use spacers to separate the unit and the auxiliary drain pan to prevent sweating on the outside of the pan.

**FIGURE 7-5** Crawl Space Installation of Indoor Unit

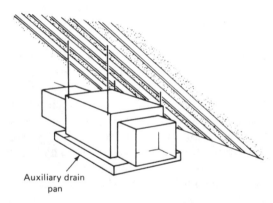

Auxiliary drain
pan

**FIGURE 7-6** Attic Installation of Indoor Unit with Auxiliary Drain Pan

The treated air is generally supplied to the room with high sidewall or ceiling diffusers in attic installations. The return air grills should be kept as low as possible. Crawl space installations generally use floor supply grills and floor return grills.

*Basement:* In the north where split systems are usually the most practical, the basement is generally the best location for the indoor unit, especially in homes with full basements.

When installing the indoor unit in a basement, the ductwork is usually designed so that it will run parallel to or between the floor joists. Floor supply and return grills are used in this type of installation.

The unit should be placed on a stand to allow a sheet-metal return plenum to be placed underneath (Figure 7-7). Some units have a knockout on the side of the blower compartment to allow connecting the return duct to the side of the unit. (Figure 7-8). When installing

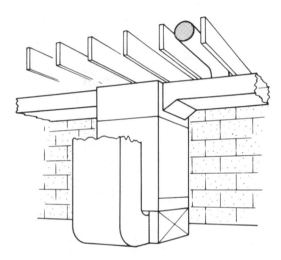

**FIGURE 7-7** Basement Installation of Indoor Upflow Unit

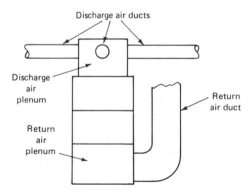

Discharge air ducts

Discharge
air
plenum

Return
air
plenum

Return
air duct

**FIGURE 7-8** Indoor Unit with Side Return Duct Connection

units that allow either vertical or horizontal air flow, the unit may be suspended from the floor joists like the crawl-space type of installation.

*Closet:* Some buildings do not lend themselves to satisfactory location of the equipment in the basement, crawl space, or attic for the indoor unit. In some buildings only a closet or utility room is available. Whether the unit is to be installed vertically or horizontally depends on the amount of room in the closet and unit flexibility (Figure 7–9). The horizontal unit will require that an auxiliary drain pan be added.

Horizontal

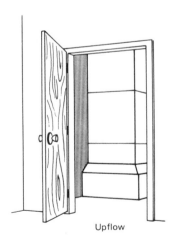

Upflow

**FIGURE 7-9** Closet Installation of Indoor Unit

The ductwork will generally be in a furred-in place overhead using individual high-sidewall supply grills or ceiling diffusers. A central high-sidewall return grill is usually satisfactory.

## INSTALLING THE AIR-DISTRIBUTION SYSTEM _____

Regardless of the type of air-distribution system selected, certain procedures must be followed. In a new home, all the concealed ductwork, such as in the wall or slab, must be in place at the proper time during the construction phase. When the building is built on a slab-type foundation, any ductwork required to be in the slab must be in place before the concrete is poured. When ducts are to be installed in the slab, capacity losses due to groundwater and heat conduction of the slab must be taken into consideration. The ducts must be installed to provide proper drainage of surface water. Installation of vapor barriers will help reduce condensation in the ducts. The supply ducts should slope toward the plenum to allow for proper drainage. All ducts embedded in the slab foundation should be covered with a minimum of $2\frac{1}{2}$ in. (63.50 mm) of concrete.

Some contractors use sheet-metal ducts in the slab while others prefer to use the cement-asbestos ducts. The one used is usually a matter of personal preference and, of course, cost. Check the local codes to be sure that all regulations are complied with.

When the indoor unit is located in the basement, the supply and return air ducts to and from the upper-floor rooms must be installed before the inside walls are sealed with sheet rock. The spaces between the wall studs are usually used for concealing the ducts. There must be openings through the floors to allow the duct to pass through from unit to grill.

The types of air-distribution system most often used for residential heat pump applications are the reducing trunk and the extended plenum types. The reasons for this are simpler design, ease of installation, and lower cost. The use of high-sidewall grills is most popular because they are more practical, efficient, and economical.

Proper air distribution and temperature balance are important in any type of installation; however, they are extremely critical in multistory homes. These homes have open stairways or upper-floor balconies and are subject to noticeable temperature variations and air currents caused by the interchange of air between the floors. The warm air from the lower floors tends to rise to the upper floors, while the cool air from the upper floors tends to fall to the lower floors. Thus it is much more difficult to heat the lower-level floors during the winter and cool the upper-level floors during the summer. To alleviate this situation, the temperatures must be carefully balanced by adjusting the air flow to these areas and by constant operation of the indoor fan.

When installing a heat pump in an existing home, certain challenges are encountered that are not present in new construction. There is very little problem in installing the ducts in new construction because the stud spaces are open and the installer is in command of the job. However, in an existing home, the installer must improvise and plan the installation around the limitations of the present structure, allowing only minor changes in the structure. Some older homes may have an existing air-distribution system, and it should be checked to be certain that it meets present-day standards. If not, it must be replaced or the required changes made to bring the system up to standards.

There are certain acceptable sound levels and comfortable air motion that depend upon air velocity. Air velocities of less than 15 feet per minute (fpm; 4.5720 m/min) in an occupied room will usually cause a feeling of stagnation. On the other hand, velocities greater than 65 fpm (19.8120 m/min) may cause uncomfortable drafts. Therefore, air motion be-

tween 20 and 50 fpm (6.0960 and 15.2400 m/min) will generally be acceptable. Therefore, when a forced-air heating system in an older home is being replaced, the air-distribution system will probably be too small. It then becomes necessary to install larger ducts or additional ducts to maintain an acceptable air velocity.

Many types of duct material are available: rectangular, shop-fabricated sheet-metal ducts; fiberglass duct that can be custom fabricated in the shop to fit the installation; round duct available from most air-conditioning equipment suppliers; and flexible round duct that can be used without many of the fittings required on the other types of materials. The type of material used will depend on the installation. The fiberglass material would not be satisfactory installed in a concrete slab. The type used will also depend upon availability, cost and building construction and codes.

All metal ducts that are not placed in the slab foundation should be insulated with a minimum 2-in. (50.80-mm) blanket of insulation with a vapor barrier on the outer surface. It is best not to pull the insulation very tight during installation because some of the insulation value will be reduced. This is because of reduced thickness, and consequently fewer air spaces, of the insulation.

A generally accepted practice in running the individual duct runs is to set the equipment, install the plenum, and start each duct run from the plenum. To start from the grill end of the run would cause extreme difficulty in making the connection into the plenum. Duct runs should be adequately supported to prevent sagging.

## INSTALLING THE REFRIGERANT LINES

The location of the indoor unit with respect to the outdoor unit will determine the length and complexity of the refrigerant lines. The tubing should be run in as short a path as possible. The line length and diameter must be sized according to the manufacturer's recommendations for the particular unit being installed. The manufacturer's pipe sizing table should be consulted to ensure proper sizing (Figure 7–10). Refrigerant lines may be run in several places. They may be run under the floor on a pier and beam construction, in or under a concrete slab, or up an outside wall and across the ceiling joists. The type of construction will usually dictate where the lines will be run. Also, whether the installation is being made in a new or existing building will dictate some installation procedures. However, if the refrigerant lines are to be buried in a concrete slab, a chase should be installed to allow removal of the lines for repairs.

Another item of concern in the installation of a heat pump is the vapor line on a split system. This line between the compressor and the indoor unit and outdoor unit serves alternately as the suction line or the hot vapor line, depending on whether the unit is heating or cooling. This line should be insulated. The insulation should be capable of withstanding the high temperatures of the discharge vapor. If precharged lines from the factory are used, the insulation will already be on the tubing. The line should not be rigidly attached to the building in any way to avoid the possibility of vibra-

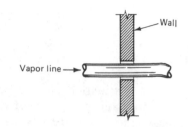

**FIGURE 7-11** Vapor Line installation

tions being transmitted to the living area. (Figure 7–11). The liquid line should not come into contact with the vapor line, which would allow an undesirable exchange of heat to occur.

If at all possible, refrigerant lines should be purchased in sets from the manufacturer in the lengths required. When tubing is to be purchased locally, it should comply with code requirements as to wall thickness. It should be insulated with $\frac{1}{2}$-in. (12.70-mm)-thick foam-rubber-type insulation. An approved type of solder should be used in making any joints or connections with the tubing.

*Purging the Lines:* Some units are shipped with the correct amount of refrigerant in the system. However, some manufacturers ship their split systems with only a holding charge. This is a very small amount of refrigerant vapor and serves only to keep air and moisture out of the system. It may be only a few ounces, whereas the operating charge may be approximately 5 lb (2.2680 kg) of refrigerant. The holding charge is usually purged, the system evacuated, and the proper amount of refrigerant charged into the system. The proper operating charge must be added according to the manufacturer's recommendations for the unit and the tube length. However, some manufacturers provide data indicating the suction and discharge pressures at given temperatures (Figure 7–12).

| SPLIT SYSTEM HEAT PUMP FIELD PIPING | | | | | | | | | | | |
|---|---|---|---|---|---|---|---|---|---|---|---|
| UNIT TONS | MAXIMUM VERTICAL SEPARATION ABOVE/BELOW OUTDOOR (FT) | | REFRIG. | REFRIGERANT LINE LENGTH (FT.) | | | | | | | |
| | | | | 10-25 | | 26-50 | | 51-75 | | 76-100 | |
| | | | | VAP | LIQ | VAP | LIQ | VAP | LIQ | VAP | LIQ |
| | | | | LINE O.D. (IN) | | | | | | | |
| 2 | 50 | 50 | R-22 | 5/8 | 3/8 | 5/8 | 3/8 | 5/8 | 3/8 | 3/4 | 3/8 |
| 2-1/2 | 40 | 60 | R-22 | 3/4 | 3/8 | 3/4 | 3/8 | 7/8 | 3/8 | 7/ | |
| 3 | 40 | 60 | R-22 | 3/4 | 3/ | | | | | | |

**FIGURE 7-10** Typical Unit Line Sizing Chart (Courtesy of Carrier Air Conditioning)

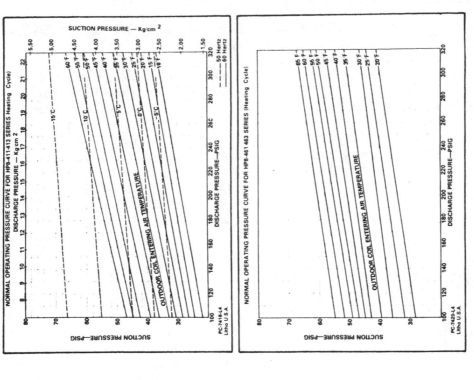

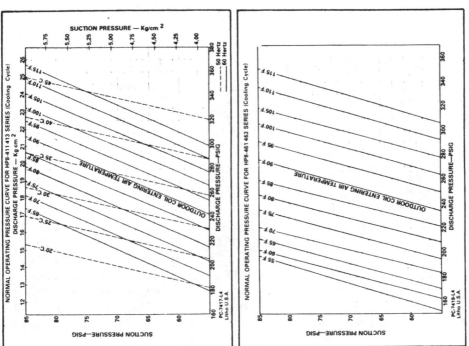

**FIGURE 7-12** Typical normal operating refrigerant pressures indicating proper charge (Courtesy of Lennox Industries, Inc.)

*Leak Testing:* All tubing connections must be tested for leakage before starting the unit to be certain that there are no refrigerant leaks. There are several methods for testing for leaks. The one used is usually a matter of personal preference. After leak testing, the system should be evacuated to remove any air and moisture. The vacuum may be broken by opening the system service valves on a fully charged unit or by adding the proper amount of refrigerant into the system from a refrigerant cylinder.

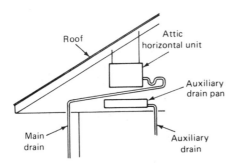

Attic unit

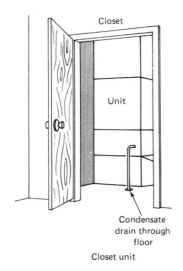

Closet unit

**FIGURE 7-13** Typical drainpipe installations

## INSTALLING THE CONDENSATE DRAIN

When installing a unit in which the coil will be operating at temperatures below the dew point, a condensate drain must be provided to carry away the accumulated water. The drain pipe should be a minimum of $\frac{3}{4}$ in. (19.05 mm) in diameter. It should have sufficient pitch downward to allow free drainage of the coil (Figure 7-13). The drain should incorporate a trap and be installed and designed in accordance with local codes.

When the indoor unit is installed above a ceiling, adequate protection from water leaks should be provided. This will normally require the use of an auxiliary drain pan below the unit. The auxiliary drain pipe should terminate in a conspicuous location such as above a back door, over a bath tub, or over a window (Figure 7-13). Dripping at these points would be noticeable and would alert the user that the primary drain is not draining properly and that service is required.

## ELECTRIC POWER SUPPLY

It is necessary that all electrical wiring comply with all local and national electrical code requirements. Both indoor and outdoor units and self-contained packaged units are factory assembled and wired internally for the specified nameplate voltage and ratings. The electrical requirements of the unit being installed must be the same as the supply voltage to the building.

The main electrical service must have enough capacity to supply the heat pump, auxiliary resistance heaters, and building load requirements. Obtain the total full-load amperes and wire sizing from the unit installation instructions (Figure 7-14).

| Style | Voltage | Comp. Motor | | Cond. Motor | | Evap. Motor | | Minimum Circuit Ampacity | Max. Fusetron Size* | No. Wires and Size |
|---|---|---|---|---|---|---|---|---|---|---|
| | | FLA | LRA | FLA | LRA | FLA | LRA | | | |
| PH2-1E | 208-230/1/60 | 14.7 | 60 | 1.6 | 4.6 | 2.2 | 4.4 | 24.8 | 30 | 3-#10 |
| PH2.5-1E | 208-230/1/60 | 19.8 | 76 | 1.9 | 4.6 | 3.0 | 6.4 | 31.9 | 35 | 3-#8 |
| PH3-1E | 230/1/60 | 23.6 | 100 | 1.9 | 4.6 | 3.5 | 9.0 | 37.3 | 40 | 3-#8 |

* Local Code takes precedence over the recommended fuse and wire sizes shown in the table.

**FIGURE 7-14** Typical wire sizing chart (Courtesy of Amana Refrigeration, Inc.)

A separate electrical disconnect should be provided for the indoor unit, outdoor unit, and auxiliary heaters. Many manufacturers provide built-in fuses for the indoor unit for each individual circuit when the auxiliary heaters exceed 10 kW. The location and fusing must conform to local and national electrical code requirements.

Fused disconnect switches provide both wire protection and a means of turning off the power supply for servicing the unit or in case of an emergency. The disconnect to the outdoor unit should be weatherproof and located within 5 ft (1.524 m) of the outdoor unit. All outdoor wiring should be run in a weathertight conduit to resist moisture and reduce corrosion and abrasion of the wiring insulation.

When wire sizes are not given by the manufacturer, the wire sizes and fuses should be sized for the ampacity of the wire to the unit. When using aluminum wire, as opposed to copper, a larger-diameter wire is required. Also, all connectors not marked AL-CU (aluminum-copper) must be coated with a corrosion inhibitor, such as Burndy-Penetrox A or Alcoa #2 EJC, and wrapped with electrical tape to reduce galvanic corrosion and loosening of the connection.

Most codes require that the unit be mechanically grounded. In such cases, a ground wire of the proper size must be connected to a ground lug on the unit chassis and to a grounded connection in the main electrical service panel.

## CONTROL WIRING

The control wiring and its function are probably the most difficult to understand parts of the heat pump. Some control functions, such as the indoor and outdoor thermostats, are quite obvious. However, many devices have been wired in by the manufacturer to perform other needed functions, such as relays, transformers, and capacitors. The majority of the controls and the circuit consist of components that are energized and operate on low voltage (24 V), which is provided by the transformer secondary winding. When these components are energized by the low voltage, they directly control the operation of some line voltage components. It is the responsibility of the installation and service technician to learn how these controls function and are wired into the overall control circuit.

## LOCATING AND INSTALLING THE PACKAGED HEAT PUMP

The installation procedures that apply to the outdoor unit on split systems will generally apply to packaged units also. However, a few additional items must be considered.

Since the packaged unit contains both the indoor and outdoor sections, the additional weight and larger dimensions require stronger mounting suppports. This is especially so in the snowbelt area where the unit must be higher above the ground.

In the south, however, roof mounting is frequently preferred (Figure 7–15). The unit mounting feet must be properly fastened to a roof that has been designed to carry the extra weight of the unit. Vibration eliminators are used in both the mounting frame and the ductwork to prevent vibration being transmitted into the living area.

In this type of installation, both the supply and return air ducts must be insulated. The insulation must be on the inside of the duct when it is exposed.

Adequate weatherstripping and a duct sealer must be used where the ductwork goes through the roof of the building. Special metal flashings are used to aid in weatherstripping these points.

When mounting the unit on the ground level, free drainage of the condensate and defrost water must be provided for. The ductwork will usually be located in a crawl space or in a basement (Figure 7–16). Some buildings may be designed with a chase on the outside wall to allow the ductwork to be installed in the attic. In this case, individual ceiling diffusers may be used. Otherwise, floor diffusers are required.

## POSTINSTALLATION AND STARTUP CHECKS _____

The following are some points that should be checked after the installation is complete and during the initial startup.

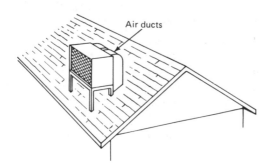

**FIGURE 7-15** Roof-mounted self-contained unit

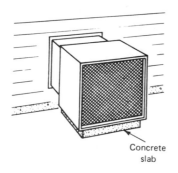

**FIGURE 7-16** Packaged heat pump ground-level installation

1. Energize the crankcase heater, if equipped, in accordance with the unit installation instructions prior to the initial startup.
2. Has the duct system been properly installed and approved?
3. Have both the indoor and outdoor units, or packaged unit, been checked for proper mounting and installation according to the manufacturer's specifications?
4. Has the refrigeration system been properly leak tested?
5. Is the complete vapor line insulated with the proper insulation?
6. Check the condensate drains for proper drainage by pouring water into the drain pan.
7. Have vibration eliminators been applied to the mounting supports and the ductwork?
8. Does the electrical wiring comply with all the codes?

9. Is there a filter in the return air duct-work to the indoor unit?

10. Open all duct dampers and supply air diffusers.

11. Turn the thermostat off.

12. Turn on the required voltage disconnects.

13. Rotate the fans to check for free rotation.

14. Open liquid and vapor line service valves.

15. Set the heat anticipator.

16. Set the thermostat to demand system operation.

17. Check the voltage readings and amperage draw to all motors.

18. Check the compressor operation in both heating and cooling modes.

19. Check the system refrigerant charge and pressures.

20. Check the indoor blower for the proper CFM requirement.

21. Balance the supply air system.

22. Jumper the outdoor thermostats to ensure auxiliary heater operation and that the emergency heat system is working.

23. Instruct the homeowner about system operation and things to check before calling a service technician.

——————— **SUMMARY** ———————

No system is better than the installation.

The proper installation of the air-distribution system will provide air circulation in the desired place and quantities.

The installation of the refrigerant lines is critical because of oil return problems and refrigerant flooding the compressor.

The wind direction, sound, closeness to the building, electrical power location, drainage of the defrost water, simple installation of the refrigerant lines, and air restrictions are important considerations in the placement of the outdoor unit.

Certain precautions should be taken to prevent snow from hindering the heat pump operation, such as raising the unit higher to be above the anticipated snow level.

The outdoor unit is heavy and should be level and properly supported to maintain this position.

There should be proper clearance around the unit for proper air flow and service access.

The defrost drain holes in the unit base pan should be clear for free water drainage.

Indoor units may be installed in a utility room, basement, attached garage, attic, or closet. Be sure to allow the proper clearances around the unit as specified in the manufacturer's instructions.

The refrigerant piping should be kept as short and as simple as possible.

The indoor unit should be located as close to the electrical service panel as possible.

Use canvas connectors when installing supply and return ducts to prevent vibration transmission to the living area.

On horizontal indoor unit installations, an auxiliary condensate drain pan must be installed under the unit.

In a basement-type installation, the unit should be placed on a stand to allow a sheet-metal return air plenum to be placed underneath.

In a new home installation, all concealed ductwork, such as in the wall or slab, must be in place at the proper time during the construction phase.

When ducts are to be installed in the slab, the capacity losses due to groundwater and heat conduction of the slab must be taken into consideration.

All ducts embedded in the slab foundation should be covered with a minimum of $2\frac{1}{2}$ in. (63.50 mm) of concrete.

Either sheet-metal ducts or cement-asbestos ducts may be installed in a slab foundation.

Proper air distribution and temperature balance are important in any type of installation; however, they are extremely critical in multistory homes. The temperatures must be carefully balanced by adjusting the air flow and constant fan operation.

Existing duct systems should be carefully checked to see that they conform to present-day standards.

Air motion between 20 and 50 fpm (6.0960 and 15.2400 m/min) will generally be acceptable.

When a forced-air heating system in an older home is being replaced, the air-distribution system will probably be too small. It then becomes necessary to install additional or larger ducts to maintain an acceptable air velocity.

All metal ducts that are not placed in the slab foundation should be insulated with a minimum 2-in. (50.80-mm) blanket of insulation with a vapor barrier on the outer surface.

The indoor and outdoor units should be located as close together as possible to reduce the length of the refrigerant lines. The diameter of the lines should meet the manufacturer's recommendations for the particular unit being installed.

The vapor line on a heat pump unit should be insulated completely with $\frac{1}{2}$ -in (12.70-mm)-thick foam-rubber-type insulation. The insulation should be capable of withstanding the high temperatures of the discharge vapor.

A holding charge is a very small amount of refrigerant vapor and serves only to keep air and moisture out of the system. The holding charge is usually purged, the system evacuated and the proper amount of refrigerant charged into the system, according to the manufacturer's recommendations.

All tubing connections must be tested for leakage before starting the unit to be certain that there are no refrigerant leaks.

When installing a unit in which the coil will be operating at temperatures below the dew point, a condensate drain must be provided to carry away the accumulated water. The drainpipe should be a minimum of $\frac{3}{4}$ in. (19.05 mm) in diameter.

It is necessary that all electrical wiring comply with all local and national electrical code requirements. The unit should be supplied with the type of electrical power called for on the equipment nameplate. There should be a separate electrical disconnect for the indoor unit, outdoor unit, and auxiliary heaters. The disconnect to the outdoor unit should be weatherproof and be located within 5 ft (1.524 m) of the outdoor unit.

Aluminum wiring requires a larger diameter than copper wiring. Also, all connectors not marked AL-CU (aluminum-copper) must be coated with an approved corrosion inhibitor.

Most electrical codes require that the unit be mechanically grounded.

The control wiring and its function are probably the most difficult to understand parts of the heat pump. The majority of the controls and the circuit consist of components that are energized and operate on low voltage (24 V), which is provided by the transformer secondary winding.

The packaged unit, being heavier in weight, requires a strong mounting and supports.

When a packaged unit is installed, both the supply and return air ducts must be insulated. The insulation must be on the inside of the duct when it is exposed.

Adequate weatherstripping and a duct sealer must be used where the ductwork goes through the building.

—— **REVIEW QUESTIONS** ——

1. What will provide air circulation in the desired place and quantity?
2. Why is the proper installation of the refrigerant lines critical?
3. What could happen if the drain holes in the outdoor unit base pan were stopped up?
4. What precaution should be observed when installing the outdoor unit in an area where snowfall is likely?
5. Name some of the places where an indoor unit may be located.
6. Where should the indoor unit be located in relation to the outdoor unit?
7. What should be used when connecting ductwork to the unit to prevent vibration?
8. What should be done to comply with FHA requirements on a horizontally mounted indoor unit?
9. Why should an indoor unit installed in a basement be placed on a stand?
10. What should be considered, in reference to unit capacity, when installing the ductwork in a slab foundation?
11. How should the supply air ducts be installed in a slab foundation?
12. What is the minimum allowed thickness of concrete covering ductwork in a slab foundation?
13. What two types of material are used to make ducts for installation in a slab?
14. What is extremely critical in multistory homes?
15. What will too low an air velocity in a room cause?
16. What is the minimum thickness of insulation placed on ductwork?
17. From where should each individual duct run be started?
18. What should be done to the vapor line on a heat pump system?
19. Is it good practice to run the vapor and liquid refrigerant lines so that they touch?
20. What should the electrical power supply match?
21. How much capacity must the electrical service have?
22. Where should the disconnect for the outdoor unit be located?
23. What precautions should be taken when connecting aluminum wire to copper wire?
24. What must be done to the supply and return air ducts used on packaged heat pump installations?
25. What precautions must be taken when installing a packaged unit on the roof?

# 8

## Troubleshooting Heat Pump Systems

The job of making a service call is challenging. The service technician should approach each problem with an open mind and no preconceived ideas about what the problem may be. Through knowledge and correct observation, a properly trained service technician can quickly and economically determine and correct any malfunctions.

Efficient servicing of the equipment is an asset to all those involved. It leads to a satisfied customer, the dealer makes a fair profit, and the technician is well paid because of his knowledge and skills.

To provide efficient service, the technician must use a systematic approach to troubleshooting the equipment. In most cases, a thorough questioning of the customer and close observation of the equipment will allow for an early solution to the problem. In this section we will outline and explain the steps that should be followed when troubleshooting a heat pump system. The primary intention of this section is to assist the service technician in diagnosing problems and repairing heat pump equipment. The use of this section should allow more competent and economical service of the equipment.

The problems encountered in servicing heat pump equipment can be divided into the following categories:

1. When the heat pump unit does not operate at all, the problem is in the electrical circuit.
2. When the system will run but not heat or cool, the problem is in the mechanical components.
3. The problem can also be the result of a combination of electrical and mechanical malfunctions.

176

## ELECTRIC HEAT

| Condition | Possible Cause | Corrective Action | Reference |
|---|---|---|---|
| Unit will not run | 1. Blown fuse | 1. Replace fuse and correct cause | 3-1 |
| | 2. Burned transformer | 2. Replace transformer and correct cause | 23-1, 23-1A |
| | 3. Thermostat not calling for heat | 3. Set thermostat | 6-1, 6-1A |
| | 4. Defective thermostat | 4. Replace thermostat | 6-1, 6-1A, B |
| | 5. Defective heating relay | 5. Replace relay | 40-1 |
| | 6. Lockout relay energized | 6. Reset lockout relay | 43–1 |
| Fan will not run | 1. Burned fan motor | 1. Repair or replace fan motor | 22-1, 22-1A, B, C, D, E, F |
| | 2. Broken fan belt | 2. Replace fan belt | 15-1E |
| | 3. Burned contacts in fan relay | 3. Replace fan relay | 25-1 |
| | 4. Defective fan control | 4. Replace fan control | 38-1, 38-1A, B |
| | 5. Defective wiring or connections | 5. Repair wiring or connections | 8-1, 9-1 |
| Fan motor hums but will not start | 1. Defective fan motor bearings | 1. Replace bearings or fan motor | 22-1, 22-1A, B, C, D, E, F |
| | 2. Defective fan motor starting switch | 2. Repair starting switch or replace motor | 22-1, 22-1D |
| | 3. Defective starting capacitor | 3. Replace capacitor | 11-1, 11-1A |
| | 4. Burned start windings in motor | 4. Repair or replace motor | 4-2, 4-2A, B, C |
| | 5. Defective blower bearings | 5. Replace bearings | 41-1, 41-1B |
| | 6. Loose wiring connections in motor starting circuit | 6. Repair wiring | 8-1, 9-1, 10-1 |
| Fan motor cycles | 1. Defective motor bearings | 1. Replace bearings or motor | 22-1, 22-1E |
| | 2. Defective blower bearings | 2. Replace blower bearings | 41-1, 41-1B |
| | 3. Defective run capacitor | 3. Replace capacitor | 11-1, 11-1B |
| | 4. Defective fan control | 4. Replace fan control | 38-1, 38-1A, B |
| | 5. Defective fan relay | 5. Replace relay | 28-1, 25-1 |
| | 6. Defective motor windings | 6. Repair or replace motor | 22-1, 22-1A, B, C |

## ELECTRIC HEAT

| Condition | Possible Cause | Corrective Action | Reference |
|---|---|---|---|
| Fan blows cold air | 1. Defective heat sequencing relays | 1. Replace relays | 32-1, 37-1A, B, C |
| | 2. Burned heat elements | 2. Replace elements | 36-1, 36-1A, B, C |
| | 3. Loose wiring connections | 3. Repair wiring | 8-1 |
| | 4. Defective thermostat | 4. Replace thermostat | 6-1, 6-1A, B, C, D |
| | 5. Fan set to "on" position | 5. Set to "auto" position | 6-1, 6-1D |
| | 6. Defective fan control | 6. Replace fan control | 38-1, 38-1A, B |
| Not enough heat | 1. Dirty air filters | 1. Clean or replace filters | 15-1B |
| | 2. Unit too small | 2. Install more elements | 27-1 |
| | 3. Too little air flow through furnace | 3. Increase air flow; remove restrictions | 15-1B, C, D, E |
| | 4. Thermostat heat anticipator not properly set | 4. Set heat anticipator | 6-1, 6-1B |
| | 5. Defective fan motor | 5. Repair or replace fan motor | 22-1, 22-1A, B, C, D, E, F |
| | 6. Air conditioning evaporator dirty | 6. Clean evaporator | 15-1C |
| | 7. Thermostat not properly located | 7. Relocate thermostat | 6-1, 6-1C |
| | 8. Thermostat set too low | 8. Set thermostat | 6-1, 6-1D |
| | 9. Thermostat out of calibration | 9. Calibrate thermostat | 6-1, 6-1A, B |
| | 10. Low voltage | 10. Correct cause | 10-1 |
| | 11. Air ducts not insulated | 11. Insulate ducts | 26-1 |
| | 12. Burned elements | 12. Replace elements | 36-1, 36-1A, B, C |
| | 13. Defective heat sequencing relays | 13. Replace relays | 37-1, 37-1A, B, C |
| | 14. Defective thermostat | 14. Replace thermostat | 6-1, 6-1A, B, C, D |
| | 15. Defective element limits | 15. Replace limits | 36-1, 36-1A, B |
| | 16. Outdoor thermostat set too low | 16. Reset thermostat | 6-1, 6-10 |
| Too much heat | 1. Unit too large | 1. Reduce BTU input | 39-1 |
| | 2. Thermostat heat anticipator not properly set | 2. Set heat anticipator | 6-1, 6-1B |
| | 3. Thermostat not properly located | 3. Relocate thermostat | 6-1, 6-1C |

# ELECTRIC HEAT

| Condition | Possible Cause | Corrective Action | Reference |
|---|---|---|---|
| Too much heat | 4. Thermostat set too high | 4. Set thermostat | 6-1, 6-1A |
| | 5. Thermostat out of calibration | 5. Calibrate thermostat | 6-1, 6-1A, B |
| High humidity in building | 1. Humidity due to cooking | 1. Vent cookstove | 35-1 |
| | 2. Humidity due to bathing | 2. Vent bathroom | 35-1 |
| | 3. Humidity due to rain | 3. Increase temperature rise through furnace | 42-1 |
| Blown element limits | 1. Shorted heating element | 1. Replace element and correct cause | 36-1, 36-1A, B, C |
| | 2. Dirty filters | 2. Clean or replace filters | 15-1B |
| | 3. Dirty blower | 3. Clean blower | 15-1D |
| | 4. Broken or slipping fan belt | 4. Adjust or replace belt | 15-1E |
| | 5. High or low voltage | 5. Notify power company | 1-1, 10-1 |
| | 6. Defective blower motor | 6. Replace motor | 22-1, 22-1A, B, C, D, E, F |
| | 7. Not enough air through furnace | 7. Remove restriction | 15-1B, D, E |
| | 8. Loose electrical connections | 8. Repair connections | 8-1, 9-1 |
| High operating costs | 1. Unit too small | 1. Increase number of elements | 22-1 |
| | 2. Dirty air filters | 2. Clean or replace filters | 15-1B |
| | 3. Dirty air conditioning evaporator | 3. Clean evaporator | 15-1C |
| | 4. Air ducts not insulated | 4. Insulate ducts | 26-1 |
| | 5. Thermostat in wrong location | 5. Relocate thermostat | 6-1, 6-1C |
| | 6. Dirty blower | 6. Clean blower | 15-1D |
| | 7. Defective thermostat | 7. Replace thermostat | 6-1, 6-1A, B, C, D |
| | 8. Fan belt slipping | 8. Replace or adjust fan belt | 15-1E |
| | 9. Low or high voltage | 9. Notify power company | 1-1, 10-1 |
| | 10. Thermostat setting too high | 10. Set to lower setting | 6-1, 6-1D |

179

## HEAT PUMP (COOLING CYCLE)

| Condition | Possible Cause | Corrective Action | Reference |
|---|---|---|---|
| No cooling, but compressor runs continuously | 1. Defective compressor valves | 1. Replace valves and valve plate or compressor | 4-1, 4-3, 4-3C |
| | 2. Shortage of refrigerant | 2. Repair leak and recharge | 15-1, 15-1A |
| | 3. Defective reversing valve | 3. Replace reversing valve | 30-1 |
| | 4. Air or noncondensables | 4. Purge noncondensables | 14-1, 14-1E |
| | 5. Wrong superheat setting on indoor expansion valve | 5. Adjust superheat setting | 16-1 |
| | 6. Loose thermal bulb on indoor expansion valve | 6. Tighten thermal bulb | 16-1, 16-1A |
| | 7. Dirty indoor coil | 7. Clean coil | 15-1, 15-1C |
| | 8. Dirty indoor filters | 8. Clean or replace filters | 15-1, 15-1B |
| | 9. Indoor blower belt slipping | 9. Replace or adjust belt | 15-1E |
| | 10. Restriction in refrigerant system | 10. Locate and remove restriction | 17-1 |
| Too much cooling; compressor runs continuously | 1. Faulty wiring | 1. Repair wiring | 8-1, 9-1, 10-1 |
| | 2. Faulty thermostat | 2. Replace thermostat | 6-1, 6-1A, B, C, D |
| | 3. Wrong thermostat location | 3. Relocate thermostat | 6-1, 6-1C |
| Liquid refrigerant flooding compressor (TXV system) | 1. Wrong superheat setting on indoor expansion valve | 1. Adjust superheat | 16-1 |
| | 2. Loose thermal bulb on indoor expansion valve | 2. Tighten thermal bulb | 16-1, 16-1A |
| | 3. Faulty indoor expansion valve | 3. Replace expansion valve | 16-1, 16-1A, B, C, D, E |
| | 4. Defective indoor check valve | 4. Replace check valve | 29-1 |
| | 5. Refrigerant overcharge | 5. Purge overcharge | 14-1, 14-1D |
| Liquid refrigerant flooding compressor (capillary tube system) | 1. Refrigerant overcharge | 1. Purge overcharge | 14-1, 14-1D |
| | 2. High head pressure | 2. See entry "High head pressure" | |
| | 3. Dirty indoor filter | 3. Clean or replace filter | 15-1, 15-1B |

## HEAT PUMP (HEATING CYCLE)

| Condition | Possible Cause | Corrective Action | Reference |
|---|---|---|---|
| Liquid refrigerant flooding compressor (capillary tube system) | 4. Dirty indoor coil<br>5. Indoor blower belt slipping<br>6. Indoor check valve defective | 4. Clean coil<br>5. Replace or adjust belt<br>6. Replace check valve | 15-1, 15-1C<br>15-1, 15-1E<br>29-1 |
| No heating, but compressor runs continuously | 1. Refrigerant shortage<br>2. Compressor valves defective<br>3. Leaking reversing valve<br>4. Defective defrost control, time clock, or relay | 1. Repair leak and recharge<br>2. Replace valves and valve plate or compressor<br>3. Replace reversing valve<br>4. Replace defrost control, time clock, or relay | 15-1, 15-1A<br>4-3, 4-3C<br>30-1<br>6-1E, 31-1 |
| Too much heat; compressor runs continuously | 1. Faulty wiring<br>2. Faulty thermostat<br>3. Wrong thermostat location | 1. Repair wiring<br>2. Replace thermostat<br>3. Relocate thermostat | 8-1, 9-1<br>6-1, 6-1A, B<br>6-1, 6-1C |
| Compressor cycles on low pressure control at end of defrost cycle | 1. Defective reversing valve<br>2. Defective power element on indoor expansion valve<br>3. Shortage of refrigerant | 1. Replace reversing valve<br>2. Replace power element<br>3. Repair leak and recharge | 30-1<br>16-1, 16-1E<br>15-1, 15-1A |
| Unit runs in cooling cycle but pumps down in cooling cycle | 1. Faulty outdoor expansion valve<br>2. Defective power element on outdoor expansion valve<br>3. Defective reversing valve<br>4. Dirty outdoor coil<br>5. Belt slipping on outdoor blower<br>6. Defective indoor check valve<br>7. Restriction in refrigerant circuit | 1. Clean or replace expansion valve<br>2. Replace power element<br>3. Replace reversing valve<br>4. Clean coil<br>5. Replace or adjust belt<br>6. Replace check valve<br>7. Locate and remove restriction | 16-1, 16-1A, B, C, D, E<br>16-1, 16-1E<br>30-1<br>14-1, 14-1B, C<br>15-1E<br>29-1<br>17-1, 17-1A, B, C, D, E |

## HEAT PUMP (HEATING CYCLE)

| Condition | Possible Cause | Corrective Action | Reference |
|---|---|---|---|
| Defrost cycle will not terminate | 1. Shortage of refrigerant | 1. Repair leak and recharge | 15-1, 15-1A |
| | 2. Defrost control out of adjustment | 2. Adjust control | 6-1E, 31-1 |
| | 3. Defective defrost control, time clock, or relay | 3. Replace defrost control, time clock, or relay | 31-1 |
| | 4. Defective reversing valve | 4. Replace reversing valve | 30-1 |
| | 5. Defective compressor valves | 5. Replace valves and valve plate or compressor | 4-3, 4-3C |
| | 6. Faulty electrical wiring | 6. Repair wiring | 8-1, 9-1, 10-1 |
| Defrost cycle initiates without ice on coil | 1. Shortage of refrigerant | 1. Repair leak and recharge | 15-1, 15-1A |
| | 2. Defrost control out of adjustment | 2. Adjust control | 6-1E, 31-1 |
| | 3. Defective defrost control, time clock, or relay | 3. Replace defrost control, time clock, or relay | 31-1 |
| | 4. Defrost control sensing element not making proper contact | 4. Improve contact | 6-1E, 31-1 |
| | 5. Outdoor coil dirty | 5. Clean coil | 14-1, 14-1B, C, 15-1 |
| | 6. Outdoor fan belt slipping | 6. Replace belt or adjust | 15-1E |
| Reversing valve will not shift | 1. Defective reversing valve | 1. Replace reversing valve | 30-1 |
| | 2. Defective compressor valves | 2. Replace valves and valve plate or compressor | 4-3, 4-3C |
| | 3. Faulty fan relay on either indoor or outdoor section | 3. Replace relay | 28-1, 25-1 |
| | 4. Burned transformer | 4. Replace transformer | 23-1, 23-1A |
| Indoor blower off with auxiliary heat on | 1. Defective indoor fan relay | 1. Replace fan relay | 25-1 |
| | 2. Defective indoor fan motor | 2. Repair or replace motor | 22-1, 22-1A, B, C, D, E, F |
| | 3. Faulty wiring or loose terminals | 3. Repair wiring or terminals | 8-1, 9-1, 10-1 |
| | 4. Faulty thermostat | 4. Replace thermostat | 6-1, 6-1A, B, C, D |

## HEAT PUMP (HEATING CYCLE)

| Condition | Possible Cause | Corrective Action | Reference |
|---|---|---|---|
| Outdoor blower runs during defrost cycle | 1. Faulty outdoor fan relay | 1. Replace fan relay | 28-1 |
| Compressor short cycles on defrost control | 1. Shortage of refrigerant | 1. Repair leak and recharge | 15-1, 15-1A |
| | 2. Defrost control out of adjustment | 2. Adjust defrost control | 6-1E, 31-1 |
| | 3. Defective defrost control, time clock, or relay | 3. Replace defrost control, time clock, or relay | 6-1E, 31-1 |
| | 4. Defective power element on outdoor expansion valve | 4. Replace power element | 16-1, 16-1E |
| | 5. Fan belt slipping on outdoor blower | 5. Replace or adjust belt | 15-1E |
| Excessive ice build-up on indoor coil | 1. Defective defrost relay | 1. Replace defrost relay | 34-1 |
| | 2. Defective compressor valves | 2. Replace valves and valve plate or compressor | 4-3, 4-3C |
| | 3. Shortage of refrigerant | 3. Repair leak and recharge | 15-1, 15-1A |
| | 4. Defrost control out of adjustment | 4. Adjust defrost control | 6-1E, 31-1 |
| | 5. Defrost control sensing element not making proper contact | 5. Improve contact | 6-1E, 31-1 |
| | 6. Defective defrost control, time clock, or relay | 6. Replace control, time clock, or relay | 31-1, 34-1 |
| | 7. Defective reversing valve | 7. Replace reversing valve | 30-1 |
| | 8. Wrong superheat setting on outdoor expansion valve | 8. Adjust superheat | 16-1, 16-1A |
| | 9. Defective power element on outdoor expansion valve | 9. Replace power element | 16-1, 16-1E |
| | 10. Plugged outdoor expansion valve | 10. Clean or replace expansion valve | 16-1, 17-1B |

## HEAT PUMP (HEATING CYCLE)

| Condition | Possible Cause | Corrective Action | Reference |
|---|---|---|---|
| Ice build-up on lower section of outdoor coil | 1. Defective defrost relay | 1. Replace defrost relay | 34-1 |
| | 2. Defective compressor valves | 2. Replace valves and valve plate or compressor | 4-3, 4-3C |
| | 3. Shortage of refrigerant | 3. Repair leak and recharge | 15-1, 15-1A |
| | 4. Defrost control out of adjustment | 4. Adjust defrost control | 6-1E, 31-1 |
| | 5. Defrost sensing element not making proper contact | 5. Improve contact | 6-1E, 31-1 |
| | 6. Defective reversing valve | 6. Replace reversing valve | 30-1 |
| | 7. Wrong superheat setting on outdoor expansion valve | 7. Adjust superheat | 16-1 |
| Liquid refrigerant flooding compressor on heating cycle (TXV system) | 1. Wrong superheat setting on outdoor expansion valve | 1. Adjust superheat | 16-1 |
| | 2. Outdoor expansion valve thermal bulb not making proper contact | 2. Improve contact | 16-1, 16-1A |
| | 3. Outdoor expansion valve stuck open | 3. Clean or replace expansion valve | 16-1, 16-1B |
| | 4. Leaking outdoor check valve | 4. Replace check valve | 29-1 |
| Liquid refrigerant flooding compressor on heating cycle (capillary tube system) | 1. Refrigerant overcharge | 1. Purge overcharge | 14-1, 14-1D |
| | 2. High head pressure | 2. See entry "High head pressure" | |
| | 3. Defective outdoor check valve | 3. Replace check valve | 29-1 |
| Excessive operating costs | 1. Refrigerant shortage | 1. Repair leak and recharge | 15-1, 15-1A |
| | 2. Defective reversing valve | 2. Replace reversing valve | 30-1 |
| | 3. Defrost control out of adjustment | 3. Adjust defrost control | 6-1E, 30-1 |
| | 4. Refrigerant overcharge | 4. Purge overcharge | 14-1, 14-1D |

## HEAT PUMP (HEATING OR COOLING CYCLE)

| Condition | Possible Cause | Corrective Action | Reference |
|---|---|---|---|
| Excessive operating costs | 5. Dirty indoor or outdoor coil | 5. Clean coil | 14-1, 14-1B, C, 15-1, 15-1C |
| | 6. Blower belt slipping on indoor or outdoor blower | 6. Replace or adjust belt | 15-1E |
| | 7. Dirty indoor air filters | 7. Clean or replace filters | 15-1, 15-1B |
| | 8. Wrong thermostat location | 8. Relocate thermostat | 6-1, 6-1C |
| | 9. Ducts not insulated | 9. Insulate ducts | 26-1 |
| | 10. Wrong size unit | 10. Replace with proper size | 27-1 |
| | 11. Outdoor thermostat not controlling auxiliary heat | 11. Adjust, relocate, or provide shield | 6-1, 6-1F |
| Compressor hums but will not start | 1. Faulty fuse | 1. Replace fuse and correct cause | 3-1 |
| | 2. Faulty wiring | 2. Repair wiring | 8-1, 9-1, 10-1 |
| | 3. Loose electrical terminals | 3. Repair loose connections | 8-1, 9-1 |
| | 4. Compressor overloaded | 4. Locate and remove overload | 4-1, 4-2, 4-3, 4-3A, B, 4-4, 4-4A, B, 27-1 |
| | 5. Faulty starting capacitor | 5. Replace capacitor | 11-1, 11-1A |
| | 6. Faulty starting relay | 6. Replace relay | 12-1, 12-1A, B, C, D, E |
| | 7. Burned compressor motor | 7. Replace compressor | 4-2, 4-2A, B, C, D |
| | 8. Defective compressor bearings | 8. Replace bearings or compressor | 4-3, 4-3A, B |
| | 9. Stuck compressor | 9. Replace compressor | 4-3, 4-3A, B |
| Compressor cycling on overload | 1. Low voltage | 1. Determine reason and repair | 10-1 |
| | 2. Loose electrical terminals | 2. Repair terminals | 8-1, 9-1, 10-1 |
| | 3. Single-phasing of phase power | 3. Replace fuse or repair wiring; or notify power company | 3-1, 1-1 |
| | 4. Defective contactor contacts | 4. Replace contacts or contactor | 5-1, 5-1C |
| | 5. Defective compressor overload | 5. Replace overload | 4-2, 4-2E, F, G |
| | 6. Compressor overloaded | 6. Locate and remove overload | 4-1, 4-2, 4-3, 4-3A, B, 4-4, 4-4A, B, 27-1 |

## HEAT PUMP (HEATING OR COOLING CYCLE)

| Condition | Possible Cause | Corrective Action | Reference |
|---|---|---|---|
| Compressor cycling on overload | 7. Defective start capacitor | 7. Replace capacitor | 11-1, 11-1A |
| | 8. Defective run capacitor | 8. Replace capacitor | 11-1, 11-1B |
| | 9. Defective starting relay | 9. Replace starting relay | 12-1, 12-1A, B, C, D, E |
| | 10. Refrigerant overcharge | 10. Purge overcharge | 14-1, 14-1D |
| | 11. Defective compressor bearings | 11. Replace bearings or compressor | 4-3, 4-3A, B |
| | 12. Air or noncondensables in system (high head pressure) | 12. Purge noncondensables from system | 14-1, 14-1E |
| | 13. Defective reversing valve | 13. Replace reversing valve | 30-1 |
| Compressor off on high pressure control | 1. Refrigerant overcharge | 1 Purge overcharge | 14-1, 14-1E |
| | 2. Control out of adjustment | 2. Adjust control | 7-1, 7-1B |
| | 3. Defective indoor fan motor | 3. Repair or replace motor | 22-1, 22-1A, B, C, D, E, F |
| | 4. Defective outdoor fan motor | 4. Repair or replace motor | 22-1, 22-1A, B, C, D, E, F |
| | 5. Defective fan relay on either indoor or outdoor section | 5. Repair or replace motor | 28-1, 25-1 |
| | 6. Too long defrost cycle | 6. Replace time clock, defrost relay, or termination thermostat | 6-1E, 34-1, 31-1 |
| | 7. Defective reversing valve | 7. Replace reversing valve | 30-1 |
| | 8. Blower belt slipping on indoor or outdoor coil | 8. Adjust or replace belt | 15-1E |
| | 9. Indoor or outdoor coil dirty | 9. Clean proper coil | 14-1, 14-1B, C, 15-1, 15-1C |
| | 10. Dirty indoor air filters | 10. Replace or clean filters | 14-1, 14-1B, 15-1, 15-1B |
| | 11. Air bypassing indoor or outdoor coil | 11. Prevent air bypassing | 14-1, 14-1B, 15-1, 15-1C, 32-1 |
| | 12. Air volume too low over indoor or outdoor coil | 12. Increase indoor ductwork or remove restriction from coils | 26-1 |
| | 13. Auxiliary heat strips ahead of indoor coil | 13. Locate heat strips downstream of indoor coil | 33-1 |

## HEAT PUMP (HEATING OR COOLING CYCLE)

| Condition | Possible Cause | Corrective Action | Reference |
|---|---|---|---|
| Compressor cycles on low pressure control | 1. Refrigerant shortage | 1. Repair leak and recharge | 15-1, 15-1A |
| | 2. Low suction pressure | 2. Increase load (see Suction Pressure Low) | |
| | 3. Defective expansion valve | 3. Repair or replace expansion valve | 16-1, 16-1A, B, C, D, E |
| | 4. Dirty indoor or outdoor coil | 4. Clean coil | 14-1, 14-1B, 15-1, 15-1C |
| | 5. Slipping blower belt | 5. Replace or adjust blower belt | 15-1E |
| | 6. Dirty air indoor filter | 6. Clean or replace filter | 14-1, 14-1B, 15-1, 15-1B |
| | 7. Ductwork restriction | 7. Increase ductwork | 26-1 |
| | 8. Liquid drier or suction strainer restricted | 8. Replace drier or strainer | 17-1, 17-1B, C |
| | 9. Defrost thermostat element loose or making poor contact | 9. Tighten or increase contact | 6-1E, 31-1 |
| | 10. Air temperature too low for evaporation | 10. Relocate unit or provide adequate air temperature | 6-1, 6-1A, B |
| | 11. Defrost cycle too long | 11. Replace time clock, defrost relay, or termination thermostat | 6-1E, 31-1, 34-1 |
| | 12. Defective evaporator fan motor | 12. Repair or replace fan motor or relay | 22-1, 22-1A, B, C, D, E, F, 25-1 |
| Outdoor fan runs, but compressor will not | 1. Faulty electrical wiring or loose connections | 1. Repair wiring or connections | 8-1, 9-1, 10-1 |
| | 2. Defective starting capacitor | 2. Replace starting capacitor | 11-1, 11-1A |
| | 3. Defective starting relay | 3. Replace starting relay | 12-1, 12-1A, B, C, D, E |
| | 4. Defective run capacitor | 4. Replace run capacitor | 11-1, 11-1B |
| | 5. Shorted or grounded compressor motor | 5. Replace compressor | 4-2, 4-2B, C |
| | 6. Stuck compressor | 6. Replace compressor | 4-3, 4-3A, B |
| | 7. Compressor overloaded | 7. Determine and remove overload | 4-1, 4-2, 4-3, 4-3A, B, 4-4, 4-4A, B, 27-1 |
| | 8. Defective contactor contacts | 8. Replace contactor or contacts | 5-1, 5-1C |

## HEAT PUMP (HEATING OR COOLING CYCLE)

| Condition | Possible Cause | Corrective Action | Reference |
|---|---|---|---|
| Outdoor fan runs, but compressor will not | 9. Single-phasing of three-phase power | 9. Locate problem and repair or contact power company | 1-1, 3-1 |
| | 10. Low voltage | 10. Locate and correct cause | 1-1, 10-1 |
| Outdoor fan motor will not start | 1. Faulty electrical wiring or loose connections | 1. Repair wiring or connections | 8-1 |
| | 2. Defective outdoor fan motor | 2. Repair or replace motor | 22-1, 22-1A, B, C, D, E, F |
| | 3. Defective outdoor fan relay | 3. Replace fan relay | 28-1 |
| | 4. Defective defrost control, timer, or relay | 4. Replace control, timer, or relay | 31-1, 34-1 |
| Outdoor section does not run | 1. No electrical power | 1. Inform power company | 1-1 |
| | 2. Blown fuse | 2. Replace fuse and correct fault | 3-1 |
| | 3. Faulty electrical wiring or loose terminals | 3. Repair wiring or terminals | 8-1, 9-1 |
| | 4. Compressor overloaded | 4. Determine overload and correct | 4-2, 4-2E, |
| | 5. Defective transformer | 5. Replace transformer | 23-1, 23-1A |
| | 6. Burned contactor coil | 6. Replace contactor coil | 5-1, 5-1A |
| | 7. Compressor overload open | 7. Determine cause and correct | 4-2, 4-2E, F, G |
| | 8. High pressure control open | 8. Determine cause and correct | 7-1, 7-1B |
| | 9. Low pressure control open | 9. Determine cause and correct | 7-1, 7-1A |
| | 10. Thermostat off | 10. Turn thermostat on and set | 6-1, 6-1D |
| Indoor blower will not run | 1. Blown fuse | 1. Replace fuse and correct cause | 3-1 |
| | 2. Faulty electrical wiring or loose connections | 2. Repair wiring or connections | 8-1 |
| | 3. Burned transformer | 3. Replace transformer | 23-1, 23-1A |

## HEAT PUMP (HEATING OR COOLING CYCLE)

| Condition | Possible Cause | Corrective Action | Reference |
|---|---|---|---|
| Indoor blower will not run | 4. Indoor fan relay defective | 4. Replace fan relay | 29-1 |
| | 5. Faulty indoor fan motor | 5. Repair or replace motor | 22-1, 22-1A, B, C, D, E, F |
| | 6. Faulty thermostat | 6. Replace thermostat | 6-1, 6-1A, B, C, D |
| Indoor coil iced over | 1. Dirty filters | 1. Clean or replace filters | 15-1, 15-1A, B, C, D, E, F |
| | 2. Dirty coil | 2. Clean coil | 15-1, 15-1C |
| | 3. Blower fan belt slipping | 3. Replace or adjust belt | 15-1, 15-1E |
| | 4. Outdoor check valve sticking closed | 4. Replace check valve | 29-1 |
| | 5. Defective indoor expansion valve | 5. Clean or replace expansion valve | 16-1, 16-1A, B, C, D, E |
| | 6. Low indoor air temperature | 6. Increase temperature | 6-1, 6-1A, B |
| | 7. Shortage of refrigerant | 7. Repair leak and recharge | 15-1, 15-1A |
| Noisy compressor | 1. Low oil level in compressor | 1. Determine reason for loss of oil and correct. Replace oil. | 15-1, 15-1A, B, C, D, E, F, 16-1, 16-1D, E, 17-1 |
| | 2. Defective suction and/or discharge valves | 2. Replace valves and plate or compressor | 4-3, 4-3C |
| | 3. Loose hold-down bolts | 3. Tighten | 19-1 |
| | 4. Broken internal springs | 4. Replace compressor | 19-1 |
| | 5. Inoperative check valves | 5. Repair or replace check valve | 29-1 |
| | 6. Loose thermal bulb on indoor expansion valve | 6. Tighten thermal bulb | 16-1, 16-1A |
| | 7. Improper superheat setting on indoor expansion valve | 7. Adjust superheat | 16-1 |
| | 8. Stuck open indoor expansion valve | 8. Clean or replace valve | 16-1, 16-1B |
| Compressor loses oil | 1. Refrigerant shortage | 1. Repair leak and recharge | 15-1, 15-1A |
| | 2. Low suction pressure | 2. Increase load on evaporator | 15-1, 15-1A |

## HEAT PUMP (HEATING OR COOLING CYCLE)

| Condition | Possible Cause | Corrective Action | Reference |
|---|---|---|---|
| Compressor loses oil | 3. Restriction in refrigerant circuit | 3. Remove restriction | 17-1, 17-1A, B, C, D, 21-1A |
| | 4. Indoor expansion valve stuck open | 4. Clean or replace expansion valve | 16-1, 16-1B |
| Unit operates normally in one cycle, but high suction pressure on other cycle | 1. Leaking check valve | 1. Replace check valve | 29-1 |
| | 2. Loose thermal bulb on outdoor or indoor expansion valve | 2. Tighten thermal bulb | 16-1, 16-1A |
| | 3. Leaking reversing valve | 3. Replace reversing valve | 30-1 |
| | 4. Expansion valve stuck open on indoor or outdoor | 4. Repair or replace expansion valve | 16-1, 16-1B |
| Unit pumps down in cool or defrost cycle but operates normally in heat cycle | 1. Defective reversing valve | 1. Replace reversing valve | 30-1 |
| | 2. Defective power element on indoor expansion valve | 2. Replace power element | 16-1, 16-1E |
| | 3. Restriction in refrigerant circuit | 3. Locate and remove restriction | 17-1, 17-1A, B, C, D, E, 21-1A |
| | 4. Clogged indoor expansion valve | 4. Clean or replace expansion valve | 16-1, 17-1, 17-1B |
| | 5. Check valve in outdoor section sticking closed | 5. Replace check valve | 29-1 |
| Head pressure high | 1. Overcharge of refrigerant | 1. Purge overcharge | 14-1, 14-1D |
| | 2. Air or noncondensables in system | 2. Purge noncondensables | 14-1, 14-1E |
| | 3. High air temperature supplied to condenser | 3. Reduce air temperature | 14-1, 14-1B, 24-1 |
| | 4. Dirty indoor or outdoor coil | 4. Clean coil | 14-1, 14-1B, C, 15-1, 15-1C, F |
| | 5. Dirty indoor air filters | 5. Clean or replace filters | 14-1, 14-1B, 15-1, 15-1B, F |
| | 6. Indoor or outdoor blower belt slipping | 6. Replace or adjust blower belt | 14-1, 14-1B, 15-1, 15-1E, F |
| | 7. Air bypassing indoor or outdoor coil | 7. Prevent air bypassing | 14-1, 15-1, 32-1 |

## HEAT PUMP (HEATING OR COOLING CYCLE)

| Condition | Possible Cause | Corrective Action | Reference |
|---|---|---|---|
| Suction pressure high | 1. Defective compressor suction valves | 1. Replace valves and valve plate or compressor | 4-2, 4-3, 4-3C |
| | 2. High head pressure | 2. See previous entry "Head pressure high" | |
| | 3. Excessive load on cooling | 3. Determine cause and correct | 4-1, 4-2, 4-3, 4-3A, B, 4-4, 4-4A, B, 27-1 |
| | 4. Leaking reversing valve | 4. Replace reversing valve | 30-1 |
| | 5. Leaking check valves | 5. Replace check valve | 29-1 |
| | 6. Indoor or outdoor expansion valve stuck open | 6. Clean or replace expansion valve | 16-1, 16-1B |
| | 7. Loose thermal bulb on indoor or outdoor expansion valve | 7. Tighten bulb | 16-1, 16-1A |
| Suction pressure low | 1. Shortage of refrigerant | 1. Repair leak and recharge | 15-1, 15-1A |
| | 2. Blower belt slipping on indoor or outdoor blower | 2. Replace belt or adjust | 14-1, 14-1B, 15-1, 15-1E |
| | 3. Dirty indoor air filters | 3. Clean or replace | 15-1, 15-1B |
| | 4. Defective check valves | 4. Replace check valves | 29-1 |
| | 5. Restriction in refrigerant circuit | 5. Locate and remove restriction | 17-1, 17-1A, B, C, D, E, 21-1, 21-1A |
| | 6. Ductwork small or restricted | 6. Repair or replace ductwork | 26-1 |
| | 7. Defective expansion valve power element on indoor or outdoor coil | 7. Replace power element | 16-1, 16-1E |
| | 8. Clogged indoor or outdoor expansion valve | 8. Clean or replace valve | 16-1, 17-1, 17-1B |
| | 9. Wrong superheat setting on indoor or outdoor expansion valve | 9. Adjust superheat setting | 16-1 |
| | 10. Dirty indoor or outdoor coil | 10. Clean coil | 15-1, 15-1C |
| | 11. Bad contactor contacts | 11. Replace contactor or contacts | 5-1, 5-1C |
| | 12. Low refrigerant charge | 12. Repair leak and recharge | 15-1, 15-1A |

Therefore, if the service technician can determine whether the trouble is electrical or mechanical, he has eliminated half of the possible causes of the problem. An example of a combination of electrical and mechanical problems would be if the bearings in a compressor became stuck and caused the compressor motor to burn out.

## HOW TO USE THE TROUBLESHOOTING CHARTS _____

When trouble is experienced with a unit in a particular cycle, turn to the chart for that cycle in the troubleshooting charts. The action of the unit can then be found in the column labeled condition. The next column lists the possible causes for the problem encountered. The third column lists the corrective action that may be taken to correct the problem. The reference column directs the reader to the specific sections of the following text material where the component is introduced and checkout procedures for the component are given. Specific procedures described under Standard Service Procedures in Chapter 9 will help the technician to accomplish many of the tasks that may be unfamiliar to him.

### EXAMPLE _____

We are having trouble with a heat pump unit cooling too much. After checking the unit we find that the compressor runs continuously. When we refer to the troubleshooting chart for Heat Pump (Cooling Cycle), the first possible cause is faulty wiring. The corrective action for this problem is to repair the wiring. The reference column indicates the specific sections in the text where the proper procedure for this job is discussed. The specific sections are: Sections 6–1, 6–1A, B, C, D, 8–1, 9–1, and 10–1. A check in Chapter 9 will provide additional information on checking compressor electrical systems.

## COMPONENT TROUBLESHOOTING _____

The primary intention of this section is to aid the service technician in checking the components indicated in the Troubleshooting Charts for possible faults. The information contained in this chapter will allow more accurate and, therefore, more economical services to be performed.

*1–1 Power Failure:* When a power failure occurs, the electric company must be contacted to make the necessary repairs. The repairs may consist of replacing a fuse or transformer on the electric pole or of correcting any one of the many other reasons for power failure. In any case, it is the electric company that must make the repairs. When the repairs have been completed, the air-conditioning service technician should check all the equipment to be certain that it is operating properly. At times, an electrical failure will cause damage to the electric motors, and they will need to be repaired or replaced.

*2–1 Disconnect Switches:* These electrical switches are used to interrupt electrical power to condensing units, large electric motors, and other equipment that require a heavy electric current flow. Disconnect switches are usually installed within an arm's reach of the equipment that they control. Some of these switches contain only a set of electrical contacts; others contain fuses to protect the equipment in addition to the contacts. Located on the outside of the switch box is a lever used to open and close the contacts (Figure 8–1). This lever may be accidentally bumped, causing the contacts to open, or it may be accidentally left in the OFF position. In either case the lever must be returned to the ON position before operation can be resumed.

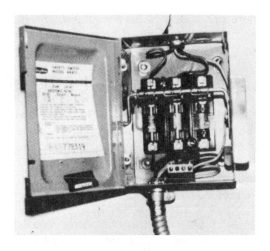

*Figure 8-1* Mounted Disconnect Box

*3-1 Fuse:* An electrical fuse is a protective device placed in the electrical line to an electric circuit. There are basically two types of fuses: the plug type and the cartridge type (Figure 8–2). The purpose of a fuse is to protect the electric circuit in case of an electrical overload. This overload may be due to tight bearings, shorted motor winding, breakdown of electrical insulation, fan belt too tight, burned contactor contacts, and so on. The problem that caused the blown fuse must be found and repaired before the job is complete.

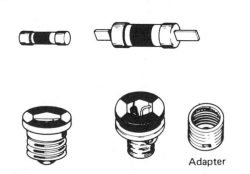

Adapter

*Figure 8-2* Plug- and Cartridge-Type Fuses (Courtesy of Bussmann Division, McGraw-Edison Company)

Defective fuses may be found with either a voltmeter or an ohmmeter. To check fuses with a voltmeter, select a scale that is high enough to prevent damage to the meter. Check the voltage across each fuse (Figure 8–3). If a voltage is found between the ends of the fuse, it is defective and must be replaced with one of the proper size. Obviously, this procedure requires the electrical power to be on. Use caution to avoid an electrical shock.

To check the fuse with an ohmmeter, remove it from the fuse holder and check for continuity between the ends of the fuse (Figure 8–4). If no continuity is found, replace the fuse with one of the proper size.

*4-1 Compressor:* The compressor is the device used to circulate the refrigerant through the system (Figure 8–5). It has two functions: (1) It draws the refrigerant vapor from the evaporator and lowers the pressure of the refrigerant in the evaporator to the desired evaporating temperature; (2) the compressor raises the pressure of the refrigerant vapor in the condenser high enough so that the satura-

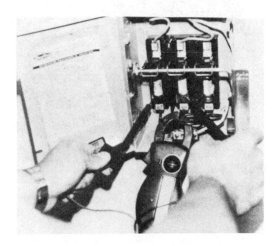

*Figure 8-3* Checking Fuses in Disconnect Box

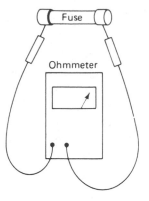

Good fuse will indicate "0" resistance

Blown fuse will indicate infinite resistance

**Figure 8-4** Checking Fuses Out of Disconnect Box with Ohmmeter

Therefore, the problems encountered could be electrical, mechanical, or a combination of the two.

*4-2 Compressor Electrical Circuit:* The problems encountered in the electrical circuit of a compressor may be divided into the following classifications: open winding, shorted winding, or grounded winding. An accurate ohmmeter is needed to check for these conditions. The following checks are good for any type of electric motor.

*4-2A Open compressor motor windings.* These problems occur when the path for electrical current is interrupted. This interruption occurs when the wire insulation becomes bad and allows the wire to overheat and burn apart.

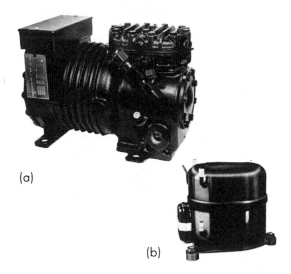

(a)

(b)

**Figure 8-5** (a) Semihermetic and (b) Hermetic Compressor (Courtesy of Copeland Corporation)

tion temperature is higher than the temperature of the cooling medium used to cool the condenser and condense the refrigerant.

The compressors in use today are usually of either the hermetic or semi-hermetic type.

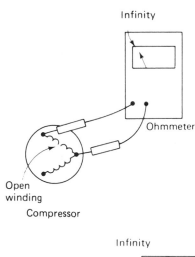

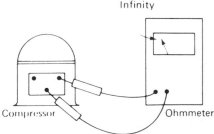

**Figure 8-6** Checking for Open Compressor Motor Winding

To check for an open winding, remove all external wiring from the motor terminals. Using the ohmmeter, check the continuity from one terminal to another terminal (Figure 8–6). Be sure to zero the ohmmeter. The open winding will be indicated by an infinity resistance reading on the ohmmeter. There should be no continuity from any terminal to the motor-compressor case.

*4–2B Shorted compressor motor windings.* This condition will occur when the insulation on the winding becomes bad and allows a shorted condition (two wires to touch), which permits the electrical current to bypass part of the winding (Figure 8–7). In some instances,

depending on how much of the winding is bypassed, the motor may continue to operate but will draw excessive amperage. To check for a shorted winding, remove all external wiring from the motor terminals. Using the ohmmeter, check the continuity from one terminal to another terminal. Be sure to zero the ohmmeter (Figure 8–8). The shorted winding will be indicated by a less than normal resistance. In some cases it will be necessary to consult the motor manufacturer's data for the particular motor to determine the correct resistance requirements. There should be no continuity from any terminal to the motor-compressor case.

*4–2C Grounded compressor motor windings.* Grounds occur when the insulation on the winding is broken down and the winding becomes shorted to the housing (Figure 8–9). In such cases the motor will rarely run and will, usually, immediately blow fuses or trip the circuit breaker. To check for a grounded winding, remove all external wiring from the motor terminals. Using the ohmmeter, check the continuity from each terminal to the motor case. Be sure to zero the ohmmeter (Figure 8–10). A grounded winding will be indicated by a low resistance reading. It may be necessary to remove some paint or scale from the motor case so that an accurate reading can be obtained.

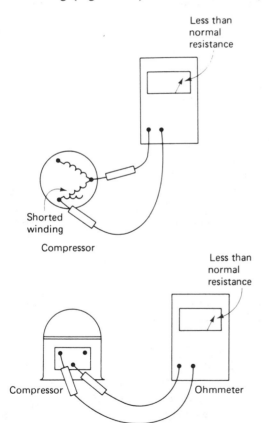

**Figure 8-7** Checking for Shorted Compressor Motor Winding

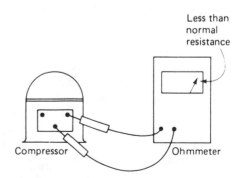

**Figure 8-8** Checking Compressor Motor Winding Resistance

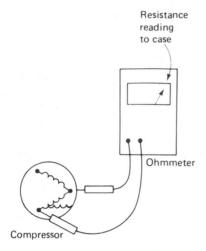

**Figure 8-9** Grounded Compressor Motor Winding

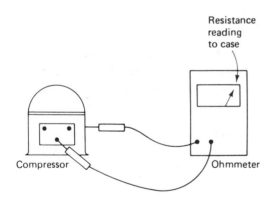

**Figure 8-10** Checking for Grounded Compressor Motor Winding

*4-2D Determining the common, run, and start terminals.* This is a relatively simple process. First, be sure that all external wiring is removed from the compressor so that no false readings are indicated. Next, draw the terminal configuration on a piece of paper. Then measure the resistance between each terminal with an ohmmeter. Be sure to zero the ohmmeter. Record the resistance found on the diagram

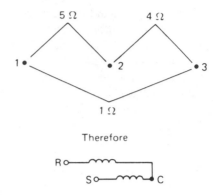

**Figure 8-11** Locating Compressor Terminals

(Figure 8-11). Apply the following rule: the least resistance indicated is between the run and common terminals, and the medium resistance indicated is between the common and start terminals, and the most resistance indicated is between the start and run terminals, then the compressor motor can now be properly wired.

*4-2E Compressor motor overloads.* These devices are used to protect the compressor motor from damage that might occur from overcurrent, overheating, or both. Motor overloads may be mounted externally or internally depending on the design of the compressor. They are generally mounted near the hottest part of the motor winding.

*4-2F Externally mounted overloads.* These types of overloads are manufactured in three configurations: (1) two-terminal, (2) three-terminal, and (3) four-terminal (Figure 8-12). To check the two-terminal overload, place an ammeter on the common electric line to the compressor. Start the compressor while observing the ammeter. The ammeter should indicate a momentary current flow of approximately six times the running amperage of the compressor motor and then drop back to the rated amperage draw of the motor or below.

**Two-terminal**          **Three-terminal**

**Four-terminal**

*Figure 8-12* Two-, Three-, and Four-Terminal Compressor Overloads (Courtesy of Control Products Division, Texas Instruments, Incorporated)

If the overload then cycles the motor, the problem is in the overload. If the amperage remains above the rated amperage of the motor, the trouble is not in the overload. To be sure that the overload is cycling the compressor motor, check across the overload terminals with a voltmeter while the compressor is off (Figure 8-13). If the overload is open, a voltage reading will be indicated. No voltage reading indicates that the overload has not opened the circuit. The trouble is elsewhere. Be sure to replace these overloads with the type recommended by the manufacturer.

*4-2G Three-terminal external overload.* These devices are used on compressor motors where it is desirable to protect the starting winding in addition to the running winding. The terminals are numbered 1, 2, and 3. Terminal 1 is connected to the electrical line that goes to the compressor. Terminal 2 is connected to the run terminal of the compressor. Terminal 3 is connected to the start capacitor. This will allow closer protection of the compressor motor in the event of a bad starting component. To check the three-terminal overload, place an ammeter on the line connected to terminal 1 and start the compressor (Figure 8-14). Start the compressor while observing the ammeter. The ammeter should indicate a momentary flow of approximately six times the running amperage of the motor and then drop back to the amperage rating of the motor or below. If the overload then cycles the motor, the problem is in the overload. If the amperage remains above the rated amperage of the motor, the trouble is not in the overload. To see if the trouble is in the starting or running circuit, measure the amperage draw through the wire connected to terminal 2, and then the wire to terminal 3, while the compressor is running. The circuit with the high amperage draw is where the fault is. The external components must be checked. To be certain that the overload is cycling the compressor motor, check the voltage between terminals 1 and 2, and then between terminals 1 and 3, while the compressor is off. If voltage is indicated, replace the overload (Figure 8-15). If the overload must be replaced, be sure to use an exact replacement to provide proper protection.

*4-3 Compressor Lubrication:* Compressors, like any moving mechanical device, require lubrication. The proper oil level should be maintained in the crankcase to provide the necessary lubrication. The oil level in the sight glass should be at or slightly above the center

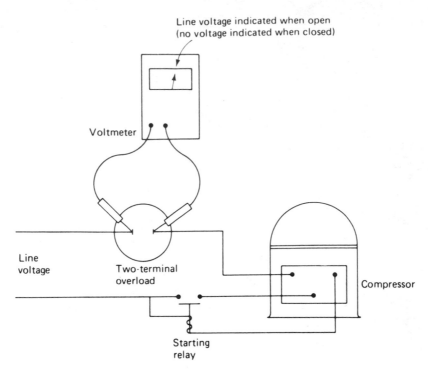

**Figure 8-13** Checking Voltage Across a Two-Terminal Overload

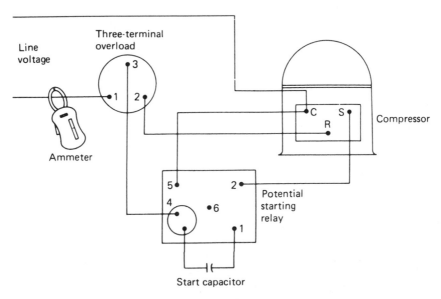

**Figure 8-14** Checking the Amperage Draw through a Three-terminal Overload

198

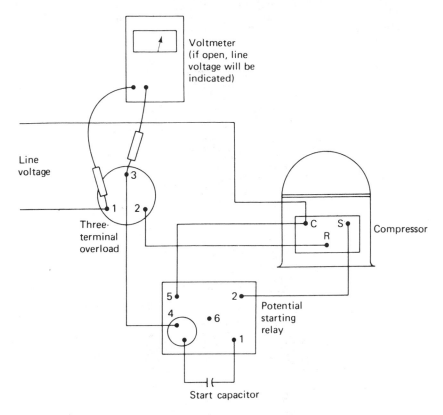

*Figure 8-15* Checking Voltage across Terminals 1 and 3 on a Three-Terminal Overload

of the sight glass (Figure 8–16). On compressors not equipped with an oil sight glass, the manufacturer's recommendations should be followed. These recommendations will generally refer to the amount of oil in ounces for a given model. When replenishing the oil in hermetic compressors, it may be necessary to remove all the oil from the housing and measure in the correct amount. The proper type of oil for the operating temperature should be used to ensure proper oil return and good compressor lubrication. Refrigeration oil containers must be kept sealed to prevent moisture and contaminants getting into the oil.

Three types of lubrication systems are used on compressors: (1) splash, (2) pressure, and (3) a combination of both. The splash system is used on 3-hp compressors and smaller. Compressors of 3-hp and larger are pressure lubricated. The oil pressure may be checked to determine if proper lubrication is being provided on forced lubrication compressors (Figure 8–17). A net oil pressure of 30 to 40 psi (206.843 to 275.790 kPa) is normal; however, adequate lubrication will be provided at pressures as low as 10 psi (68.948 kPa). To obtain the net oil

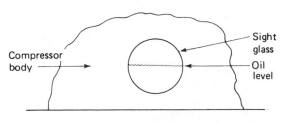

*Figure 8-16* Oil Level in Compressor Sight Glass

199

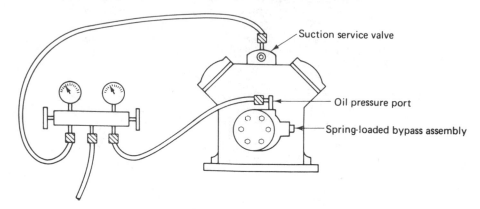

**Figure 8-17** Checking Oil Pressure

pressure, subtract the suction pressure from the oil pump pressure.

## EXAMPLE

Ninety psig (620.528 kPa) pump pressure − 50 psig (344.738 kPa) suction pressure = 40 psi (275.790 kPa).

*4-3A Compressor bearing lubrication.* Bearings that have not received proper lubrication will become tight and sometimes will freeze the shaft and prevent operation. If this condition should occur, the compressor motor will become overloaded and will trip out on the overload, the circuit breaker or will blow the electrical fuses. In any case, the amperage draw of the compressor motor will be higher than normal. When the compressor motor is not allowed to turn, it will draw locked rotor amperage. This amperage rating is indicated by LR on the motor nameplate. The compressor must be replaced when an LR amperage condition is found. The compressor oil should be checked, and if it is found insufficient, the system must be checked to determine the reason for the lack of oil. The loss of oil may be due to refrigerant leaks, oil logging of the evaporator, low refrigerant charge, and so on. The reason must be corrected before the new compressor is placed in operation or it may soon

fail. Do not confuse this condition with faulty starting components or a faulty running capacitor.

A malfunctioning oil pump will generally go undetected until the compressor bearings are damaged enough to cause knocking or until the compressor is frozen mechanically. A malfunctioning oil pump may be due to mechanical wear of the pump. It may also become vapor locked with refrigerant, or the inlet screen may become plugged with dirt or sludge. Should a worn pump be the culprit, it must be repaired or replaced at the same time that the compressor is repaired. A vapor-locked oil pump will produce no oil pressure. The vapor lock must be removed by bleeding off the vapor through the gauge connection. Use the necessary precautions to prevent oil being blown onto and damaging personnel or surroundings.

When the oil inlet screen becomes clogged, the oil flow will be restricted and perhaps completely stopped. In this case the screen must either be cleaned or replaced, along with complete cleaning of the compressor crankcase, replacement of the oil, and replacement of the refrigerant filter-drier.

*4-3B Worn compressor bearings.* Compressor bearings that are worn become loose and noisy. The compressor loses its efficiency, and poor refrigeration results. Loose bearings will be indicated by more noise than usual, and sometimes excessive vibration will occur. When the compressor is equipped with a forced lubrication system, the oil pressure will be low. The amperage draw will be from normal to low. The suction pressure may be high and the discharge pressure low. When this condition occurs, the compressor must be replaced or overhauled depending on the type. This condition is generally a result of age and usage and not faulty lubrication.

*4-3C Compressor valves.* These are the components that control the flow of refrigerant through a compressor. If they are broken or leaking, the compressor will become inefficient. A broken or leaking suction valve will result in a higher than normal suction pressure. To check for a defective compressor suction valve, connect the gauge manifold to the compressor service valves. Open the service valves (Figure 8–18). Next, front-seat the suction service valve (screw all the way in) and observe the suction pressure while the compressor is operating. The suction pressure should pull down to at least 28 in. (8.534-mm) vacuum in 1 or 2 min. If it does not pump down to this reading, stop the compressor for 2 or 3 min; then restart the compressor and allow it to run for 1 or 2 min. If

the desired 28-in. (8.534-mm) vacuum is not reached, the valves must be replaced. In the case of a hermetic compressor, the compressor must be replaced.

A broken or leaking discharge valve will result in a lower than normal discharge pressure. To check for a defective compressor discharge valve, connect the gauge manifold to the service valves on the compressor and open the service valves (Figure 8–18). Next, front-seat the suction service valve (screw all the way in) and start the compressor and allow it to pump as deep a vacuum as possible. Stop the compressor and observe the compound gauge. If the pressure rises, start the compressor and pump another vacuum. Stop the compressor and again observe the compound gauge. If the pressure rises, start the compressor and pump another vacuum. Stop the compressor and again observe the compound gauge. If the pressure increases again, close off the discharge service valve. If the presure reading on the compound gauge stops rising, the discharge valve is bad and must be replaced. In the case of a hermetic compressor, the compressor must be replaced.

*4-4 Compressor Slugging:* Slugging is a noisy condition that occurs when the compressor is pumping oil or liquid refrigerant. When a com-

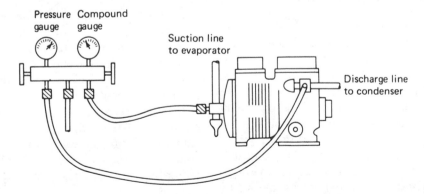

***Figure 8-18*** Gauges Connected to Compressor to Take Refrigerant Pressure Readings

pressor is slugging, it will sound like the clattering of an automobile engine that is under strain. Continued slugging will probably result in broken valves, scored pistons, and galled bearings. It should be evident that a slugging condition must be corrected.

*4-4A Oil slugging.* Oil slugging of a compressor occurs when there is too much oil in the compressor crankcase. When this occurs, some of the oil must be removed to provide the oil level recommended by the manufacturer. The excess oil may be drained through a drain plug, or some installations may require removal of the compressor so that the oil may be poured out. In either case the refrigerant must be pumped from the compressor or purged from the system. Do not attempt to remove oil from the compressor with refrigerant pressure in the crankcase.

To pump refrigerant from the compressor, connect the gauge manifold to the compressor service valves (Figure 8–18). Front-seat the suction service valve and allow the unit to run until approximately 2 psig (13.790 kPa) of pressure is shown on the compound gauge. Do not pump the unit below atmospheric pressure for this procedure. Front-seat the compressor discharge service valve and relieve any remaining pressure through the gauge manifold. The compressor may now be serviced.

*4-4B Refrigerant slugging.* Refrigerant slugging of a compressor is the result of liquid refrigerant being returned to the compressor. This can be caused by several things and will usually result in moisture condensing on the compressor housing because of the lower temperature; sometimes ice or frost will form on the compressor housing. Some of the causes are overcharge of refrigerant, especially on capillary tube systems, superheat setting too low on the thermostatic expansion valves, automatic expansion valves open too much, or a low load condition on the evaporator. The obvious solu-

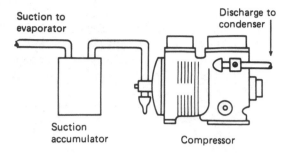

**Figure 8-19** Suction Line Accumulator Installation

tion is to correct any of these causes. However, if these conditions cannot be remedied, a suction line accumulator may be installed to prevent refrigerant slugging of the compressor (Figure 8–19). Almost all equipment manufacturers also install or recommend the installation of crankcase heaters to reduce slugging due to liquid refrigerant entering the crankcase.

*5-1 Motor Starters and Contactors:* The purpose of a motor starter or contactor is to provide the switching action of the high current and voltage required by a compressor. This is done by a signal given by the control circuit on demand from the thermostat or temperature controller. These are electromagnetic-operated devices.

*5-1A Burned coil.* A burned coil will prevent the operation of starters and contactors because there will be no electromagnetic field to operate the device. To check for a burned coil, turn off the electrical power to the unit and remove the electric wiring from the terminals of the coil. Zero an ohmmeter and check the continuity of the coil (Figure 8–20). No continuity should be indicated if the coil is open. Many times the coil will be discolored, indicating that it has been overheated.

*5-1B Sticking starter or contactor.* A sticking motor starter or contactor can cause

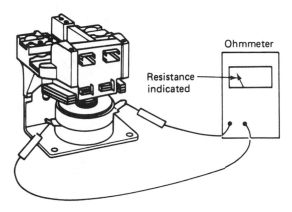

**Figure 8-20** Checking Continuity of a Contactor Coil

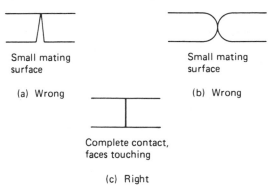

**Figure 8-21** Right and Wrong Contact Mating Surfaces

permanent damage to the motor or compressor. A starter or contactor that sticks may prevent the motor from starting or may keep it running when there is no demand for it. When the starter or contactor sticks during the initial startup, it will usually buzz and either prevent starting of the motor or cause a delayed starting of the motor. When the contactor or starter sticks closed, the compressor or motor will never stop. There are several types of sprays on the market that may be used to lubricate these troublesome and dangerous controls. However, it is generally recommended that starters and contactors that fall into this category be replaced.

*5-1C Burned contacts.* Burned starter and contactor contacts can cause permanent damage to the motor windings by preventing the proper flow of current through them. These contacts will be severely pitted and will not make good contact, thus causing a higher current draw than normal. In an emergency, these contacts may be lightly filed until the mating surfaces match (Figure 8-21). However, the damaged contacts should be replaced as soon as possible because they will again burn and become pitted in a very short time.

Sometimes these contacts may become so bad that they will not make contact at all. This can be determined by energizing the starter or contactor and checking across the contact points with a voltmeter (Figure 8-22). If the contacts are open, the applied voltage will be indicated. However, if the contacts are closed, there will be no indication of voltage.

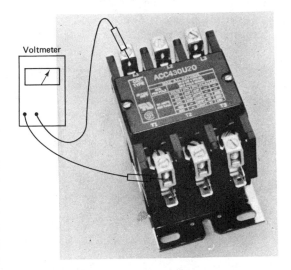

**Figure 8-22** Checking Voltage Across Contacts (Courtesy of Crouse-Hinds Co., Arrow-Hart Division)

*6-1 Thermostat:* Thermostats are temperature-sensitive devices used to control equipment in response to the demands of the space in which they are located. Thermostats may be operated by a bimetal or a "feeler" bulb filled with a fluid that expands and contracts in response to temperature changes.

*6-1A Room thermostat.* A room thermostat generally makes use of a bimetal element for its operation. This type of thermostat is the most popular for air-conditioning and heating applications (Figure 8-23). To check the thermostat, turn it below room temperature and place a reliable thermometer as close to the thermostat as possible. Allow the thermometer to remain there for 10 min. Next, turn the thermostat temperature selector up. The contacts should "make" (close) at no more than 2°F (1.12°C) above the temperature indicated by the thermometer. If not, the thermostat must be calibrated. There are several means of calibrating a thermostat. Therefore, the manufacturer's specifications should be consulted. If more than 10°F (5.6°C) calibration is needed, replace the thermostat. To check for an inoperative room thermostat, check the voltage from the red terminal to the terminal

for the section in question (heating or cooling). If the contacts are closed, no voltage will be indicated. If the contacts are open, a voltage will be indicated when the thermostat is demanding.

*6-1B Heat anticipator.* The heat anticipator incorporated in a thermostat is used during the heating cycle only. Most heat anticipators are adjustable (Figure 8-24). When adjusting a heat anticipator, the total amperage draw of the control circuit must be known. To determine the amperage draw, wrap the tong of an ammeter with one wire to the main gas valve and read the amperage draw while the circuit is in operation (Figure 8-25). Divide the amperage indicated by the number of turns taken on the ammeter tong. Set the heat anticipator to match this amperage draw.

*6-1C Thermostat location.* Location of the thermostat is important to satisfactory operation of the equipment. The thermostat should be located on an inside wall about 5 ft (1.52 m) from the floor. The thermostat should

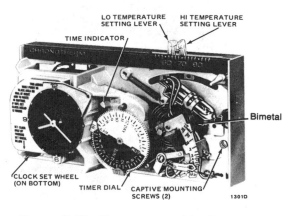

**Figure 8-23** Thermostat with Cover Removed Showing Bimetal Element (Courtesy of Honeywell, Inc.)

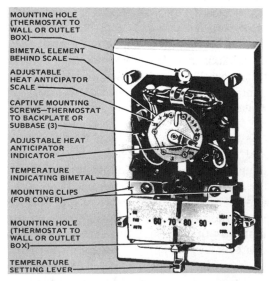

**Figure 8-24** Internal View of a Thermostat (Courtesy of Honeywell, Inc.)

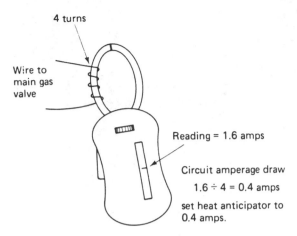

4 turns

Wire to
main gas
valve

Reading = 1.6 amps

Circuit amperage draw
1.6 ÷ 4 = 0.4 amps
set heat anticipator to
0.4 amps.

**Figure 8-25** Checking Amperage in a Temperature-Control Circuit

**Figure 8-26** Air-Conditioning Thermostat (Courtesy of White-Rodgers Division, Emerson Electric Co.)

not be affected by any external heat source such as lights, sun, or a television set. It should be located so that it will sense the average return-air temperature.

*6-1D Thermostat switches.* Thermostat switches are placed according to the type of operation desired. The system switch controls heating or cooling equipment operation. The fan switch controls the fan operation. Both switches are usually incorporated in the thermostat subbase (Figure 8-26). To check out the switches, use a jumper to jump from the R or V terminal to the Y or C terminal for cooling, or from the R or V terminal to the $W_1$ terminal for heating. The first letters of each of these sets refer to Honeywell thermostats; the second letters refer to General Control thermostats. When the fan switch is set to ON, the fan will run continously. The user can select the operation desired. Should these switches become defective, the subbase or thermostat must be replaced before normal operation can be resumed.

*6-1E Defrost thermostats.* These thermostats generally make use of a fluid-filled bulb-

sensing element (Figure 8–27). These types of thermostats are used in the defrost system circuitry on heat pump systems. They are located so that their bulb will sense the outdoor coil temperature. There is generally not much

**Figure 8-27** Refrigeration-Type Thermostat (Courtesy of Johnson Controls, Inc., Control Products Division)

calibration required on this type of thermostat. It is usually best to replace these thermostats if they give problems. To check for a faulty thermostat, place a reliable thermometer as close to the sensing element as possible and allow it to remain there for at least 10 min. The thermostat contacts should open at not more than 2 °F (1.12 °C) above the temperature indicated by the thermometer. If necessary, make only minor adjustments on the thermostat to obtain this temperature. If the thermostat requires very much calibrating, it should be replaced. To check the contacts, place a voltmeter across the contacts of the thermostat. There should be a voltage reading on the voltmeter if the contacts are open. Turn the temperature selector down below the thermometer reading. When the contacts close, the voltage on the voltmeter should drop back to zero. If these readings are not obtained, replace the thermostat.

*6-1F Outdoor thermostats.* Outdoor thermostats are remote bulb-type thermostats that are used on heat pump systems. They are sometimes mounted in the terminal box of the condensing unit. Others are mounted under the eaves of buildings. When mounted under the eaves, protection from the wind, rain, and sun must be provided. These thermostats are used to energize the auxiliary heating elements when the outdoor temperature falls below the balance point. The thermostat setting is determined by the designer to provide greater efficiency and economy. There may be more than one outdoor thermostat. Therefore, each thermostat is set at a temperature equal to the combined point after all the other heat strips plus the heat pump are operating. Electric power is provided from the second state of the indoor thermostat (Figure 8–28). To check the thermostat, the bulb must be cooled to see if the contacts open and close at the desired temperature. Insert the bulb in an ice and salt solution containing a thermometer. Adjust the

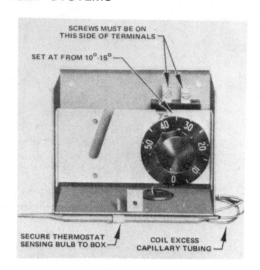

**Figure 8-28** Outdoor Thermostat (Courtesy of Lennox Industries, Inc.)

thermostat to the desired temperature if possible. If adjustment is not possible, replace the thermostat.

*7-1 Pressure Controls:* Pressure controls are designed to protect compressors and motors from damage as a result of excessive pressures, either high or low (Figure 8–29). Low-pressure

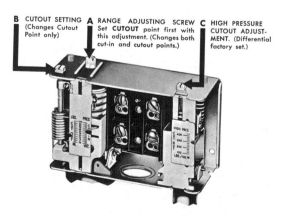

**Figure 8-29** Dual Pressure Control (Courtesy of Johnson Controls, Inc., Control Products Division)

controls are used to open the control circuit when the refrigerant pressure in the high side of the system rises to a given pressure. These pressure settings are generally recommended by the equipment manufacturer. Usually, when these controls cause the compressor to cycle, the problem is due to some cause other than the control.

*7–1A Low-pressure controls.* Low-pressure controls are designed to respond to the refrigerant pressure in the low side of the system. They will stop the compressor motor to protect it from overheating and to prevent the compressor from pumping oil out of the crankcase. On some smaller units the low-pressure control may also be used as a temperature control. To check a low-pressure control, install a compound gauge on the compressor suction service valve (Figure 8–30). Do not disconnect the low-pressure control or cause it to be inoperative. Crack the service valve off the back seat. With the compressor running, front-seat the suction service valve and observe the pressure on the compound gauge when the pressure control stops the compressor. If the actual pressure does not correspond to the control setting, adjust the control, back-seat the suction service valve, and repeat the procedure until the desired cut-out and cut-in points are obtained. Rarely

do these controls need replacement except in the case of refrigerant leakage, in which case replacement is preferred to repair.

*7–1B High-pressure controls.* High-pressure controls are designed to respond to the refrigerant pressure in the high side of the system. They will stop the compressor motor to protect it from being overloaded owing to excessive discharge pressures. To check the compressor discharge service valve, crack the service valve off the back seat (Figure 8–31). Block the air flow through the condenser, or stop the water pump if it is a water-source heat pump. Start the compressor and observe the pressure on the pressure gauge when the pressure control stops the compressor. If the actual pressure does not correspond with the control setting, adjust the control, push the reset, and repeat the procedure until the desired cut-out point is obtained. Rarely do these controls need to be replaced except in the case of refrigerant leakage, in which case replacement is preferred to repair.

*8–1 Loose Electrical Wiring:* Loose electrical wiring can cause many problems and at times be extremely difficult to locate. The problems caused by loose wiring do not follow any set pattern. Most loose wiring can be found by

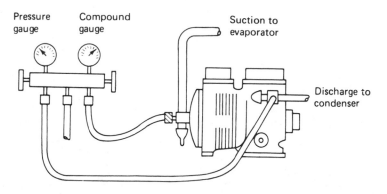

**Figure 8-30** Gauges Installed on a Compressor

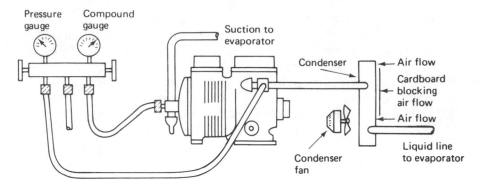

*Figure 8-31* Checking High-Pressure Control Setting

visual inspection (Figure 8–32). However, when a loose wire is suspected, it is sometimes necessary to check each wire and its connections individually. This is usually a time consuming and grueling task. However, it must be done before the unit will satisfactorily operate again. Once the bad wire or connection is found, it must be repaired or replaced. Any wire that is not properly repaired will only cause problems in the future.

*9-1 Improperly Wired Units:* Improperly wired units will operate inefficiently, if at all. They will not produce the results for which they were designed. Each manufacturer designs his own wiring diagrams for each piece of equipment to produce the desired results. If a service technician is in doubt about the wiring on a piece of equipment, the recommended wiring diagram should be consulted and the wiring changed to match the diagram.

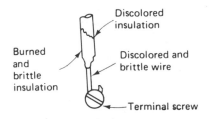

*Figure 8-32* Hot Electrical Joint

*10-1 Low Voltage:* Low voltage to an electric motor can cause it to overheat, which will damage the motor windings. This overheating is caused by excessive current draw due to the low voltage. There are several causes of low voltage, such as wire too small, loose connections, or low voltage supplied by the power company. To check for low voltage, connect a voltmeter to the common and run terminals of the motor (Figure 8–33). Start the unit and observe the voltage reading. It should not vary more than 10% from the rated voltage of the unit. If a voltage drop of more than 10% is experienced, check the size of the wire to the unit. Be sure that it is at least as big as the recommendation of the equipment manufacturer. If not, it should be replaced with the proper size. If the wire is of sufficient size and the voltage is still low, check for loose connections. These connections will usually be indicated by the wire insulation being overheated or burned. Repair these connections. If there are no loose connections, check the voltage at the electric meter. If found to be low here, contact the power company.

*11-1 Starting and running capacitors.* Capacitors are used by many manufacturers to improve the starting and running characteristics of their motors. Capacitors are manufactured

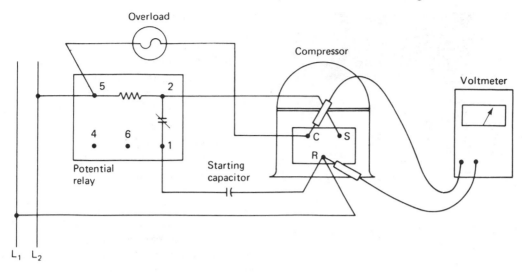

**Figure 8-33** Checking Starting and Running Voltage to a Compressor

both for use in starting and in running motors. Each manufacturer determines the proper size for his motor. His recommendations should be followed.

*11–1A Starting capacitors.* Starting capacitors are used in the starting circuit of a motor to improve its starting capabilities. They are generally round, encased in a plastic casing, and have a relatively high microfarad ($\mu$F) rating (Figure 8–34). These capacitors are designed for short periods of use only. Prolonged use will usually result in damage to the capacitor. If a starting capacitor is found to be defective, be sure to check the starting relay before the unit is placed back in service or the new capacitor may also be damaged. The best way to check a capacitor is by the use of a capacitor analyzer. This type of meter provides a direct reading of the microfarad output of a capacitor without the use of bulky equipment and mathematical formulas (Figure 8–35). Starting capacitors that are found to be out of the range of 0 to +20% of the microfarad rating of the capacitor must be replaced with

**STARTING CAPACITOR WITH BLEED-RESISTOR**

**Figure 8-34** Starting Capacitor with a Bleed-Resistor (Courtesy of Copeland Corporation)

the proper size. Replacement capacitors must have the same voltage rating as those being replaced. Also, the voltage rating of any capacitor must be equal to or greater than the capacitor being replaced. When replacing start-

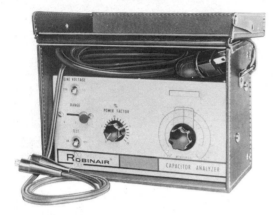

**Figure 8-35** Capacitor Analyzer (Courtesy of Robinair Manufacturing Co.)

ing capacitors, be sure that a 15,000- to 18,000-Ω, 2-W resistor is soldered across the terminals of the capacitor (Figure 8-34). This is a precautionary measure to prevent the arcing and burning of the starting relay contacts. To wire a starting capacitor into the circuit, see Figure 8-36.

*11-1B  Running capacitors.* Running capacitors are in the operating circuit continuously. They are normally oil-filled-type capacitors. The oil is used to prevent overheating of the capacitor. The microfarad rating of these capacitors is relatively low, even though they are larger in size. Running capacitors are used to improve the running efficiency of motors. They also provide enough torque to start the PSC (permanent split capacitor) types of motors. Running capacitors are provided with a terminal marked with a red dot (Figure 8-37). Because of the relatively high voltage generated in the starting winding, the unmarked terminal is connected to the starting terminal on the motor. The red dot indicates the terminal that is most likely to short out in case of capacitor breakdown. If the terminal with the red dot is connected to the motor starting termi-

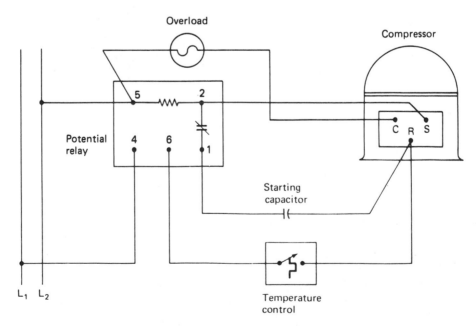

**Figure 8-36** Compressor Wiring Diagram Showing Starting Circuit.

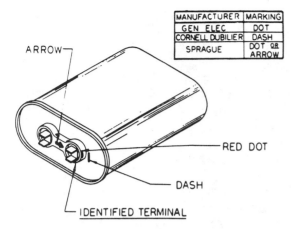

| MANUFACTURER | MARKING |
|---|---|
| GEN ELEC | DOT |
| CORNELL DUBILIER | DASH |
| SPRAGUE | DOT OR ARROW |

*Figure 8-37* Run Capacitor (Courtesy of Copeland Corporation)

nal, damage to the winding could result. The best way to check running capacitors is with a capacitor analyzer. These meters provide a direct reading of the microfarad output of the capacitor without the use of bulky equipment

and mathematical formulas (Figure 8–35). Running capacitors that are found to be out of the range of $\pm 10\%$ of the microfarad rating of the capacitor must be replaced with the proper size. The voltage rating of any capacitor must be equal to or greater than the capacitor being replaced. A higher than normal running amperage usually indicates a weak capacitor. To wire a running capacitor into the circuit, see Figure 8–38.

*12-1 Starting Relays:* Starting relays are used to remove the starting circuit from operation when the motor reaches approximately 75% of its normal running speed. Their function is basically the same as the centrifugal switch used in split-phase motors. Four types of starting relays are in use today: (1) amperage (current) relay, (2) hot-wire relay, (3) solid-state relay, and (4) potential (voltage) relay. The horsepower size and the equipment design regulate which type of starting relay is used.

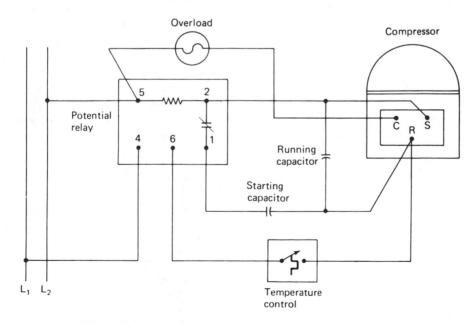

*Figure 8-38* Compressor Wiring Diagram Using Run and Start Capacitors

*12-1A Amperage (current) relay.* These relays are of the electromagnetic type, which are normally used on $\frac{1}{2}$-hp units and smaller (Figure 8-39). These relays are positional types and must be properly mounted for satisfactory operation. They must be sized for each motor horsepower and amperage rating. To check an amperage relay, turn off the electricity to the unit and remove the wire from the S terminal and touch it to the L terminal on the relay (Figure 8-40). Place an ammeter on the common wire to the compressor. Start the compressor and immediately remove the S wire from the L terminal. If the compressor continues to run and the amperage draw is within the rating of the compressor, replace the relay.

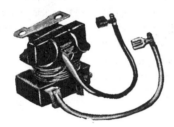

GENERAL ELECTRIC
3ARR2
BRACKET TYPE

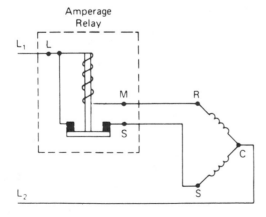

**Figure 8-40** Amperage Relay Connections

*Caution:* To prevent electrical shock, do not allow the loose wire to come in contact with anything or anyone.

An amperage relay that is too large for a motor may not allow the relay contacts to close, thus leaving out the starting circuit. The motor will not start under these conditions. A relay that is rated too small for a motor may keep the contacts closed at all times while electrical power is applied, causing the starting circuit to be engaged continuously. Damage to the starting circuit may occur under these conditions. A motor protector must be used with this type of relay.

*12-1B Hot-wire relays.* Hot-wire relays are a form of current relay, but they do not operate with an electromagnetic coil. They are designed to sense the heat produced by the flow of electrical current through a resistance wire (Figure 8-41). There are two sets of contacts in these relays, a set for starting and a set for running. To check these relays, turn off the electrical current and remove the wire from the S terminal on the relay and touch it to the L terminal on the relay (Figure 8-42). Place an ammeter on the common wire to the compressor. Start the compressor and immediately remove the S wire from the L terminal. To prevent electrical shock, do not allow the loose end of the wire to touch anything or anyone. If the

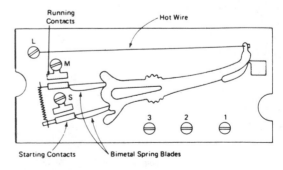

**Figure 8-41** Hot-Wire Relay

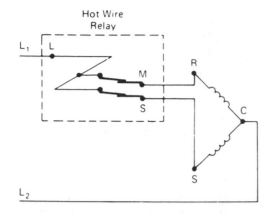

**Figure 8-42** Hot-Wire Relay Connections

compressor continues to run and the amperage draw is within the rating of the compressor, replace the relay. If the compressor operates within the amperage rating indicated by the manufacturer, but still stops within 1 or 2 minutes, the overload portion of the relay is defective. Replace the relay with one of the proper size for the horsepower and amperage rating of the compressor motor.

A hot-wire relay that is too large for the motor will not remove the starting components from the circuit, resulting in possible motor damage. One that is rated too small will stop the motor with the overload after 1 or 2 minutes of operation. An additional overload is not necessary when these relays are used. These are nonpositional relays.

*12-1C Solid-state starting relays.* These relays use a self-regulating conductive ceramic developed by Texas Instruments, Inc., which increases in electrical resistance as the compressor starts and thus quickly reduces the starting winding current flow to a milliampere level. The relay switches in approximately 0.35 second. This allows this type of relay to be applied to refrigerator compressors without being tailored to each particular system within the specialized current limitations. These relays

will start virtually all split-phase 120-V hermetic compressors up to $\frac{1}{3}$-hp. An overload must be used with these relays (Figure 8–43). Since these relays are push-on devices, the easiest method of checking their operation is to simply install a new relay. Be sure to check the amperage draw of the compressor motor.

*12-1D Potential relays.* Potential (voltage) relays operate on the electromagnetic principle. They incorporate a coil of very fine wire wound around a core. These relays are used on motors of almost any size. They are nonpositional. The contacts are normally closed and are caused to open when a plunger is pulled into the relay coil. These relays have three connections to the inside in order for the relay to perform its function. These terminals are numbered 1, 2, and 5. Other teminals numbered 4 and 6 are sometimes used as auxiliary terminals (Figure 8–44). To check a potential relay, turn off the electrical current and remove the wire from terminal 2 on the relay and touch it to terminal 1 on the relay. Place an ammeter on the common wire to the motor. Start the motor and immediately remove the terminal 2 wire from the terminal 1 wire on the relay.

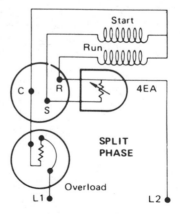

**Figure 8-43** Solid-State Relay Connections (Courtesy of Control Products Division, Texas Instruments, Incorporated)

**Figure 8-44** Potential Starting Relay

*Caution:* To prevent electrical shock, do not allow the loose wire to touch anything or anyone. If the compressor continues to run and the amperage draw is within the rating of the compressor, replace the relay.

The sizing of potential relays is not as critical as with the amperage and hot wire relays. A good way to determine the required relay is to manually start the motor and check the voltage between the start and common terminals while the motor is operating at full speed (Figure 8–45). Multiply the voltage obtained by 0.75 and this will be the pickup voltage of the required relay.

*12–1E Starting relay mounting.* An excessive amount of vibration of a starting relay will cause the relay contacts to arc excessively and become burned. If such vibration occurs, the relay must be remounted on a more solid surface. Generally, the relay will need to be replaced during the remounting process. Be sure to replace the relay with the proper type and size, and if it is a positional relay, it must be mounted to satisfy the manufacturer's recommendations.

*13–1 Crankcase Heaters:* Crankcase heaters are electrical resistors designed to provide just enough heat to boil off any liquid refrigerant that might enter the crankcase. There are three

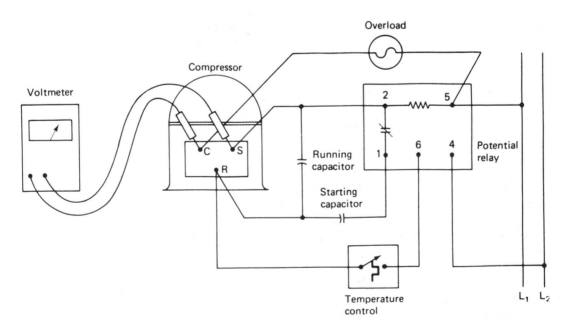

**Figure 8-45** Checking Voltage between Start and Run Terminals

different designs: (1) externally mounted, (2) internally mounted, and (3) motor winding heater (Figure 8–46). Externally mounted crankcase heaters are manufactured by several manufacturers and are easily installed on most compressors. The internally mounted heaters are usually designed for a specific model of compressor. The motor winding heater is specially designed for a specific model of compressor. The motor winding heater is specially designed with the proper resistance placed in the circuit to prevent overheating of the winding. Precaution should be taken to ensure that the manufacturer's specifications are maintained when this type of heater is encountered. To check crankcase heaters, dampen a finger and gently touch the heater. If it is working, it will be hot to the touch. The electrical power must be on for several hours before this test can be performed. Another method of checking is to disconnect both electrical wires and check the resistance of the heater (Figure 8–47). Crankcase heaters are generally energized continuously and are designed to prevent overheating of the compressor oil.

*14-1 High Discharge Pressure:* A high discharge pressure can be the result of one or a combination of several things. It is a condition

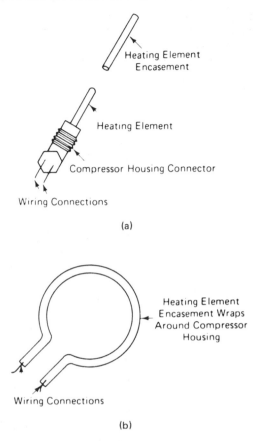

(a)

(b)

*Figure 8-46* (a) Internal Crankcase Heater; (b) External Crankcase Heater

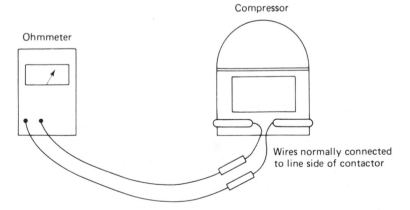

*Figure 8-47* Checking Resistance of Crankcase Heater

that causes an overload on the motor and decreases the efficiency of the compressor and the refrigeration system. The most common causes are (1) compressor discharge service valve front-seated, (2) lack of cooling air, (3) lack of cooling water, (4) an overcharge of refrigerant, or (5) noncondensable gases.

*14–1A Front-seated discharge service valve.* The compressor discharge service valve when front-seated will reduce or completely stop the flow of refrigerant from the compressor. Caution should be exercised to prevent this condition because damage to the compressor or motor is likely. The rapid buildup of pressure within the compressor cylinder head is tremendous and increases rapidly as the piston completes its upward stroke. Never front-seat the compressor discharge service valve while the compressor is running or start the compressor with the valve front-seated.

*14–1B Lack of cooling air.* The lack of cooling air over an outdoor unit will cause the discharge pressure to increase. This is because the higher pressure is required to condense the vapor to a liquid. The higher refrigerant temperature also reduces the unit efficiency because of increased flash gas, as well as reducing the compressor efficiency. This condition is generally caused by a dirty outdoor coil, loose fan belt, or bad fan motor bearings.

A dirty outdoor coil can be cleaned by using a garden hose with a high-pressure nozzle. The water should be forced through the coil from both sides. Be sure to prevent water from entering the fan motor, which might cause an electrical short. This can be done by wrapping a piece of sheet plastic around the motor. Be sure to remove the plastic before starting the unit. Turn the unit off during this procedure.

A loose or broken fan belt will prevent the blower moving air to cool the coil. This condition is generally obvious and is easily corrected. To adjust the belt tension, turn the adjustment screw until the belt can be flexed about 1 in. (25.40 mm) with one finger using moderate pressure (Figure 8–48). If the belt is frayed, has wear grooves on the sides, or has become hard, it should be replaced. Be sure to use the proper size of belt. A belt that is too narrow will ride the bottom of the pulley and slip, which results in decreased efficiency (Figure 8–49). A belt that is too wide will ride too high in the pulley; it will not maintain the desired efficiency and will probably overload the fan motor.

Bad fan motor bearings will cause the motor to overheat and cut out on the overload. When the fan motor stops, the outdoor coil overheats and the compressor will cut out on high pressure. To check for bad bearings, stop the unit, remove the belt, and move the motor shaft from side to side. Any movement of the shaft in a sideways direction indicates bad bearings. The bearings must be replaced or the motor replaced, depending on the motor size.

*14–1C Lack of cooling water.* The lack of cooling water on a water-source heat pump will cause the discharge pressure to increase, because higher pressure is required to condense the vapor to a liquid. The higher refrigerant temperature also reduces the unit efficiency because of increased flash gas, as well as reducing compressor efficiency. This condition is generally caused by plugged strainers, pumps, or spray nozzles. When this condition is suspected, check the temperature rise of the water through the condenser. This rise should not be

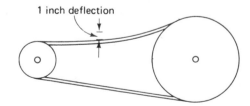

1 inch deflection

*Figure 8-48* Proper Belt Tension Adjustment

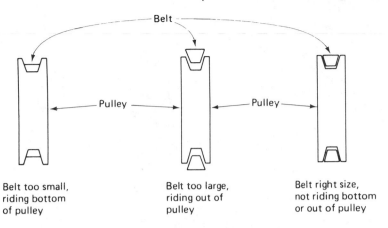

Belt too small, riding bottom of pulley

Belt too large, riding out of pulley

Belt right size, not riding bottom or out of pulley

*Figure 8-49* Comparison of Belt and Pulley Fitting

more than 10°F (5.56°C). If a temperature rise of more than 10°F is found, a strainer, pump, or spray nozzle is stopped up or there is not enough water in the tower water sump. The necessary steps must be taken to relieve this condition.

A temperature rise of less than 10°F (5.56°C) indicates that the tubes in the coil are scaled and must be cleaned. Several commercial cleaners are available for cleaning (acidizing) water coils. The amount used is recommended by the manufacturer of the cleaner. Caution should be used to prevent damage to equipment, personnel, and surrounding vegetation when acidizing a unit. The most common method of acidizing a unit is first to be sure that the strainers, pump, and spray nozzles are clean. Then slowly dump the cleaner into the tower sump and check the mixture with pH strips; add cleaner until the desired pH is indicated, and allow the unit to run until all the scale has been removed. The cleaner and sediment must be completely removed from the system. To do this, drain and flush the tower, coil, and water lines with fresh water. Then add a neutralizer and allow the unit to run until the neutralizer has had time to neutralize the cleaner; then drain and refill the system with fresh water. Never leave the cleaner in the

system because of the possible damage that might be done to the equipment.

Another method is to disconnect the water lines from the condenser and circulate the cleaner through the coil with a pump. This method is usually expensive and, therefore, not popular. However, this method must be used if even the slightest possibility of contaminating the ground water is possible.

*14–1D Refrigerant overcharge.* An overcharge of refrigerant will cause a high discharge pressure because the extra refrigerant will take up space in the condensing coil that is needed to condense the vapor to a liquid. An overcharge of refrigerant will cause at least one-half of the condenser tubes to be cooler than the rest. The cool tubes are the ones that are full of liquid refrigerant (Figure 8–50). When a capillary tube is used, the suction pressure will also be higher than normal. The suction line will also be cooler than normal and may be frosted over, depending on the amount of overcharge.

The excess of refrigerant must be removed from the system. When removing refrigerant, remove only a small amount at a time. This procedure is recommended to prevent removing too much refrigerant from the system,

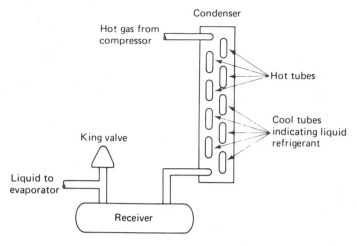

**Figure 8-50** Checking Liquid Level in Condensing Unit

which would require the addition of refrigerant back into the unit. Also, it prevents the removal of lubricating oil from the compressor crankcase.

*14-1E Noncondensable gases.* The presence of noncondensable gases in a refrigeration system will cause a high discharge pressure because the noncondensables will not condense under the normal pressures encountered in refrigeration systems and will take up space needed by the refrigerant in the condenser. To determine if noncondensables are present in the system, pump the system down; that is, pump all the refrigerant into the receiver or condensing coil by front-seating the liquid line service (King) valve (Figure 8–51). Operate the compressor until the low side has been pumped

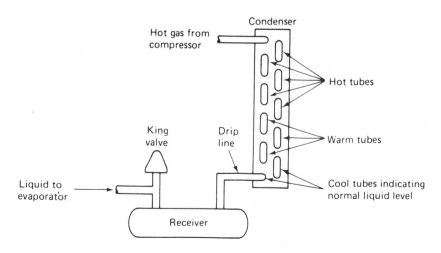

**Figure 8-51** Location of King Valve

down to approximately 5 psig (34.474 kPa). Stop the unit and allow it to stand idle until it has cooled to the ambient temperature. Compare the idle discharge pressure to the pressure indicated on a pressure-temperature chart at that temperature. If the discharge pressure is higher than that indicated by the chart, slowly purge the noncondensables from the highest point in the system. Purge the air slowly to prevent purging an excess of refrigerant from the system. If the system contains only a small charge of refrigerant, it will probably be better to remove the entire charge, evacuate the system, and add a complete new charge. Noncondensables should be kept from a system because they not only reduce the system efficiency, but they also contain moisture, which is extremely dangerous to a refrigeration system.

*15–1 Low Suction Pressure:* A suction pressure that is lower than normal may be due to any or all of the following reasons: (1) shortage of refrigerant, (2) dirty air filter, (3) dirty evaporating coil, (4) dirty blower, (5) loose fan belt, (6) iced-over coil, (7) TXV (thermostatic expansion valve) superheat set too low, (8) AXV (automatic expansion valve) set too low, or (9) restriction in the refrigerant system. The problem or problems causing a low suction pressure should be found and corrected, because the compressor may pump all the lubricating oil out of the crankcase and into the system, resulting in possible damage to the compressor, as well as inefficient operation.

*15–1A Refrigerant shortage.* A shortage of refrigerant is usually due to a leak that has developed in the refrigerant circuit. The leak should be found and repaired and the system recharged with the proper charge of refrigerant. If the leak is not found and repaired, the refrigerant will escape from the system, which will require more service. A refrigerant leak is

generally indicated by oil at the place where the leak has occurred. Many times a visual inspection will locate the problem area (Figure 8–52). If the problem area is not found by visual inspection, an electronic or halide torch leak detector should be used. Either of these will pinpoint a leak under most conditions. Two conditions that make leak detection difficult with these devices are when the wind is blowing outdoors or in an enclosed place where the refrigerant concentration is high. When these conditions are encountered, a soap and water solution or one of the liquid plastic gas-leak detectors must be used. The leak may be found by applying the solution to the suspected area and watching for bubbles to appear (Figure 8–53). The bubbles will appear within 5 seconds. When repairing leaks that require the tub-

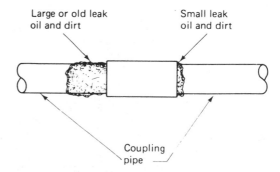

**Figure 8-52** Oil around Refrigerant Leaks

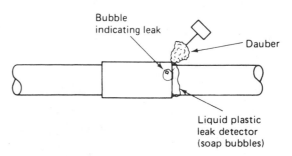

**Figure 8-53** Leak Testing with Liquid Plastic Leak Detector (Soap Bubbles)

ing to be heated, be sure to relieve any pressure from inside the tubing that is being heated. This will prevent a blowout, which can result in damage to the equipment and injury to the service technician.

*15–1B Dirty air filters.* This condition is frequently the cause of low suction pressures in heat pump systems. Dirty air filters restrict the flow of air over the indoor coil and result in a low load on the refrigeration system. The air filter is located on the air inlet side of the blower (Figure 8–54). The filters should be changed or cleaned on a regular basis to ensure peak efficiency from the unit. Throwaway-type filters should not be cleaned. They should be replaced. When cleanable-type filters are used, they should be coated with a filter coater after cleaning to increase the effectiveness of the filter.

*15–1C Dirty evaporating coil.* Dirty evaporating coils are often the cause of low suction pressures on heat pump systems. A dirty coil restricts the air flow through the unit and the insulating value, and the dirt prevents the proper transfer of heat to the refrigerant. A dirty indoor coil is the result of a dirty air filter or one that does not fit properly. When an air filter does not fit properly, the dust-laden air will bypass it and the dust will stick to the coil

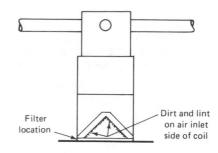

**Figure 8-55** Dirty Coil because of Filter Air Bypass

fins (Figure 8–55). Precautions must be taken to ensure proper fit of the air filter, and the coil must be cleaned before the system will function properly. Many times the coil will need to be removed and be steam cleaned. Be sure to prevent moisture entering the refrigerant system during this process.

*15–1D Dirty blower.* A dirty blower will not deliver enough air to place the proper load on the coil and will result in low suction pressure. A dirty indoor blower is the result of improperly fitting air filters or a unit that has been operated with an excessively dirty filter. The blower must be removed and the vanes cleaned so that only the metal can be seen. The filter must be cleaned or replaced, and any air bypass must be prevented. Otherwise, the problem will reoccur. When steam cleaning or washing the blower with water, take precautions to prevent moisture entering the electric blower motor.

*15–1E Loose or broken fan belt.* A loose or broken fan belt will prevent proper operation of the blower and reduce the flow of air over the evaporating coil, resulting in low or no load on the unit and a low suction pressure. A loose belt that is in good condition can be properly adjusted and the unit put back in service. A broken or frayed belt or one that has become hard on the sides must be replaced and

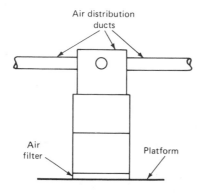

**Figure 8-54** Air Filter Location

properly adjusted. To properly adjust a belt, tighten the adjustment screw until the belt can be flexed about 1 in. (25.40 mm) with one finger using moderate pressure (Figure 8–48). Be sure to use the proper size of belt. A belt that is too narrow will ride the bottom of the pulley and will slip, which causes decreased efficiency (Figure 8–49). A belt that is too wide will ride too high in the pulley, will not maintain the desired efficiency, and will probably overload the motor.

*15–1F  Iced  or  frosted  evaporating coil.*  An iced or frosted evaporating coil will cause a low suction pressure because the insulating function of the ice or frost will prevent proper heat transfer from the air to the refrigerant. An iced coil is usually caused by an insufficient load on the refrigeration system or a shortage of refrigerant. A reduced load can be the result of (1) a dirty air filter, (2) a dirty evaporating coil, (3) a dirty blower, or (4) a loose or broken fan belt. A shortage of refrigerant is due to a refrigerant leak. Any of these conditions must be corrected and the coil de-iced before the system will operate satisfactorily. De-icing can be accomplished by turning on the blower and turning off the compressor. Allow the fan to operate until the ice has been melted. A faster method is to apply a small amount of heat to the ice. Caution should be exercised to prevent overheating of the coil or fins.

*16–1 Thermostatic Expansion Valves:*  Thermostatic expansion valves are the most common type of flow-control devices used on units of 5-ton capacity and above. They operate as a result of pressure and temperature inside the evaporating coil. The proper superheat adjustment of these valves will have a tremendous effect on the efficiency of the equipment.

A thermostatic expansion valve that is adjusted for too great a superheat or an im-

properly located feeler bulb will result in a low suction pressure. If the superheat is found improperly adjusted, it may be adjusted by the following procedure:

1.  Measure the temperature of the suction line at the point where the bulb is clamped.
2.  Obtain the suction pressure that exists in the suction line at the bulb location by either of the following methods.
    a.  If the valve is externally equalized, a pressure gauge in the external equalizer line will indicate the pressure directly and accurately.
    b.  Read the gauge pressure at the suction service valve of the compressor. To the reading, add the estimated pressure drop through the suction line between the bulb location and the compressor suction service valve. The sum of the gauge reading and the estimated pressure drop will equal the approximate line pressure at the bulb.
3.  Convert the pressure obtained in step 2 to the saturated evaporating coil temperature by using a temperature-pressure chart as shown in Table 8–1.
4.  Subtract the two temperatures obtained in steps 1 and 3. The difference is the superheat setting of the valve.

Figure 8–56 illustrates a typical example of superheat measurement on a system using refrigerant 22 as the refrigerant. The temperature of the suction line at the bulb location is read at 52 °F (11.11 °C). The suction pressure at the compressor suction valve is 66 psi (455.054 kPa), and the estimated pressure drop is 2 psi (13.790 kPa). The total suction pressure is 66 psi + 2 psi = 68 psi (468.843 kPa). This is equivalent to a 40 °F (4.4 °C) saturation

## Table 8-1
Temperature-Pressure Chart  (Courtesy of Alco Controls Division, Emerson Electric Co.)

BOLD FIGURES – INCHES MERCURY VACUUM        LIGHT FIGURES = PSIG

| °F | R-12 | R-13 | R-22 | R-500 | R-502 | R-717 Ammonia | °F | R-12 | R-13 | R-22 | R-500 | R-502 | R-717 Ammonia |
|----|------|------|------|-------|-------|---------------|----|------|------|------|-------|-------|---------------|
| -100 | **27.0** | 7.5 | **25.0** | — | **23.3** | **27.4** | 16 | 18.4 | 211.9 | 38.7 | 24.2 | 47.8 | 29.4 |
| -95 | **26.4** | 10.9 | **24.1** | — | **22.1** | **26.8** | 18 | 19.7 | 218.8 | 40.9 | 25.7 | 50.1 | 31.4 |
| -90 | **25.7** | 14.2 | **23.0** | — | **20.7** | **26.1** | 20 | 21.0 | 225.7 | 43.0 | 27.3 | 52.5 | 33.5 |
| -85 | **25.0** | 18.2 | **21.7** | — | **19.0** | **25.3** | 22 | 22.4 | 233.0 | 45.3 | 29.0 | 55.0 | 35.7 |
| -80 | **24.1** | 22.2 | **20.2** | — | **17.1** | **24.3** | 24 | 23.9 | 240.3 | 47.6 | 30.7 | 57.5 | 37.9 |
| -75 | **23.0** | 27.1 | **18.5** | — | **15.0** | **23.2** | 26 | 25.4 | 247.8 | 49.9 | 32.5 | 60.1 | 40.2 |
| -70 | **21.8** | 32.0 | **16.6** | — | **12.6** | **21.9** | 28 | 26.9 | 255.5 | 52.4 | 34.3 | 62.8 | 42.6 |
| -65 | **20.5** | 37.7 | **14.4** | — | **10.0** | **20.4** | 30 | 28.5 | 263.2 | 54.9 | 36.1 | 65.4 | 45.0 |
| -60 | **19.0** | 43.5 | **12.0** | — | **7.0** | **18.6** | 32 | 30.1 | 271.3 | 57.5 | 38.0 | 68.3 | 47.6 |
| -55 | **17.3** | 50.0 | **9.2** | — | **3.6** | **16.6** | 34 | 31.7 | 279.5 | 60.1 | 40.0 | 71.2 | 50.2 |
| -50 | **15.4** | 57.0 | **6.2** | — | **0.0** | **14.3** | 36 | 33.4 | 287.8 | 62.8 | 42.0 | 74.1 | 52.9 |
| -45 | **13.3** | 64.6 | **2.7** | **7.9** | 2.1 | **11.7** | 38 | 35.2 | 296.3 | 65.6 | 44.1 | 77.2 | 55.7 |
| -40 | **11.0** | 72.7 | **0.5** | **4.8** | 4.3 | **8.7** | 40 | 37.0 | 304.9 | 68.5 | 46.2 | 80.2 | 58.6 |
| -35 | **8.4** | 81.5 | 2.6 | **1.4** | 6.7 | **5.4** | 45 | 41.7 | 327.5 | 76.0 | 51.9 | 88.3 | 66.3 |
| -30 | **5.5** | 91.0 | 4.9 | 0.0 | 9.4 | **1.6** | 50 | 46.7 | 351.2 | 84.0 | 57.8 | 96.9 | 74.5 |
| -28 | **4.3** | 94.9 | 5.9 | 0.7 | 10.6 | 0.0 | 55 | 52.0 | 376.1 | 92.6 | 64.2 | 106.0 | 83.4 |
| -26 | **3.0** | 98.9 | 6.9 | 1.5 | 11.7 | 0.8 | 60 | 57.7 | 402.3 | 101.6 | 71.0 | 115.6 | 92.9 |
| -24 | **1.6** | 103.0 | 7.9 | 2.3 | 13.0 | 1.7 | 65 | 63.8 | 429.8 | 111.2 | 78.2 | 125.8 | 103.1 |
| -22 | **0.3** | 107.3 | 9.0 | 3.1 | 14.2 | 2.6 | 70 | 70.2 | 458.7 | 121.4 | 85.8 | 136.6 | 114.1 |
| -20 | 0.6 | 111.7 | 10.1 | 4.0 | 15.5 | 3.6 | 75 | 77.0 | 489.0 | 132.2 | 93.9 | 148.0 | 125.8 |
| -18 | 1.3 | 116.2 | 11.3 | 4.9 | 16.9 | 4.6 | 80 | 84.2 | 520.8 | 143.6 | 102.5 | 159.9 | 138.3 |
| -16 | 2.1 | 120.8 | 12.5 | 5.8 | 18.3 | 5.6 | 85 | 91.8 | — | 155.7 | 111.5 | 172.5 | 151.7 |
| -14 | 2.8 | 125.7 | 13.8 | 6.8 | 19.7 | 6.7 | 90 | 99.8 | — | 168.4 | 121.2 | 185.8 | 165.9 |
| -12 | 3.7 | 130.5 | 15.1 | 7.8 | 21.3 | 7.9 | 95 | 108.3 | — | 181.8 | 131.2 | 199.7 | 181.1 |
| -10 | 4.5 | 135.4 | 16.5 | 8.8 | 22.8 | 9.0 | 100 | 117.2 | — | 195.9 | 141.9 | 214.4 | 197.2 |
| -8 | 5.4 | 140.5 | 17.9 | 9.9 | 24.4 | 10.3 | 105 | 126.6 | — | 210.8 | 153.1 | 229.7 | 214.2 |
| -6 | 6.3 | 145.7 | 19.3 | 11.0 | 26.0 | 11.6 | 110 | 136.4 | — | 226.4 | 164.9 | 245.8 | 232.3 |
| -4 | 7.2 | 151.1 | 20.8 | 12.1 | 27.7 | 12.9 | 115 | 146.8 | — | 242.7 | 177.3 | 262.6 | 251.5 |
| -2 | 8.2 | 156.5 | 22.4 | 13.3 | 29.5 | 14.3 | 120 | 157.7 | — | 259.9 | 190.3 | 280.3 | 271.7 |
| 0 | 9.1 | 162.1 | 24.0 | 14.5 | 31.2 | 15.7 | 125 | 169.1 | — | 277.9 | 203.9 | 298.7 | 293.1 |
| 2 | 10.2 | 167.9 | 25.6 | 15.7 | 33.1 | 17.2 | 130 | 181.0 | — | 296.8 | 218.2 | 318.0 | 315.0 |
| 4 | 11.2 | 173.7 | 27.3 | 17.0 | 35.0 | 18.8 | 135 | 193.5 | — | 316.6 | 233.2 | 338.1 | 335.0 |
| 6 | 12.3 | 179.8 | 29.1 | 18.4 | 37.0 | 20.4 | 140 | 206.6 | — | 337.3 | 248.8 | 359.1 | 365.0 |
| 8 | 13.5 | 185.9 | 30.9 | 19.8 | 39.1 | 22.1 | 145 | 220.6 | — | 358.9 | 265.2 | 381.1 | 390.0 |
| 10 | 14.6 | 192.1 | 32.8 | 21.2 | 41.1 | 23.8 | 150 | 234.6 | — | 381.5 | 282.3 | 403.9 | 420.0 |
| 12 | 15.8 | 198.6 | 34.7 | 22.7 | 43.3 | 25.6 | 155 | 249.9 | — | 405.2 | 300.1 | 427.8 | 450.0 |
| 14 | 17.1 | 205.2 | 36.7 | | 45.5 | 27.5 | 160 | 265.12 | — | 429.8 | 318.7 | 452.6 | 490.0 |

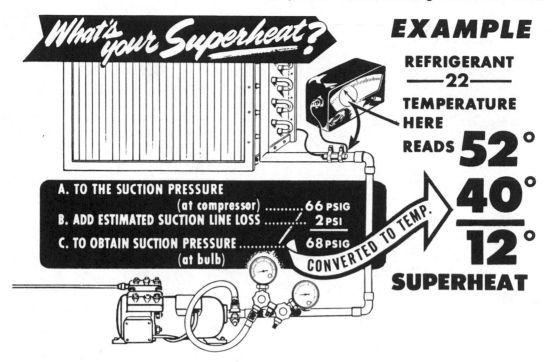

*Figure 8-56* Determination of Superheat (Courtesy of Sporlan Valve Co.)

temperature. Then 40 °F subtracted from 52 °F equals 12 °F (6.72 °C) superheat setting.

Notice that subtracting the difference between the temperature at the coil inlet and the temperature at the coil outlet is not an accurate measure of superheat. This method is not recommended because any evaporator pressure drop will result in an erroneous superheat indication.

The valve should be adjusted to provide the desired superheat setting. To determine if the expansion valve is operating, remove the bulb and warm it with the hand while observing the suction pressure. If the pressure increases, the valve needs adjusting.

*16–1A Thermostatic expansion valve feeler bulb.* The feeler bulb on a thermostatic expansion valve is one of the major forces con-

trolling the valve. If the bulb is located in a cold location, it will cause the valve to starve the evaporating coils, which will cause a lower than normal suction pressure. The location of the feeler bulb is extremely important and in some cases determines the success or failure of the heat pump. For satisfactory expansion valve control, good thermal contact between the bulb and the vapor line is essential. The bulb should be securely fastened with two bulb straps to a clean, straight section of the vapor line.

Installation of the bulb to a horizontal run of vapor line is preferred. If a vertical installation cannot be avoided, the bulb should be mounted so that the capillary tube comes from the top (Figure 8–57). To install, clean the vapor line thoroughly before clamping the remote bulb in place. When a steel vapor line is used, it is advisable to paint the line with

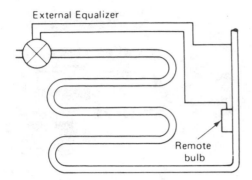

**Figure 8-57** Remote Bulb Installation on Vertical Tubing

aluminum paint to minimize future corrosion and to eliminate faulty remote bulb contact with the line. On lines smaller than $\frac{7}{8}$ in. (22.225 mm) OD (outside diameter), the remote bulb may be installed on top of the line. On $\frac{7}{8}$-in. OD and larger, the remote bulb should be installed at about the 4 or 8 o'clock position (Figure 8-58). If it is necessary to protect the remote bulb from the effects of an air stream after it is clamped to the line, use a material such as sponge rubber that will not absorb water when evaporating coil temperatures are above 32°F (0°C).

*16-1B Thermostatic expansion valve stuck open.* When a thermostatic expansion valve is stuck open, there will be an excessive amount of sweating on the vapor line. The compressor

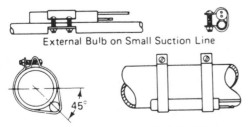

External Bulb on Small Suction Line

45°

External Bulb on Large Suction Line

**Figure 8-58** Remote Bulb Installation on Horizontal Tubing (Courtesy of Alco Controls Division, Emerson Electric Co.)

crankcase will also sweat because of the liquid refrigerant being fed back to the compressor. It is best to replace a sticking expansion valve. Be sure to check the system load and the valve superheat before replacing the valve.

*16-1C Thermostatic expansion valve too large.* If the thermostatic expansion valve is too large for the system, it will not maintain a constant suction pressure. The feeler bulb will attempt to control the flow of liquid to maintain the superheat setting, but the oversized valve will admit too much liquid too rapidly. The feeler bulb sensing this liquid will close the valve, and the pressure in the evaporating coil will drop until the valve opens and admit more liquid refrigerant. This hunting action will cause the suction pressure to fluctuate, which can be seen on the compound gauge. This variation is usually 10 to 15 psi (68.948 to 103.421 kPa). When this condition occurs, either replace the complete valve or replace the valve seat.

*16-1D Thermostatic expansion valve too small.* A thermostatic expansion valve that is too small cannot pass enough liquid refrigerant to properly refrigerate the evaporating coil. When the unit is heavily loaded, the superheat will be high and the system capacity will be low. An expansion valve that is too small will generally cause a lower than normal suction pressure. Either replace the valve or replace the valve seat with the proper size.

*16-1E Weak thermostatic expansion valve power element.* The power element of a thermostatic expansion valve contains a vapor charge. If a leak should develop in the power element assembly, the valve will tend to close and stop the flow of refrigerant. To check the power element, use the following procedure:

1. Stop the compressor.
2. Remove the feeler bulb from the line and place it in ice water.

3. Start the compressor.

4. Remove the bulb from the ice water and warm it with the hand. At the same time feel the vapor line for a drop in temperature. If liquid refrigerant floods through the valve, the power element is operating properly. Be sure not to flood liquid back through the vapor line for a long period of time because it can cause damage to the compressor.

*17-1 Restriction in a Refrigerant Circuit:* A restriction in the refrigerant circuit will reduce the flow of refrigerant and result in a lower than normal suction pressure. A restriction may be in the form of a kinked line, a plugged strainer, a plugged drier, a plugged capillary tube, or ice in the orifice of a flow-control device. The restriction must be removed before proper operation can be realized.

*17-1A Kinked line.* A kinked line occurs when the line has been bent too far without proper support, which has resulted in a flattened place. This type of restriction can be located by visual inspection. However, if the restriction is in the liquid line, there will be a temperature difference across the kink. If the tubing is flattened enough, there may be condensation or frost on the outlet side of the kink (Figure 8-59). In cases when the tube is not flattened excessively, the flattened spot may be

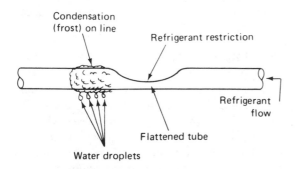

*Figure 8-59* Flattened Tube causing Refrigerant Restriction

straightened by placing a flaring block around the tube, using the proper size hole, and tightening it down. If this fails, the piece of tubing must be removed and a new piece installed. Be sure to relieve all pressure on the tubing before attempting to make repairs. Personal injury may result when repairs are attempted if there is pressure inside the tube that is being repaired.

*17-1B Strainer.* The purpose of a strainer is to trap foreign particles in the refrigeration system. A plugged strainer will reduce the flow of refrigerant and may stop any passage of refrigerant. Strainers are placed in all expansion valves, as well as suction line filter-driers and liquid line driers, and at the compressor suction inlet. All these features are used to trap foreign materials in the system to prevent damage to the particular component. A plugged strainer will develop a pressure drop on one side and usually a temperature difference on each side that can be felt with the bare hand. When a strainer becomes plugged, it is best to replace it if another is readily available. If another strainer is not available, thoroughly clean the original and replace it. Do not remove the strainer permanently. To do this will allow foreign particles to enter the protected device, resulting in possible damage.

*17-1C Drier.* The purpose of a drier is to remove moisture and trap foreign particles in the refrigeration system. A plugged drier will restrict or completely stop the flow of refrigerant through the system. All field built-up systems are equipped with driers to remove any moisture and foreign particles that may accidentally enter the system during the installation process. A plugged drier can be determined by the temperature drop through it (Figure 8-60). A plugged drier should be replaced. Never permanently remove a drier from the system. A plugged drier is an indication that there are still moisture and foreign particles in the system. When removing a drier, be sure to relieve all

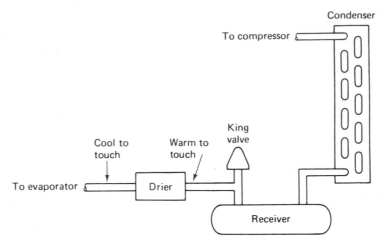

**Figure 8-60** Checking a Plugged Drier

pressure inside the tube being worked on. To open a refrigeration system while it is pressurized can result in personal injury.

*17–1D Plugged capillary tube.* A plugged capillary tube is caused by untrapped foreign particles in a refrigeration system and can result in a reduced or completely stopped refrigerant flow through the system. A plugged capillary tube will be indicated by a longer than usual pressure equalization time, accompanied by a loss of refrigeration. It is recommended that a plugged capillary line be replaced rather than trying to clean it out. When replacing it, be sure to use the same length and inside diameter tubing as the original tubing. To use a different size will result in an unbalanced refrigeration system and poor refrigeration. Always replace the filter-strainer along with the capillary tube. Be sure to relieve the pressure in the low side of the system before attempting repairs or personal injury may result.

*17–1E Iced flow-control device.* Ice in the orifice of a refrigerant flow-control device is due to free moisture in the system. This condition generally occurs when the drier has

trapped all the moisture it can absorb, and there is surplus free moisture that turns to ice in the orifice of the flow-control device. This condition is usually indicated by poor operation, a lower than normal suction, and low discharge pressures. To be sure that moisture is causing the problem, stop the unit and apply a cloth moistened with hot water to the outlet of the flow-control device. If after a few minutes a hissing sound is heard and an increase in the low-side pressure is indicated, moisture is the culprit. To correct this problem, a new drier must be installed. If the problem continues, it may be necessary to completely discharge the refrigerant from the system, triple-evacuate the system, install oversized driers, and recharge the system with dry refrigerant.

*18-1 Tubing Rattle:* A tubing rattle is the result of mishandling or misuse of the refrigeration equipment. Not only is this condition annoying, it is also damaging to the tubing. If let go long enough, a refrigerant leak will result because of a hole being rubbed in the tubing. To eliminate tubing rattle, either slightly bend the tubing apart with the hand or place a cushion between the tubes. Be careful not to apply enough pressure to the lines to break or kink them.

*19–1 Loose Mounting:* When mountings become worn or broken, an annoying noise will result, as well as possible damage to the equipment. The only proper method of correcting this situation is to replace the mounting. In cases where a bolt or nut has become loose, it can be tightened without replacement of the mounting. When springs are used as the mounting, it is usually best to replace all the springs, not just the broken one, because the rest are probably weak and may break in the near future.

*20–1 Liquid Flashing in Liquid Line:* Liquid flashing, or turning to vapor, in the liquid line may be due to a shortage of refrigerant, a liquid line that is too small, or a liquid line that has too high a vertical lift. Liquid flashing will be indicated by a hissing or gurgling noise at the expansion valve and a lower than normal suction presssure.

*20–1A Shortage of refrigerant.* When a system is short of refrigerant, the leak should be found and repaired and the system charged to the proper level with the proper refrigerant and checked for proper operation. Be sure to observe all safety precautions during this process.

*20–1B Liquid line too small.* A liquid line that is too small will offer an excessive amount of resistance to the flow of refrigerant. As a result, the refrigerant pressure will drop near the outlet of the line, which will result in the evaporation of a portion of the refrigerant and the picking up of heat outside the refrigerated space. This reduces the effectiveness of the system. The liquid line will also drop in temperature near the evaporating coil. This drop in temperature can be felt with the hand. Oil return may also be a problem when this situa-

tion is experienced. When a too small liquid line is found, it must be replaced with one of proper size. Manufacturers provide line sizing charts that can be used to properly size refrigerant lines.

*20–1C Vertical liquid lines.* When liquid lines extend vertically for more than approximately 20 ft (6.09 m), the weight of the liquid forcing downward will probably cause flash gas. This flash gas is due to the drop in pressure. For example, if liquid refrigerant 22 is forced up 30 ft (9.144 m), the pressure on top of the liquid will be 15 psig (103.42 kPa) lower than that at the bottom of the column. This pressure difference plus the heat in the liquid will cause flash gas. When this condition is encountered, a lower than normal suction pressure and a reduction in temperature of the liquid line near the evaporating coil will be present. To solve this problem, the liquid refrigerant must be subcooled. This may be accomplished by installing another condensing coil in the liquid line near the condensing coil outlet; in severe cases a vapor to liquid heat exchanger should be added to the system (Figure 8–61). To prevent overloading the compressor, use the proper size of heat exchanger.

*21–1 Restricted or Undersized Refrigerant Lines:* Restricted or undersized refrigerant lines can be a source of continual trouble in an otherwise carefully selected unit. Often the system capacity is reduced, and oil return problems are frequently experienced with undersized lines. When these two conditions are repeated, a reevaluation of the refrigerant line sizing should be done. Line sizing charts are available for almost every situation encountered. These charts should be consulted and the lines replaced with the proper size or an alternate recommended multiple set of lines installed.

*21–1A Restricted refrigerant lines.* Restricted refrigerant lines are usually due to

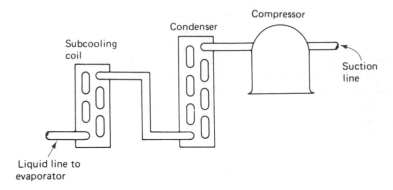

**Figure 8-61** Subcooling Coil Location

contaminants or kinks in the system. A restriction in a line can be located by the difference in temperature on each side of the restriction (Figure 8-62). A restriction may be found in a drier, filter, strainer, or some other system component. This condition will be accompanied by pressures that are lower than normal and a reduction in system capacity. The restriction must be removed before proper operation can be achieved. Be sure to practice safety precautions when removing restrictions from the system.

*22-1 Electric Motors:*  Electric motors are used

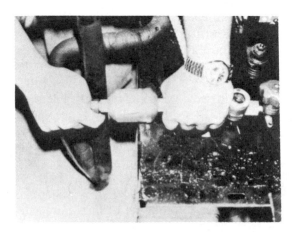

**Figure 8-62** Feeling a Line for Temperature Drop

for many purposes in refrigeration and air conditioning other than to drive the compressor. The majority of these are fan motors and pump motors. Several different types of motors are in use: (1) split-phase (SP), (2) permanent split capacitor (PSC), (3) capacitor start (CS), (4) capacitor start, capacitor run (CSCR), and (5) shaded pole. The amount of starting and running torque required to do the job will determine the type of motor used. If any of the following conditions are found, the motor must be either repaired or replaced.

*22-1A Open motor windings.* Open motor windings occur when the path for electric current is interrupted. This interruption occurs when the insulation on the wire becomes bad and allows the wire to overheat and burn apart (Figure 8-63). To check for an open winding, remove all external wiring from the motor terminals or connections. Using an ohmmeter, check the continuity from one terminal

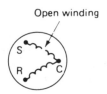

**Figure 8-63** Open Motor Winding

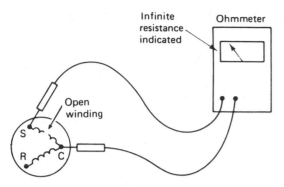

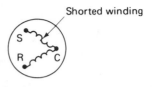

Figure 8-65 Shorted Motor Winding

**Figure 8-64** Checking Continuity of Open Motor Winding

to another terminal (Figure 8-64). Be sure to zero the ohmmeter. The open winding will be indicated by an infinity resistance reading on the ohmmeter. There should be no continuity from any terminal to the motor case. Repair or replace the motor if found to be faulty.

*22-1B Shorted motor windings.* Shorted motor windings occur when the insulation on the winding becomes bad and allows a shorted condition (two wires to touch), which causes the electric current to bypass part of the winding (Figure 8-65). In some instances, depending on how much of the winding is bypassed, the motor may continue to operate but will draw excessive amperage. To check for a shorted

winding, remove all external wiring from the motor terminals. Using an ohmmeter, check the continuity from one terminal to another terminal (Figure 8-66). Be sure to zero the ohmmeter. The shorted winding will be indicated by a less than normal resistance. In some cases, it will be necessary to consult the motor manufacturer's data for the particular motor to determine the correct resistance requirements. There should be no continuity from any terminal to the motor case. Repair or replace the motor if found to be faulty.

*22-1C Grounded motor windings.* Grounded motor windings occur when the insulation on the winding is broken down and the winding becomes shorted to the motor housing (Figure 8-67). To check for a grounded winding, remove all external wiring from the motor terminals. Using an ohmmeter, check for continuity from each terminal to the motor case (Figure 8-68). Be sure to zero the ohmmeter. The grounded winding will be indicated by a

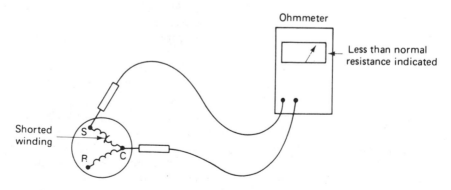

**Figure 8-66** Checking Continuity of Shorted Motor Winding

**Figure 8-67** Grounded Motor Winding

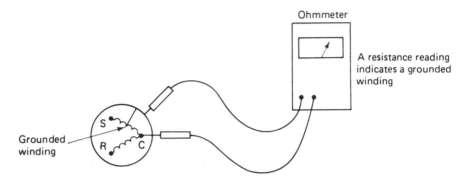

**Figure 8-68** Checking for Grounded Motor Winding

resistance reading. It may be necessary to remove some paint or scale from the motor case so that an accurate reading can be obtained. If a reading is obtained, repair or replace the motor.

*22-1D Bad starting switch.* A bad starting switch on a split-phase motor will prevent the proper starting of the motor. This switch provides a path for electrical power to flow to the starting (auxiliary) winding during the starting period, and it interrupts the electrical power when the motor has reached approximately 75% of its running speed. The contacts on these switches may become stuck closed, pitted, or stuck open.

A starting switch that is stuck closed will allow the motor to start but will cause the overload to open and stop the motor after a short period of operation. To check for a stuck-closed starting switch, start the motor and check the amperage draw. The amperage draw to a motor with a stuck-closed starting switch will not drop when 75% of the running speed

is reached. If there is any drop in amperage, check for a bad overload, bad bearing, or an overloaded condition. If the amperage draw does not drop, replace starting switch.

A starting switch that is pitted or stuck open will not allow the motor to start. A pitted or stuck-open starting switch will, however, allow the motor to hum and try to start. When this condition is encountered, start the motor turning while it is humming. If the motor comes up to speed, the amperage draw is normal, and the motor operates normally, replace the starting switch. If the motor sometimes runs in the wrong direction, the starting switch is probably the cause. Replace the starting switch.

*22-1E Bad bearings.* Bad motor bearings will cause an overloaded condition, which will cause an excessive current draw. The motor may cut off owing to the overload after operating for a while. To check for bad bearings, remove the belt, if one is used, and free the motor shaft. Move the motor shaft in a sideways direction with the hand (Figure 8-69).

Figure 8-69   Checking Motor Bearings

AT40

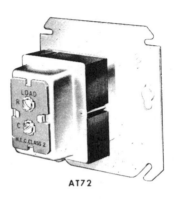

AT72

AT87

Figure 8-70   Class 2 Transformers (Courtesy of Honeywell, Inc.)

If movement in the shaft is found, either replace the bearings or the entire motor.

*22-1F Motor replacement.* Motor replacement should be done with great care. An exact replacement must be found to be certain that efficiency is maintained. Be certain that the motor manufacturer's wiring diagram is followed. After tightening all the motor mounts, check to be sure that the shaft is free to rotate. Start the motor momentarily to see that it turns in the right direction and that nothing is dragging. Then start the motor and allow it to operate under normal conditions while checking the amperage draw. If the amperage draw is excessive, the reason should be found and corrected before leaving the job.

*23-1 Transformers:* Transformers are electrical devices that are used to reduce, or increase, electrical voltage. In refrigeration and air-conditioning work, they are used to reduce line voltage to low voltage (24 V; Figure 8-70). The 24 V is used in the control circuit because it is safer, the controls are cheaper to manufacture, and the controls are more responsive to temperature change than line voltage controls. To check a transformer, use a voltmeter and check the input voltage to the primary side of the transformer (Figure 8-71). If voltage is

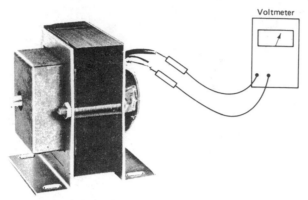

**Figure 8-71** Checking Voltage on a Transformer Primary (Courtesy of Honeywell, Inc.)

found, check the output voltage from the secondary side of the transformer (Figure 8-72). If no secondary voltage is found and primary voltage is found, the transformer is bad and must be replaced. If no voltage is found to the primary side of the transformer, the trouble is elsewhere. An alternative method of checking a transformer is to first disconnect all external

wiring to the transformer. Then check the primary and secondary windings, in turn, with an ohmmeter. If either winding is open, the transformer is bad and must be replaced.

Some transformers incorporate a fuse in the secondary winding. If this fuse should blow, the transformer is rendered inoperative and usually must be replaced. Fuses can be replaced in only a few transformers.

*23-1A Replacing a transformer.* When replacing a transformer, be sure to use one with at least an equal VA (voltampere) rating as the one being replaced. Always use a replacement designed for the same electrical characteristics as the one being replaced. Never use one with a smaller VA rating unless it is known that the replacement will have sufficient capacity. A transformer with too small a VA rating will only burn out because of an overloaded condition. Always check to be sure that there is no short to cause the replacement transformer to burn out. To check for an overload or short, measure the amperage draw in the low-voltage circuit. Then multiply the amperage times the

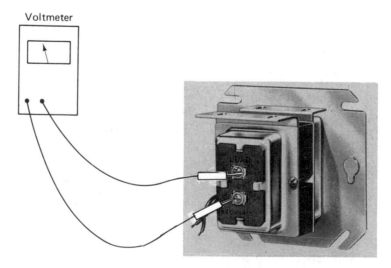

**Figure 8-72** Checking Voltage on a Transformer Secondary (Courtesy of Honeywell, Inc.)

voltage to obtain the VA draw of the circuit. The circuit VA must be equal to or smaller than the transformer rating. This check must be made with the unit running. Because of the small current draw in the low-voltage circuit, it may be necessary to use a multiplier. These multipliers may be purchased or may be hand made. To make one, simply coil a piece of wire around the tong of an ammeter (Figure 8–73). Connect the multiplier into the circuit and check the current draw by the number of turns in the handmade coil. For example, a coil has 10 turns and the current reading is 4.5 A. Therefore, $4.5 \div 10 = 0.45$ A. The VA would be $0.45 \times 24 = 10.8$ VA.

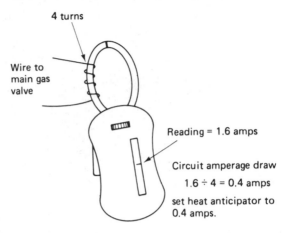

**Figure 8-73** Checking Amperage in a Temperature-Control Circuit

*24–1 High Ambient Temperatures:* High ambient temperatures are usually encountered in the summertime, especially in the cooling season. High ambient temperatures result in a high discharge pressure and poor operation of the unit. The best method of alleviating this situation is to provide a shade of some type over the outdoor unit. The shade should extend several feet past the unit toward the direction of air flow (Figure 8–74).

An alternative method is to provide a spray of water on the outdoor coil. This can involve an intricately designed unit or a lawn sprinkler, depending on the efficiency required (Figure 8–75). Care should be taken to protect the elec-

tric motors and components from becoming electrically shorted to the unit.

*25–1 Indoor Fan Relay:* The purpose of an indoor fan relay is to allow the user to select continuous or intermittent operation of the indoor fan motor at the thermostat and to allow the fan to operate on a certain speed for each season. Some heat pump systems require a higher fan speed for heating than for cooling, and vice versa. This type of relay is normally equipped with one set of open and one set of closed contacts. There are two general types of

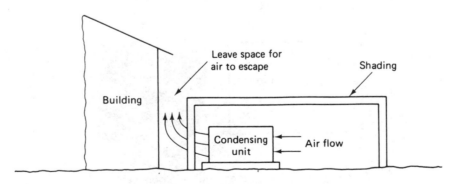

**Figure 8-74** Shading for an Outdoor Unit

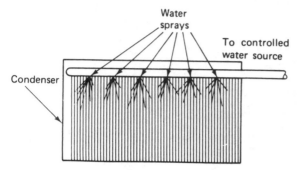

**Figure 8-75** Water Spray on Air-Cooled Condenser

should click. If there is no click, check the voltage on the two coil conenctions (Figure 8–77). If voltage is indicated and no click is heard when the relay is energized, the relay is sticking or burned out and should be replaced. If a click is heard and the fan still does not start, check the line voltage across the common connection and the other two connections in turn (Figure 8–78). With the relay de-energized, a voltage reading should be obtained between the common and low-speed connections. If these two checks do not prove out, the relay is bad and should be replaced. However, if these two checks prove the relay is good, the problem is elsewhere and should be found. When the relay must be replaced, be sure the replacement has the same voltage and amperage ratings as the relay being replaced.

fan relays (Figure 8–76). The open type has marked terminals, and the shrouded type has colored wires. To check out a fan relay, first turn the thermostat fan switch to ON. The relay

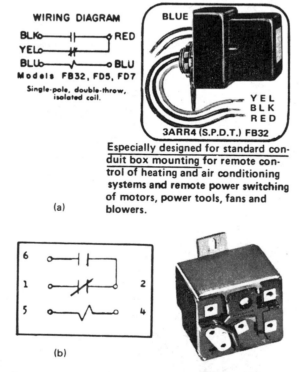

**Figure 8-76** (a) Shrouded and (b) Open-Type Fan Relays and Corresponding Schematics

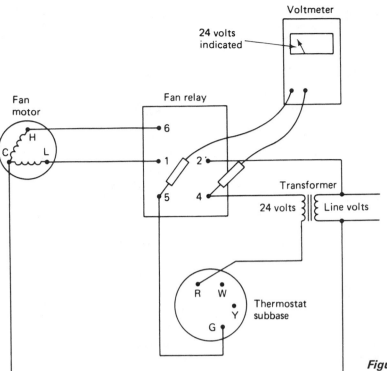

**Figure 8-77** Checking Voltage on Fan Relay Coil

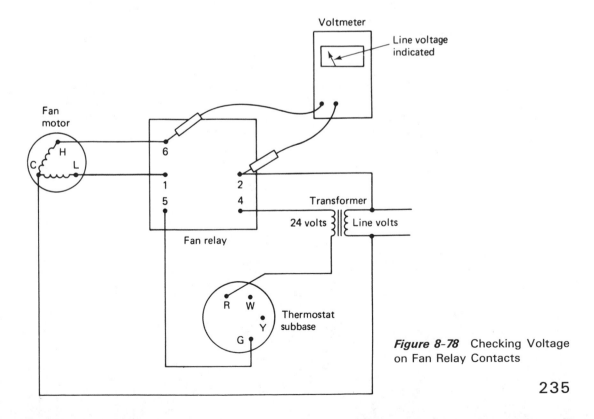

**Figure 8-78** Checking Voltage on Fan Relay Contacts

*26-1 Air Ducts:*   Air ducts are the hollow tubes that direct the flow of air from the air handler to the conditioned space. These ducts must be properly sized to direct the desired amount of air to the required space. To design a practical and efficient duct system requires much time, effort, and calculation. The proper amount of insulation around the ducts is an important feature that requires careful calculation. There should be a minimum 2-in. (25.4-mm) thickness of insulation with a vapor barrier. There are times when more insulation may be required. Enough insulation should be applied to prevent a heat loss or gain of more than about 2 °F (1.11 °C; Figure 8–79).

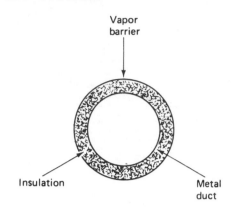

**Figure 8-79**   Cross Section of Metal Duct with Insulation

*27-1 Unit Too Small:*   Occasionally, a designer will miscalculate and the wrong-sized unit will be installed. When this occurs, the only remedy is to install a unit of the proper size. A unit that is too small will operate properly, but the space temperature will be higher than desired on cooling and lower than desired on heating.

*28-1 Outdoor Fan Relay:*   Outdoor fan relays are used on heat pump systems to aid in the starting and stopping of the outdoor fan in the various cycles involved in this type of system. The relays are generally SPST-type (single pole, single throw) switches, which are actuated by a 24-V coil. The contacts are in the line voltage circuit to the fan motor (Figure 8–80). To check

an outdoor fan relay, set the thermostat to a temperature that will cause the unit to run. With a voltmeter, check the voltage across the relay coil. A reading of 24 V should be indicated. Cycle the unit and listen for the relay to click. No click with voltage present indicates a faulty relay. If a click and voltage to the coil are present, check the contacts. With the relay energized and the system turned on, there should be no voltage indicated across the contacts. The relay is bad if a voltage is indicated. If no voltage is indicated and the fan does not run, the trouble is not in the relay and further checking is necessary. When replacing a relay, be sure that the replacement has the correct voltage and amperage rating.

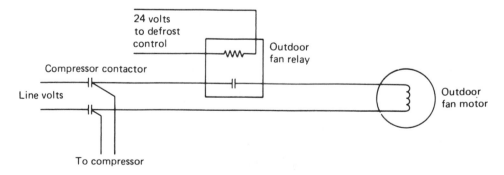

**Figure 8-80**   Outdoor Fan Relay Wiring Connections

*29-1 Check Valves:*  Check valves are used on heat pump systems to help in directing the flow of refrigerant through the proper path for the desired cycles of heating, cooling, or defrosting (Figure 8–81). The check valve will allow the passage of refrigerant either around or through the flow-control device. Use the following suggestions to check the operation of a check valve.

1.  A check valve sticking closed in the outdoor section during the cooling cycle will cause the suction pressure to be low. The superheat will be high on the indoor coil.

2.  A check valve sticking open in the indoor section during the cooling cycle will cause a high suction pressure, flooding of the compressor, and low superheat on the indoor coil.

3.  A check valve sticking closed in the indoor section during the heating cycle will cause the suction pressure to be low and the superheat to be high on the outdoor coil.

4.  A check valve sticking open in the outdoor section during the heating cycle will cause a high suction pressure, flooding of the compressor, and a low superheat on the outdoor coil.

If any of these conditions are found, the check valve should be replaced. Be sure to remove the refrigerant charge or pump the system down to prevent personal injury. Be sure to install a check valve with the arrow pointing in the direction of refrigerant flow at that point (Figure 8–82).

*30-1 Reversing Valve:*  The purpose of the

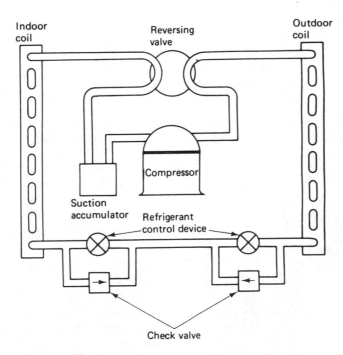

**Figure 8-81**  Check Valve Installation on a Heat Pump System

**Type 1160**
Check Valve

**Type 119**
Check Valve

*Figure 8-82* Check Valves: (a) Ball Check; (b) Swing Check (Courtesy of Henry Valve Company)

reversing valve is to reverse the refrigerant direction on a heat pump and reverse cycle refrigeration systems (Figure 8–83). Some systems energize the coil on heating and some energize the coil on cooling. Therefore, the system in question will determine the energized condition of the coil. Any troubles that occur in a heat pump system that will affect the normal operating pressures could possibly affect proper shifting of the valve. Some examples of problems are a refrigerant leak in the system resulting in a shortage, a compressor that is not

operating at full capacity, a defective check valve, a damaged valve, or a defective electrical system. Any of these conditions will indicate an apparent malfunctioning valve. The following checks should be made on a system and its components before attempting to diagnose any valve troubles by using the touch-test method.

1. Inspect the electrical system. This is best done by having the system in operation so that the reversing valve solenoid is energized. While the unit is operating, remove the locknut from the solenoid cover. Pull the solenoid coil partway off the stem. There should be a magnetic force trying to hold the coil on the stem. Be sure that electric power is applied to the solenoid. If the coil is moved off the stem, the valve will return to the deenergized position, which will be indicated by a clicking sound. A clicking sound will also be heard when the coil is replaced on the stem.

2. Inspect the reversing valve for physical damage. Check the solenoid coil for deep scratches, dents, and cracks.

3. Check the operation of the system against the equipment manufacturer's recommendations. Make any corrections indicated.

If all these checks prove that the system is operating properly, perform the touch test on the reversing valve, using Table 8–2. The touch test is performed simply by feeling the temperature differences on the six tubes on the valve and comparing these differences. Refer to the table for possible causes and corrections.

When replacing the reversing valve, the refrigerant will need to be purged. Follow the manufacturer's recommendations when installing a new valve. Evacuate the system, install new filter-driers, and recharge the system according to the manufacturer's recommendations.

*Figure 8-83* Reversing Valve (Courtesy of Ranco Controls Division)

**TABLE 8-2**
Touch Test Chart

| Valve operating condition | Discharge tube from compressor | Suction tube to compressor | Tube to inside coil | Tube to outside coil | Left pilot back capillary tube [a] | Right pilot front capillary tube [a] | Possible causes | Corrections |
|---|---|---|---|---|---|---|---|---|
| | 1 | 2 | 3 | 4 | 5 | 6 | | |
| | Normal Operation of Valve | | | | | | | |
| Normal cooling | Hot | Cool | Cool, as (2) | Hot, as (1) | TVB | TVB | | |
| Normal heating | Hot | Cool | Hot, as (1) | Cool, as (2) | TVB | TVB | | |
| | Malfunction of Valve | | | | | | | |
| Valve will not shift from cool to heat | Check electrical circuit and coil | | | | | | No voltage to coil Defective coil | Repair electrical circuit. Replace coil. |
| | Check refrigeration charge | | | | | | Low charge Pressure differential too high | Repair leak, recharge system. Recheck system. |
| | Hot | Cool | Cool, as (2) | Hot, as (1) | TVB | Hot | Pilot valve okay, dirt in one bleeder hole | Deenergize solenoid, raise head pressure, reenergize solenoid to break dirt loose. If unsuccessful, remove valve, wash out. Check on air before installing. If no movement, replace valve, add strainer to discharge tube, mount valve horizontally. |

239

**Table 8-2 (cont.)** Touch Test Chart

| Valve operating condition | Discharge tube from compressor (1) | Suction tube to compressor (2) | Tube to inside coil (3) | Tube to outside coil (4) | Left pilot back capillary tube [a] (5) | Right pilot front capillary tube [a] (6) | Possible causes | Corrections |
|---|---|---|---|---|---|---|---|---|
| | | | | | | *Malfunction of Valve (cont.)* | | |
| Valve will not shift from cool to heat | Hot | Cool | Cool, as (2) | Hot, as (1) | TVB | Hot | Piston cup leak | Stop unit. After pressures equalize, restart with solenoid energized. If valve shifts, re-attempt with compressor running. If still no shift, replace valve. |
| | Hot | Cool | Cool, as (2) | Hot, as (1) | TVB | TVB | Clogged pilot tubes | Raise head pressure, operate solenoid to free. If still no shift, replace valve. |
| | Hot | Cool | Cool, as (2) | Hot, as (1) | Hot | Hot | Both ports of pilot open. (back seat port did not close) | Raise head pressure, operate solenoid to free partially clogged port. If still no shift, replace valve. |
| | Warm | Cool | Cool, as (2) | Warm as (1) | TVB | Warm | Defective compressor | |
| Start to shift but does not complete reversal | Hot | Warm | Warm | Hot | TVB | Hot | Not enough pressure differential at start of stroke or not enough flow to maintain pressure differential | Check unit for correct operating pressure and charge. Raise head pressure. If no shift, use valve with smaller ports. |
| | | | | | | | Body damage | Replace valve. |

| Symptom | | | | | | | Cause | Correction |
|---|---|---|---|---|---|---|---|---|
| | Hot | Warm | Warm | Hot | Hot | Hot | Both ports of pilot open | Raise head pressure, operate solenoid. If no shift, replace valve. |
| | Hot | Hot | Hot | Hot | TVB | Hot | Body damage | Replace valve. |
| | Hot | | | | | | Valve hung up at mid-stroke; Pumping volume of compressor not sufficient to maintain reversal | Raise head pressure, operate solenoid. If no shift, use valve with smaller ports. |
| | Hot | Hot | Hot | Hot | Hot | Hot | Both ports of pilot open | Raise head pressure, operate solenoid. If no shift, replace valve. |
| Apparent leak in heating | Hot | Cool | Hot, as (1) | Cool, as (2) | TVB | WVB | Piston needle on end of slide leaking | Operate valve several times then recheck. If excessive leak, replace valve. |
| | Hot | Cool | Hot, as (1) | Cool, as (2) | WVB | WVB | Pilot needle and piston needle leaking | Operate valve several times then recheck. If excessive leak, replace valve. |
| Will not shift from heat to cool | Hot | Cool | Hot, as (1) | Cool, as (2) | TVB | TVB | Pressure differential too high | Stop unit. Will reverse during equalization period. Recheck system. |
| | | | | | | | Clogged pilot tube | Raise head pressure, operate solenoid to free dirt. If still no shift, replace valve. |
| | Hot | Cool | Hot, as (1) | Cool, as (2) | Hot | TVB | Dirt in bleeder hole. | Raise head pressure, operate solenoid. Remove valve and wash out. Check on air before reinstalling, if no movement, replace valve. Add strainer to discharge tube. Mount valve horizontally. |

**Table 8-2 (cont.)**
Touch Test Chart

| Valve operating condition | Discharge tube from compressor | Suction tube to compressor | Tube to inside coil | Tube to outside coil | Left pilot back capillary tube [a] | Right pilot front capillary tube [a] | Possible causes | Corrections |
|---|---|---|---|---|---|---|---|---|
| | 1 | 2 | 3 | 4 | 5 | 6 | | |
| *Malfunction of Valve (cont.)* | | | | | | | | |
| Will not shift from heat to cool | Hot | Cool | Hot, as (1) | Cool, as (2) | Hot | TVB | Piston cup leak. | Stop unit, after pressures equalize, restart with solenoid deenergized. If valve shifts, reattempt with compressor running. If it still will not reverse while running, replace valve. |
| | Hot | Cool | Hot, as (1) | Cool, as (2) | Hot | Hot | Defective pilot | Replace valve. |
| | Warm | Cool | Warm, as (1) | Cool, as (2) | Warm | TVB | Defective compressor | |
| Valve operated satisfactorily prior to compressor motor burnout | | | | | | | Dirt and small greasy particles inside the valve | Remove valve, thoroughly wash it out. Check on air before reinstalling, or replace valve. Add strainer and filter-dryer to discharge tube between valve and compressor. |

[a]TVB, temperature of valve body; WVB, warmer than valve body.

*Source:* Ranco Controls Division.

242

*31-1 Defrost Controls:* Heat pump defrost controls are used to detect ice and frost on the outdoor coil during the heating cycle. The most popular type of defrost control is the time–temperature method (Figure 8–84). With this method, both the time and temperature sections must demand defrost before the defrost cycle can be initiated. When the defrost cycle is initiated, the reversing valve changes and the outdoor fan motor stops running. The clock motor operates only when the compressor is running. The timer can be set to initiate the defrost cycle on 30-, 60-, and 90-minute intervals. If it is found that the 90-minute cycle allows too much frost on the outdoor coil, then the timer can be adjusted to provide the proper amount of defrost time. The defrost can be initiated only by demand from both the timer and the temperature sections of the defrost control. The defrost control temperature will initiate the defrost cycle at a temperature of approximately 30 °F (−1.1 °C) and will terminate the defrost cycle at approximately 60 °F (15.6 °C).

The timer motor is operated by 240 V and operates only when the compressor is operating. The temperature portion of the control is attached to a return bend on the outdoor coil (Figure 8–85). The bulb is insulated to ensure that the ambient air does not affect the bulb. To check a time–temperature defrost control, be sure that the outdoor coil is cold enough to need defrosting; then use the following steps:

1. Place a jumper between terminals 2 and 3 on the clock timer (Figure 8–86). If the reversing valve changes, the clock is not operating. Check to be certain that power is supplied to the clock motor terminals. If power is found, the motor is defective and must be replaced.
2. If the reversing valve does not change, jump the terminals of the defrost thermostat while the jumper remains on terminals 3 and 2. If the reversing valve changes, the defrost overload is defective and must be replaced.

WIRING DIAGRAM

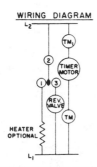

*Figure 8-84* Time–Temperature Defrost Control and Schematic (Courtesy of Ranco Controls Division)

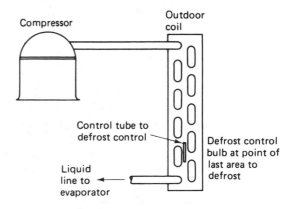

*Figure 8-85* Installation of Temperature Bulb for a Defrost Control

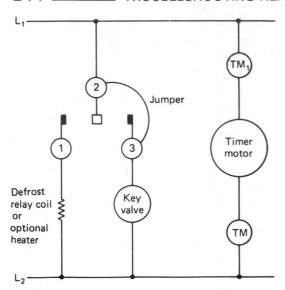

**Figure 8-86** Jumper between Terminals 2 and 3 on a Defrost Timer

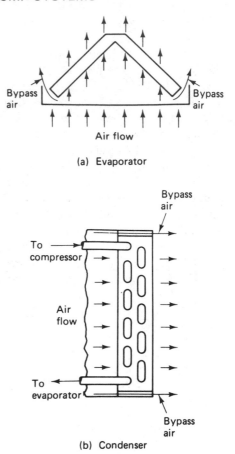

(a) Evaporator

(b) Condenser

**Figure 8-87** Air Bypassing Coils

If the additional jumper does not cause the reversing valve to change, the trouble is not in the defrost control. The reversing valve solenoid coil, the reversing valve, and the defrost relay must be checked to find the trouble. Be sure that the specified voltage is applied to these controls.

*32-1 Air Bypassing Coil:* Air bypassing a coil will reduce the efficiency of the unit (Figure 8–87). The greater the amount of air bypassing the coil, the greater the reduction in efficiency. The opening through which the air bypasses must be closed so that peak efficiency can be obtained. Air bypassing an evaporating coil causes a low suction pressure. Air bypassing a condensing coil causes a high discharge pressure.

*33-1 Auxiliary Heat Strip Location:* The auxiliary heat strips must be installed downstream of the indoor coil (Figure 8–88). If the heat strips are installed upstream of the indoor coil, the heat pump will act as a reheat system.

This hot air entering the indoor coil during the heating cycle will cause the discharge pressure and temperature to be excessively high, which will result in high compression ratio on the compressor, one that will probably cause permanent damage and require compressor replacement.

*34-1 Defrost Relay:* The purpose of the defrost relay is to cause the reversing valve to change to the cooling position, stop the outdoor fan motor, and energize the auxiliary heaters during the defrost cycle. The coil voltage is usually 240 V (Figure 8–89). This relay is ener-

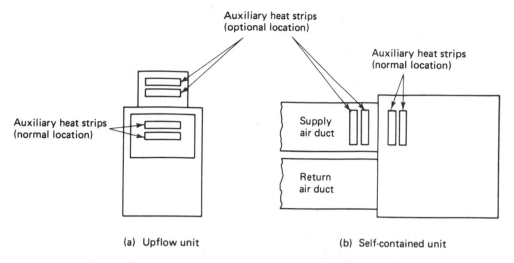

**Figure 8-88** Location of Auxillary Heat Strips on Heat Pump System

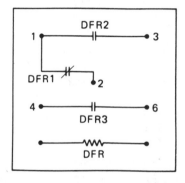

**Figure 8-89** Internal Schematic of Defrost Relay

gized by the defrost control. Because some manufacturers prefer to energize the reversing valve during the heating cycle and some during the cooling cycle, the wiring diagram for the particular unit in question should be used.

If the relay is energized and a click is heard when the defrost control demands defrost but the system does not shift, check the voltage across each set of contacts with a voltmeter. Open contacts will show full line voltage across them. Closed contacts will show no voltage across them. If conditions other than these are found, replace the relay.

*35–1 Humidity:* During the winter months, a desired relative humidity inside a conditioned space will be between 40 and 60%. However, when the outdoor temperature is very low, condensation on doors, walls, and windows may be experienced. This is an indication of excessive relative humidity, which must be lowered to prevent damage. This humidity may be the result of cooking with open pots and pans or bathing. When cooking with open pots and pans, the pans should be covered to prevent the escape of moisture into the conditioned space. If covering the pots is not desirable, a vent system may be installed above the stove. When bathing causes the problem, install a vent system in the bathroom. Thus vents can be used when cooking or bathing is in progress.

*36–1 Heating Elements:* There are several types of heating elements available. However, the open-wire element is the most popular. Heating elements generally are equipped with protective devices in case of excessive current or overheating of the element (Figure 8–90). These heating elements produce heat when electricity is applied to them. Sometimes they will develop hot spots and the wire will separate, causing an open circuit, and the heating will stop (Figure 8–91). A visual inspection of the element will show this condition. A separated element must be replaced before normal operation can be resumed.

*36–1A Overtemperature protection.* Overtemperature protective devices are required by

Heating elements are staggered for more uniform heat transfer, eliminates hot spots, and insures black heat. Each element is protected by an over temperature disc in one end and a fuse link in the other.

**Figure 8–90** Open-Wire-Type Heating Element (Courtesy of Square D Company)

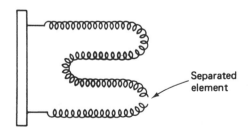

Separated element

**Figure 8–91** Separated Electric Heating Element

both Underwriters' Laboratories (UL) and the National Electrical Code® (NEC). Two types of protection are required by both agencies. The primary system is automatically reset and is designed to interrupt the flow of electrical current to the heater if insufficient air-flow, air blockage, or other causes result in an overheating condition. The secondary or backup system is designed to operate at a higher temperature and interrupt the flow of electrical current to the heater in the event of failure of the primary system or continued operation under unsafe conditions.

When standard slip-in heaters are used, either a disk-type or a bulb- and capillary-type thermal cutout may be used for both the primary and secondary system of temperature protection (Figure 8–90). While this type of cutout is standard in the field for primary protection, some manufacturers also use them for control of the secondary system. The fusible links require field replacement, but the manual reset controls can immediately be put back in operation by simply pressing the reset button located in the heater terminal box. Be sure to determine and correct the cause of the trouble before leaving the job.

*36–1B Overcurrent protection.* Over-current protection is required on all electrical devices by UL and NEC. These agencies require that an electric heater drawing more than 48 A total be subdivided into circuits that draw no more than 48 A each. Each circuit must be protected by fuses or circuit breakers. Because electric heaters are regarded as a continuous load, overcurrent protective devices must be rated for at least 125% of the circuit load. Thus, while the circuits are limited to 48 A, the fuses are limited to 60 A. These fuses must be supplied by the heater manufacturer. When a fuse is blown, there is an electrical overload, which must be found and removed before the service job is complete.

*37-1 Heat-Sequencing Relays:* Heat-sequencing relays are used on electric furnaces to prevent all the heating elements coming on or off at the same time. Therefore, these relays may be used to vary the capacity of the unit. Three basic types of relays are used to accomplish sequencing: (1) thermal element switches, also known as bimetal time-delay switches, which heat up and close circuits, (2) relays that activate another relay along with the heating element, and (3) modulating motors that rotate to close contacts at various points in the rotation.

*37-1A Thermal element switches.* These switches make use of a bimetal blade heated by an electric current to actuate the contacts (Figure 8–92). These switches are generally termed stack relays. They have one heated bimetal that closes more than one set of contacts. Use of the different contacts will vary depending on the equipment design. One set of contacts may complete the electrical circuit to

**(a) SPST**

**(b) DPST**

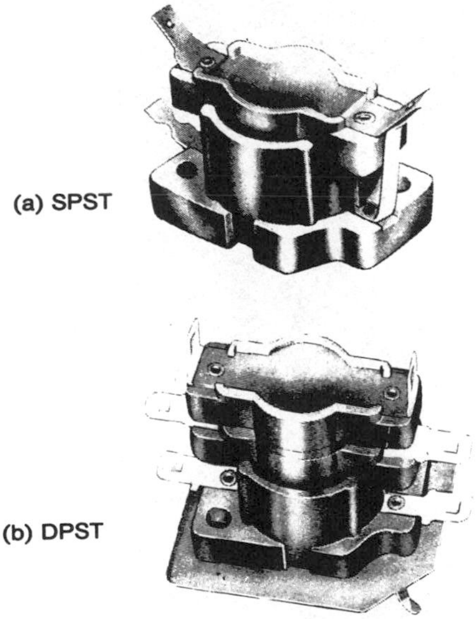

*Figure 8-92* Bimetal Heating Relays (Courtesy of Control Products Division, Texas Instruments, Incorporated)

the fan motor; another set may complete the electrical circuit to the heating element; and yet another set may complete the electrical circuit to the heating element of another relay. The specific use will need to be determined on each unit.

To check out these switches, first set the thermostat to demand heating. Then listen to the switch. If a click is heard, the heating element is working. If no click is heard, check the voltage across the heating coil (Figure 8–93). If voltage is indicated, the heating element is bad and the switch must be replaced. If a click is heard, indicating that the heating element is working, check the voltage across each set of contacts. If a voltage is indicated, the contacts are bad and the switch must be replaced. Be sure that full electrical power is applied to the unit. The replacement relay must be an exact replacement or the unit will not function as desired.

*37-1B Time-delay relays.* These single contact relays, which actuate relays along with the heating element, are also sometimes used (Figure 8–94). When these relays are used, the relays energized by these devices must be equipped with heating elements of the same voltage as the heating element energized by the original relays. The checkout procedure is the same as that described in Section 37–1A.

*37-1C Modulating motors.* Modulating motors are generally used on more sophisticated units requiring more exact control than other units (Figure 8–95). These motors will operate in any position within the number of degrees of rotation for which they are designed. The motor is controlled by use of a Wheatstone bridge (Figure 8–96). The motor shaft is extended through the housing. The switches are placed on a shaft to complete the desired circuit at a given degree of rotation. The thermostat controls the motor by the use of a po-

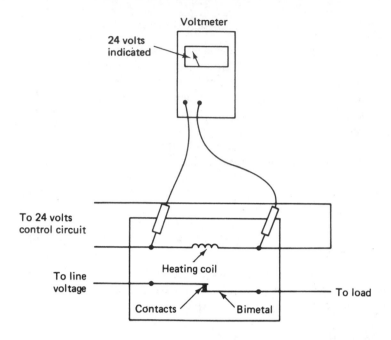

Voltmeter

24 volts
indicated

To 24 volts
control circuit

To line
voltage

Heating coil

Contacts       Bimetal

To load

*Figure 8-93* Checking Voltage across Thermal Element on a Time-Delay Relay

*Figure 8-94* Single-Contact Time-Delay Relay (Courtesy of Cam-Stat, Inc.)

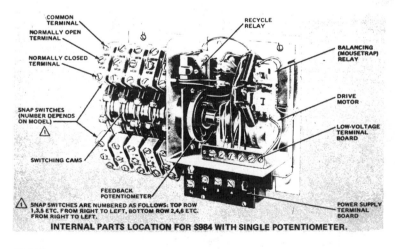

COMMON
TERMINAL

NORMALLY OPEN
TERMINAL

NORMALLY CLOSED
TERMINAL

SNAP SWITCHES
(NUMBER DEPENDS
ON MODEL)

SWITCHING CAMS

FEEDBACK
POTENTIOMETER

RECYCLE
RELAY

BALANCING
(MOUSETRAP)
RELAY

DRIVE
MOTOR

LOW-VOLTAGE
TERMINAL
BOARD

POWER SUPPLY
TERMINAL
BOARD

⚠ SNAP SWITCHES ARE NUMBERED AS FOLLOWS: TOP ROW
1,3,5 ETC. FROM RIGHT TO LEFT, BOTTOM ROW 2,4,6 ETC.
FROM RIGHT TO LEFT.

INTERNAL PARTS LOCATION FOR S984 WITH SINGLE POTENTIOMETER.

*Figure 8-95* Modulating Motor with End Switches (Courtesy of Honeywell, Inc.)

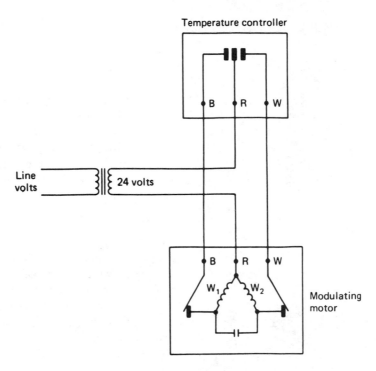

**Figure 8-96** Wheatstone Bridge Connections for Modulating Motor

tentiometer, which produces a balanced voltage effect and controls the balancing relay in the motor. To check out these motors, first be sure that the required voltage is being supplied to the motor and controls. Next, set the thermostat to demand heating. The motor should start turning. If not, clean the potentiometers and the balancing relay contacts with a cleaning solvent. It is not generally recommended to burnish the potentiometer because of possible damage. If the motor does not operate, check to see that voltage is supplied to the motor. The manufacturer's wiring diagram may be required to determine the motor terminals. If voltage is indicated at these terminals and the motor still does not operate, replace the motor. If the motor operates but the heating elements do not function, check the voltage across the switch after making certain that the correct voltage is available. If voltage is indicated across the contacts, the switch is defective and must be re-

placed. Be sure to install the replacement switch in the exact position as the replaced switch to maintain equipment efficiency.

*38-1 Fan Control:* The fan control is used on heating systems to start and stop the fan motor on an increase in temperature inside the furnace. This control is manufactured in adjustable and nonadjustable types, as well as in extended element and bimetal disk types (Figure 8-97). Some manufacturers are using electric time-delay relays for fan controls on some units. The bimetal types are set to bring the fan on when the temperature at the sensing element reaches a temperature of 135 to 150°F (57 to 66°C) and stop the fan when the temperature drops to approximately 100°F. The electric types are preset for a specified time to lapse for the ON and OFF settings. To check out a fan control, insert a thermometer in the unit as close as possible to the sensing element (Figure

**(a) Standard**

**(b) Electric**

*Figure 8-97* Fan Controls (Courtesy of Honeywell, Inc.)

8-91). Start the heating elements and observe the thermometer. When the desired temperature is reached, or the specified time has lapsed, the fan should start. If not, adjust the control if it is of the adjustable type. Next, stop the elements and observe the thermometer. When approximately 100°F (37.8°C) is reached, the fan should stop. If it does not stop, adjust the control if it is adjustable. If the control is not adjustable, or more than about 15°F (8.33°C) of adjustment is needed, replace the control. Be sure to use an exact replacement or the unit will not operate as designed. Never attempt to make internal adjustments on these controls because of the dangers of overheating and per-

manently damaging the equipment, possible fire, and personal injury.

*38–1A Unnecessary fan operation.* If the fan should run when the elements are off, check to see that the air flow is not restricted by a dirty filter or dirty indoor coil. When no air restriction is found, widening the differential between the fan ON and OFF settings will usually solve the problem.

*38–1B Fan cycling.* If the fan cycles on and off when the elements remain on, the air temperature may be too low or too much air may be blown through the furnace. Allow the room temperature to increase, and reduce the air flow to provide the desired temperature rise, as indicated on the furnace nameplate. Sometimes, widening the differential between the fan ON and OFF settings will solve the problem.

*39-1 Btu Input:* The Btu input to an electric heating unit can be changed in the field without much problem. When more or less heat is required, change the number of heating elements that are in operation. If more heat is required, add enough heating elements to supply the demand. When less heat is required, remove the extra heating elements. The Btu input of a gas furnace cannot be increased above the input rating of the furnace. To do so would cause overheating of the heat exchanger, and this would upset the vent system. The Btu input to

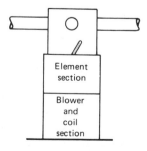

*Figure 8-98* Checking Temperature with Fan On

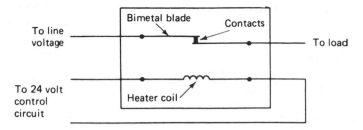

**Figure 8-99** Inside of a Bimetal Heating Relay

a gas furnace can be reduced by only 20% of the input rating. To reduce the input Btu more would upset the vent system. An improperly operating vent system is very dangerous and must be avoided. When changing the Btu input of any furnace, be sure to check the temperature rise through the furnace. Make any adjustments on the blower speed to maintain the recommended temperature rise. In some cases a larger blower motor will be required.

*40-1 Heating Relay:*   Heating relays are used on electric furnaces to complete the electrical circuit to the heating elements on demand from the thermostat. These relays operate by an electrically operated bimetal warpswitch (Figure 8-98). The bimetal heater is in the 24-V circuit, and the contacts are in the line voltage to the heating elements (Figure 8-99). To check the heating relay, use a voltmeter to check the voltage across the contacts. If voltage is indicated here, the contacts are open. Next, jumper the contacts. If the heating element is energized, the trouble is in the relay. An energized heating element will be indicated by a current flow to the element. Replace the relay with an exact replacement or the unit will not operate as designed.

*41-1 Blowers:*   Blowers are used to force air through the equipment and into the conditioned space (Figure 8-100). Blowers are generally located at the air inlet to the furnace or to the air-distribution system. A blower will not move a sufficient amount of air if the vanes become

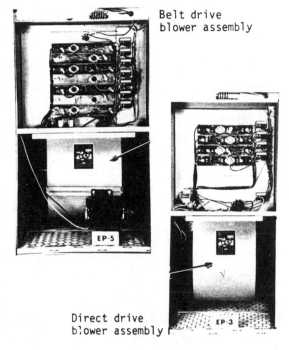

**Figure 8-100** Furnace Blower Assembly (Courtesy of Square D Company)

clogged with dirt, if the belt becomes loose and slipping, if the bearings become worn or out of lubricant, or if the motor develops a problem.

*41-1A Cleaning.*   The blower should be removed from the unit before it can be effectively cleaned. Sometimes a brush and soapy water are required to clean the blower. In such cases the blower should be taken out of the building and the motor protected from the water spray.

*41–1B Defective bearings.* Defective bearings will overload the motor and cause the blower to turn more slowly than is required. Therefore, air flow will be reduced and equipment operation adversely affected. To check for defective bearings, remove the fan belt and grasp the blower shaft in the hand. Move the shaft from side to side (Figure 8–101). If sideways movement is detected, replace the bearings. Be sure to check the blower shaft for wear or scoring. If wear or scoring is indicated, replace the shaft also. If no sideways movement is detected, spin the blower and observe if the blower turns free or if it is stiff. If stiffness is indicated, lubricate the bearings. If lubrication does not solve the problem, replace the bearings as described previously. Do not overtighten the belt or excessive bearing wear will result.

*42–1 Temperature Rise through a Furnace:* The amount of temperature rise through a furnace is recommended by the manufacturer. The minimum and maximum limits are designated on the furnace nameplate. When these limits are exceeded, the unit will not perform as designed. The temperature rise may be determined by subtracting the inlet dry bulb temperature from the outlet dry bulb temperature. To obtain these temperatures, insert thermometers as close as possible to the inlet and outlet of the furnace (Figure 8–102). When the

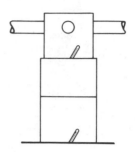

**Figure 8-102** Checking Temperature Drop or Rise through an Indoor Unit

maximum temperature rise is reached, the minimum of air is flowing through the furnace. If the limit is exceeded, damage to the heater may occur. When the minimum temperature rise is reached, the maximum amount of air is flowing through the furnace. If this limit is reached, cold drafts may be experienced. When operating on a maximum temperature rise, the relative humidity will be lowered. When operating on a minimum temperature rise, the relative humidity will be raised. To change the temperature rise, adjust the fan speed to deliver the amount of air required.

*43–1 Lockout Relay:* The purpose of the lockout relay is to interrupt the electrical power to the compressor contactor coil in the event that an overload occurs in the refrigerant circuit. The coil of the lockout relay is a high-impedance type which requires full circuit voltage on the coil terminals before it will activate the contacts. Once the NC contacts have opened, the electrical power to the lockout relay coil must be manually interrupted. When the power is interrupted the contacts will close again, resetting the system. Should the unit not start again after resetting, and all the other controls are demanding operation, turn off the electrical power and remove one wire from either terminal 2 or 3 of the lockout relay. Then check the continuity between terminals 2 and 3 with an ohmmeter (Figure 8–103). No continuity indicates that the contacts are open. The

**Figure 8-101**  Checking Fan Bearings

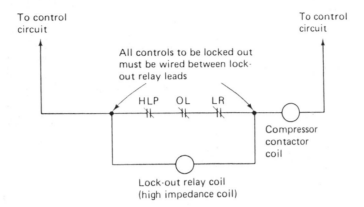

**Figure 8-103** Lockout Relay Connections.

relay must be replaced with an exact replacement. If continuity is indicated, apply control circuit voltage to the relay coil terminals. If continuity still remains between terminals 2 and 3, the contacts are stuck closed. Replace the relay with an exact replacement.

## ELECTRICAL TROUBLESHOOTING THE COLEMAN D.E.S. HEAT PUMP _____

It is important that the following steps be followed for satisfactory troubleshooting procedures:

1. When troubleshooting the Coleman D.E.S. heat pump the electrical power to the indoor unit must be on. Otherwise, no power will be available to the transformer and no secondary voltage (24 V) will be available to check the relays and circuits.

2. All terminal locations are numbered and all factory-installed wiring is color coded.

3. When using these charts, be sure that the temperature lever on the thermostat is set all the way down with the thermo-

stat set to cool, or all the way up with the thermostat set to the heating position.

4. All high voltage must be 240 V ± 10% (207 to 253 V).

5. All service charts cover the 3300 series indoor units with the greatest number of heating strips. When checking an indoor unit with fewer strips (such as the 10-kW model), disregard the service chart steps dealing with any additional heat strips, sequencers, fuses, or outdoor thermostats not on the unit being checked.

*Danger*: Remember that electricity, which works for us so faithfully, can be deadly dangerous unless it is handled carefully and intelligently. Always turn off the power and test with a voltmeter before touching any conductor. Keep in mind that you are working with hot circuits when using a voltmeter or an ammeter. Carelessness or disregard for safety can kill you in a fraction of a second. *Think safety all the time.*

The charts in Figure 8-104. have been designed to help isolate electrical service problems. Use a good-quality meter which will measure volts, amps, and ohms. A representative wiring diagram is also provided.

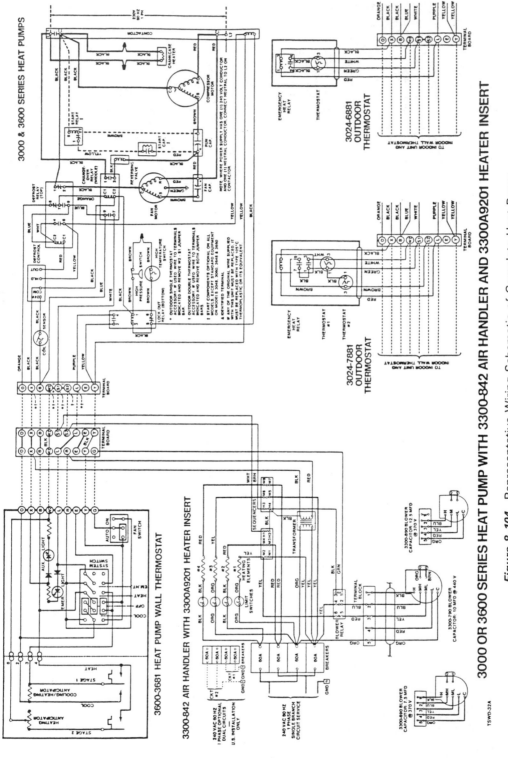

**Figure 8-104** Representative Wiring Schematic of Complete Heat Pump System (Courtesy of The Coleman Company.)

254

# INDOOR UNIT
## COOLING
## LOW VOLTAGE

| WALL THERMOSTAT SETTINGS | | |
|---|---|---|
| System Switch | Fan Switch | E.H. Light |
| COOL | AUTO | OFF |

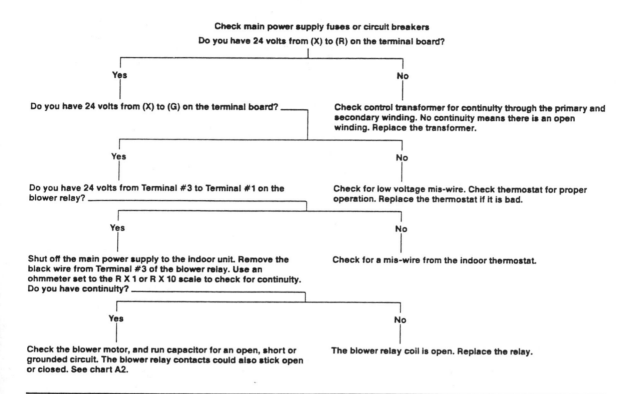

Check main power supply fuses or circuit breakers

Do you have 24 volts from (X) to (R) on the terminal board?

**Yes**

Do you have 24 volts from (X) to (G) on the terminal board?

**No**

Check control transformer for continuity through the primary and secondary winding. No continuity means there is an open winding. Replace the transformer.

**Yes**

Do you have 24 volts from Terminal #3 to Terminal #1 on the blower relay?

**No**

Check for low voltage mis-wire. Check thermostat for proper operation. Replace the thermostat if it is bad.

**Yes**

Shut off the main power supply to the indoor unit. Remove the black wire from Terminal #3 of the blower relay. Use an ohmmeter set to the R X 1 or R X 10 scale to check for continuity. Do you have continuity?

**No**

Check for a mis-wire from the indoor thermostat.

**Yes**

Check the blower motor, and run capacitor for an open, short or grounded circuit. The blower relay contacts could also stick open or closed. See chart A2.

**No**

The blower relay coil is open. Replace the relay.

---

# INDOOR UNIT
## HIGH VOLTAGE
## COOLING

| WALL THERMOSTAT SETTINGS | | |
|---|---|---|
| System Switch | Fan Switch | E.H. Heat |
| COOL | AUTO | OFF/ON |

Check the main power supply, be sure that circuit breakers are on.

Do you have 230 volts ± 10% from terminal 5 of blower motor speed selection terminal block to Terminal #4 on the blower relay?

**Yes**

Do you have 230 volts from L1 to Terminal #2 of blower relay?

**No**

Check main power for open fuse or circuit breaker.

Continued
Next Page

*Figure 8-104* (cont.)

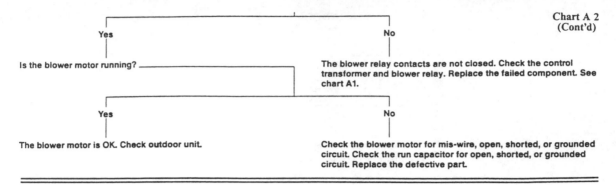

Yes                  No

Is the blower motor running? _____

The blower relay contacts are not closed. Check the control transformer and blower relay. Replace the failed component. See chart A1.

Yes                  No

The blower motor is OK. Check outdoor unit.

Check the blower motor for mis-wire, open, shorted, or grounded circuit. Check the run capacitor for open, shorted, or grounded circuit. Replace the defective part.

---

OUTDOOR UNIT
LOW VOLTAGE — COOLING

Chart A 3

| WALL THERMOSTAT SETTING | | |
|---|---|---|
| System Switch | Fan Switch | E.H. Light |
| COOL | AUTO | OFF |

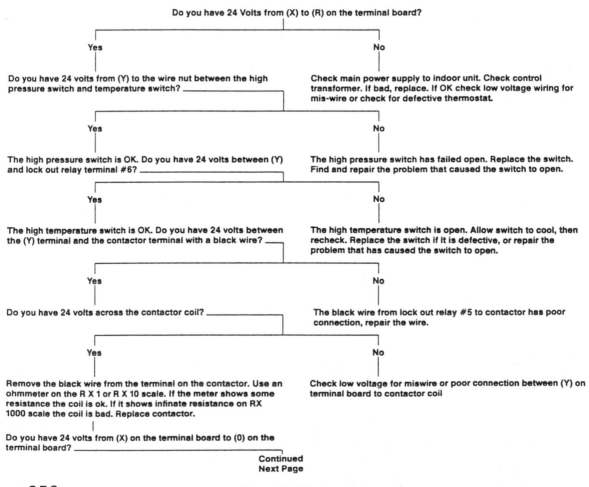

Do you have 24 Volts from (X) to (R) on the terminal board?

Yes                  No

Do you have 24 volts from (Y) to the wire nut between the high pressure switch and temperature switch? _____

Check main power supply to indoor unit. Check control transformer. If bad, replace. If OK check low voltage wiring for mis-wire or check for defective thermostat.

Yes                  No

The high pressure switch is OK. Do you have 24 volts between (Y) and lock out relay terminal #6? _____

The high pressure switch has failed open. Replace the switch. Find and repair the problem that caused the switch to open.

Yes                  No

The high temperature switch is OK. Do you have 24 volts between the (Y) terminal and the contactor terminal with a black wire? ____

The high temperature switch is open. Allow switch to cool, then recheck. Replace the switch if it is defective, or repair the problem that has caused the switch to open.

Yes                  No

Do you have 24 volts across the contactor coil? _____

The black wire from lock out relay #5 to contactor has poor connection, repair the wire.

Yes                  No

Remove the black wire from the terminal on the contactor. Use an ohmmeter on the R X 1 or R X 10 scale. If the meter shows some resistance the coil is ok. If it shows infinate resistance on RX 1000 scale the coil is bad. Replace contactor.

Check low voltage for miswire or poor connection between (Y) on terminal board to contactor coil

Do you have 24 volts from (X) on the terminal board to (0) on the terminal board? _____

Continued
Next Page

256

*Figure 8-104* (cont.)

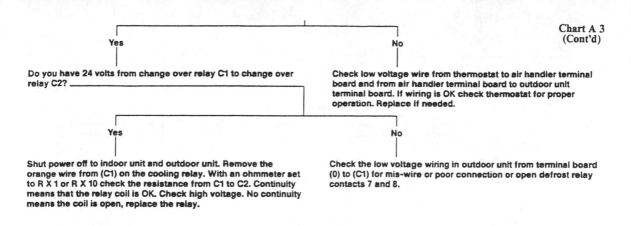

**Yes**

Do you have 24 volts from change over relay C1 to change over relay C2?

**No**

Check low voltage wire from thermostat to air handler terminal board and from air handler terminal board to outdoor unit terminal board. If wiring is OK check thermostat for proper operation. Replace if needed.

**Yes**

Shut power off to indoor unit and outdoor unit. Remove the orange wire from (C1) on the cooling relay. With an ohmmeter set to R X 1 or R X 10 check the resistance from C1 to C2. Continuity means that the relay coil is OK. Check high voltage. No continuity means the coil is open, replace the relay.

**No**

Check the low voltage wiring in outdoor unit from terminal board (0) to (C1) for mis-wire or poor connection or open defrost relay contacts 7 and 8.

---

OUTDOOR UNIT
HIGH VOLTAGE
COOLING

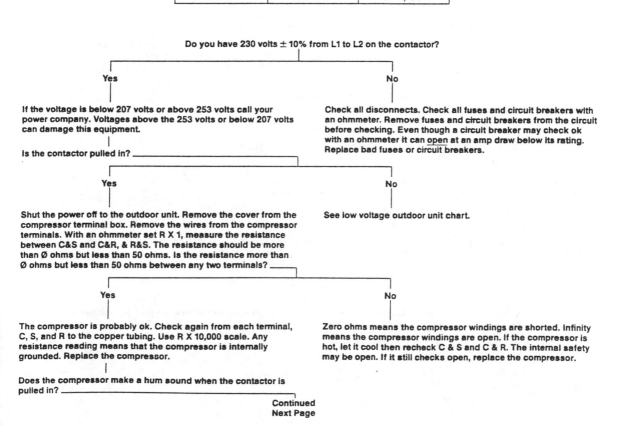

| WALL THERMOSTAT SETTINGS | | |
|---|---|---|
| System Switch | Fan Switch | E.H. Light |
| COOL | AUTO | OFF/ON |

Do you have 230 volts ± 10% from L1 to L2 on the contactor?

**Yes**

If the voltage is below 207 volts or above 253 volts call your power company. Voltages above the 253 volts or below 207 volts can damage this equipment.

Is the contactor pulled in?

**No**

Check all disconnects. Check all fuses and circuit breakers with an ohmmeter. Remove fuses and circuit breakers from the circuit before checking. Even though a circuit breaker may check ok with an ohmmeter it can open at an amp draw below its rating. Replace bad fuses or circuit breakers.

**Yes**

Shut the power off to the outdoor unit. Remove the cover from the compressor terminal box. Remove the wires from the compressor terminals. With an ohmmeter set R X 1, measure the resistance between C&S and C&R, & R&S. The resistance should be more than Ø ohms but less than 50 ohms. Is the resistance more than Ø ohms but less than 50 ohms between any two terminals?

**No**

See low voltage outdoor unit chart.

**Yes**

The compressor is probably ok. Check again from each terminal, C, S, and R to the copper tubing. Use R X 10,000 scale. Any resistance reading means that the compressor is internally grounded. Replace the compressor.

Does the compressor make a hum sound when the contactor is pulled in?

**No**

Zero ohms means the compressor windings are shorted. Infinity means the compressor windings are open. If the compressor is hot, let it cool then recheck C & S and C & R. The internal safety may be open. If it still checks open, replace the compressor.

Continued
Next Page

*Figure 8-104* (cont.)

257

Chart A 4
(Cont'd)

**Yes**

This could indicate a bad run capacitor, bad start capacitor or a bad start relay. The compressor could also be mechanically locked up. Check these components for failure.

Is the outdoor fan running? ————————————

**No**

Check compressor windings for open, shorted or grounded motor. Check disconnect or fuses or circuit breakers.

**Yes**

The fan motor is ok. Check refrigerant system pressures or low voltage electrical circuits.

**No**

With contactor pulled in, do you have 230 volts from L2 to defrost relay contact #2?

**Yes**

Turn power off, then try to spin the outdoor fan blade by hand, does it spin freely? ————————————

**No**

The defrost relay contacts 1 & 2 are open. Check defrost relay.

**Yes**

The motor bearings are ok. Check the motor for possible open, short or grounded circuit. If you find any of these, replace the motor.

Did the motor check out ok? ————————————

**No**

The bearings are dry. This can cause motor failure. Disconnect all outdoor fan motor wires from the circuit. With an ohmmeter set to the R x 1 scale, check resistance from black lead to brown lead, from black to red lead from red to brown. The resistance should be more than zero ohms. If the resistance is Ø or infinity, replace the motor. If you can measure resistance from any motor wire to the motor frame, replace the motor.

**Yes**

The outdoor fan motor run capacitor may be bad. Check the capacitor for open, shorted or grounded circuit.

Is the unit cooling? ————————————

**No**

Replace the motor.

**Yes**

If the unit does cool, there is no electrical problem in the outdoor unit. Check indoor unit and refrigeration charge.

**No**

Do you have 230 volts from contact #3 on the change over relay L2 on the contactor?

**Yes**

The change over relay is pulled in. Remove the black wire from the change over relay #3 and the compressor run capacitor. Using an ohmmeter on the R X 10 scale check the reversing valve coil. Continunity means the coil is ok. No continuity means the coil is open. If the coil is ok check for stuck reversing valve.

**No**

The change over relay is not energized. See outdoor unit cooling low voltage.

*Figure 8-104* (cont.)

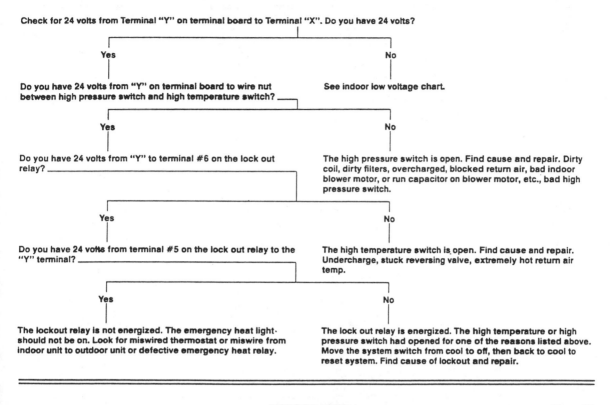

## OUTDOOR UNIT
## LOW VOLTAGE — COOLING

| WALL THERMOSTAT SETTING | | |
|---|---|---|
| System Switch | Fan Switch | E.H. Light |
| COOL | AUTO | ON |

Check for 24 volts from Terminal "Y" on terminal board to Terminal "X". Do you have 24 volts?

**Yes**

**No**

Do you have 24 volts from "Y" on terminal board to wire nut between high pressure switch and high temperature switch?

See indoor low voltage chart.

**Yes**

**No**

Do you have 24 volts from "Y" to terminal #6 on the lock out relay?

The high pressure switch is open. Find cause and repair. Dirty coil, dirty filters, overcharged, blocked return air, bad indoor blower motor, or run capacitor on blower motor, etc., bad high pressure switch.

**Yes**

**No**

Do you have 24 volts from terminal #5 on the lock out relay to the "Y" terminal?

The high temperature switch is open. Find cause and repair. Undercharge, stuck reversing valve, extremely hot return air temp.

**Yes**

**No**

The lockout relay is not energized. The emergency heat light should not be on. Look for miswired thermostat or miswire from indoor unit to outdoor unit or defective emergency heat relay.

The lock out relay is energized. The high temperature or high pressure switch had opened for one of the reasons listed above. Move the system switch from cool to off, then back to cool to reset system. Find cause of lockout and repair.

---

Chart C 1

## INDOOR UNIT
## STAGE 1 HEAT
## LOW VOLTAGE

| WALL THERMOSTAT SETTINGS | | |
|---|---|---|
| System Switch | Fan Switch | E.H. Heat |
| HEAT | AUTO | OFF |

Check main power supply fuses or circuit breakers.

Do you have 24 volts from (X) to (R) on the terminal board?

**Yes**

**No**

Do you have 24 volts from (X) to (G) on the terminal board?

Check control transformer for continuity through the primary and secondary windings. No continuity means there is an open winding. Replace the transformer.

Continued
Next Page

*Figure 8-104* (cont.)

259

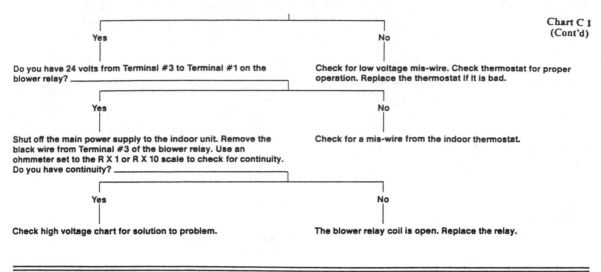

Chart C 1
(Cont'd)

Yes — No

Do you have 24 volts from Terminal #3 to Terminal #1 on the blower relay?

Check for low voltage mis-wire. Check thermostat for proper operation. Replace the thermostat if it is bad.

Yes — No

Shut off the main power supply to the indoor unit. Remove the black wire from Terminal #3 of the blower relay. Use an ohmmeter set to the R X 1 or R X 10 scale to check for continuity. Do you have continuity?

Check for a mis-wire from the indoor thermostat.

Yes — No

Check high voltage chart for solution to problem.

The blower relay coil is open. Replace the relay.

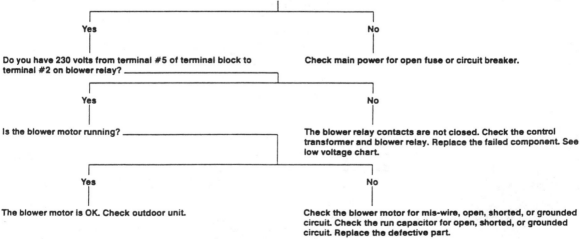

**INDOOR UNIT
HIGH VOLTAGE
STAGE 1 HEAT**

Chart C 2

| WALL THERMOSTAT SETTINGS | | |
|---|---|---|
| System Switch | Fan Switch | E.H. Heat |
| HEAT | AUTO | OFF/ON |

Check the main power supply, be sure that circuit breakers are on.
Do you have 230 volts ± 10% from terminal #5 of blower motor speed selection terminal block to Terminal #4 on the blower relay?

Yes — No

Do you have 230 volts from terminal #5 of terminal block to terminal #2 on blower relay?

Check main power for open fuse or circuit breaker.

Yes — No

Is the blower motor running?

The blower relay contacts are not closed. Check the control transformer and blower relay. Replace the failed component. See low voltage chart.

Yes — No

The blower motor is OK. Check outdoor unit.

Check the blower motor for mis-wire, open, shorted, or grounded circuit. Check the run capacitor for open, shorted, or grounded circuit. Replace the defective part.

*Figure 8-104* (cont.)

OUTDOOR UNIT
LOW VOLTAGE — STAGE 1 HEAT

Chart C 3

| WALL THERMOSTAT SETTING | | |
|---|---|---|
| System Switch | Fan Switch | E.H. Light |
| HEAT | AUTO | OFF |

Measure voltage from Terminal (X) to Terminal (Y) on the terminal board. Do you have 24 volts?

**Yes**

Check from (Y) to Terminal (1) on lock out relay. Do you have 24 volts?

**No**

Problem not in outdoor unit. See control voltage in indoor unit.

**Yes**

Check from (Y) to wire nut connection between high pressure and high temperature switch. Do you have 24 volts?

**No**

Loose connection or broken wire in control circuit from (X) on terminal strip to lockout relay terminal #1.

**Yes**

Check from (Y) to lock out relay terminal #6. Do you have 24 volts?

**No**

The high pressure switch is open. Replace switch if defective.

**Yes**

Check from (Y) to lock out relay #5 terminal. Do you have 24 volts?

**No**

The high temperature switch is open. Replace switch if defective.

**Yes**

Check from (Y) to black wire on contactor coil, 24 volts present?

**No**

Normally closed contacts on lock out relay are open.

**Yes**

Check from yellow on contactor to black on contactor, 24 volts?

**No**

Black wire from lock out relay is broken or has poor connection at spade connection at either end.

**Yes**

Is contactor pulled in?

**No**

Yellow wire from (Y) to contactor coil is broken or has poor connection at either end.

**Yes**

The control voltage is OK. Check high voltage.

**No**

Check the contactor coil with ohmmeter. Infinity - coil open, replace contactor. Resistance - contactor is mechanically stuck - replace.

*Figure 8-104* (cont.)

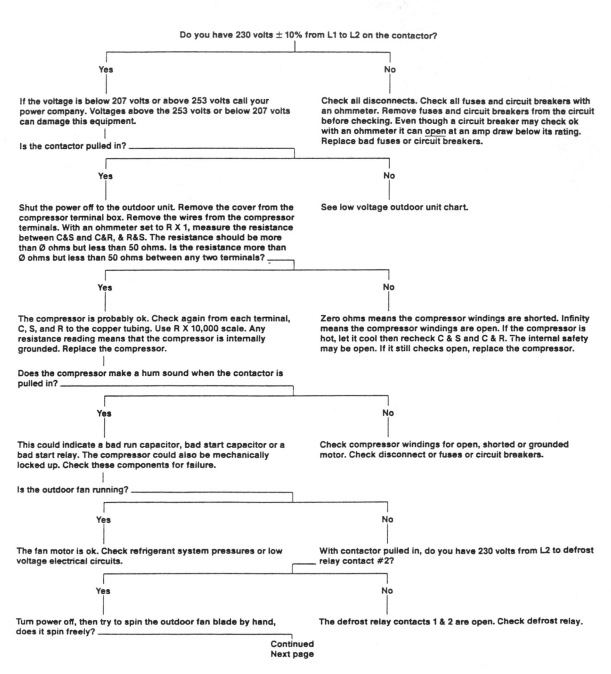

OUTDOOR UNIT
HIGH VOLTAGE
STAGE 1 HEAT

Chart C 4

| WALL THERMOSTAT SETTING | | |
|---|---|---|
| System Switch | Fan Switch | E.H. Light |
| HEAT | AUTO | OFF/ON |

Do you have 230 volts ± 10% from L1 to L2 on the contactor?

**Yes**

If the voltage is below 207 volts or above 253 volts call your power company. Voltages above the 253 volts or below 207 volts can damage this equipment.

Is the contactor pulled in?

**No**

Check all disconnects. Check all fuses and circuit breakers with an ohmmeter. Remove fuses and circuit breakers from the circuit before checking. Even though a circuit breaker may check ok with an ohmmeter it can open at an amp draw below its rating. Replace bad fuses or circuit breakers.

**Yes**

Shut the power off to the outdoor unit. Remove the cover from the compressor terminal box. Remove the wires from the compressor terminals. With an ohmmeter set to R X 1, measure the resistance between C&S and C&R, & R&S. The resistance should be more than Ø ohms but less than 50 ohms. Is the resistance more than Ø ohms but less than 50 ohms between any two terminals?

**No**

See low voltage outdoor unit chart.

**Yes**

The compressor is probably ok. Check again from each terminal, C, S, and R to the copper tubing. Use R X 10,000 scale. Any resistance reading means that the compressor is internally grounded. Replace the compressor.

Does the compressor make a hum sound when the contactor is pulled in?

**No**

Zero ohms means the compressor windings are shorted. Infinity means the compressor windings are open. If the compressor is hot, let it cool then recheck C & S and C & R. The internal safety may be open. If it still checks open, replace the compressor.

**Yes**

This could indicate a bad run capacitor, bad start capacitor or a bad start relay. The compressor could also be mechanically locked up. Check these components for failure.

Is the outdoor fan running?

**No**

Check compressor windings for open, shorted or grounded motor. Check disconnect or fuses or circuit breakers.

**Yes**

The fan motor is ok. Check refrigerant system pressures or low voltage electrical circuits.

**No**

With contactor pulled in, do you have 230 volts from L2 to defrost relay contact #2?

**Yes**

Turn power off, then try to spin the outdoor fan blade by hand, does it spin freely?

**No**

The defrost relay contacts 1 & 2 are open. Check defrost relay.

Continued
Next page

262

*Figure 8-104* (cont.)

Chart C 4
(Cont'd)

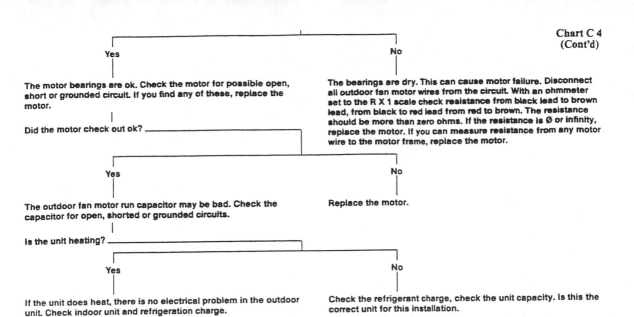

**Yes**

The motor bearings are ok. Check the motor for possible open, short or grounded circuit. If you find any of these, replace the motor.

Did the motor check out ok? _____

**Yes**

The outdoor fan motor run capacitor may be bad. Check the capacitor for open, shorted or grounded circuits.

Is the unit heating? _____

**Yes**

If the unit does heat, there is no electrical problem in the outdoor unit. Check indoor unit and refrigeration charge.

**No**

The bearings are dry. This can cause motor failure. Disconnect all outdoor fan motor wires from the circuit. With an ohmmeter set to the R X 1 scale check resistance from black lead to brown lead, from black to red lead from red to brown. The resistance should be more than zero ohms. If the resistance is Ø or infinity, replace the motor. If you can measure resistance from any motor wire to the motor frame, replace the motor.

**No**

Replace the motor.

**No**

Check the refrigerant charge, check the unit capacity. Is this the correct unit for this installation.

---

## OUTDOOR UNIT
### LOW VOLTAGE — STAGE 1 HEAT

| WALL THERMOSTAT SETTING | | |
|---|---|---|
| System Switch | Fan Switch | E.H. Light |
| HEAT | AUTO | ON |

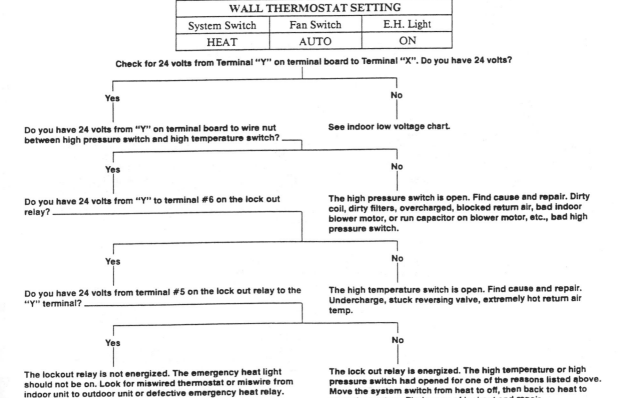

Check for 24 volts from Terminal "Y" on terminal board to Terminal "X". Do you have 24 volts?

**Yes**

Do you have 24 volts from "Y" on terminal board to wire nut between high pressure switch and high temperature switch? ___

**Yes**

Do you have 24 volts from "Y" to terminal #6 on the lock out relay? ___

**Yes**

Do you have 24 volts from terminal #5 on the lock out relay to the "Y" terminal? ___

**Yes**

The lockout relay is not energized. The emergency heat light should not be on. Look for miswired thermostat or miswire from indoor unit to outdoor unit or defective emergency heat relay.

**No**

See indoor low voltage chart.

**No**

The high pressure switch is open. Find cause and repair. Dirty coil, dirty filters, overcharged, blocked return air, bad indoor blower motor, or run capacitor on blower motor, etc., bad high pressure switch.

**No**

The high temperature switch is open. Find cause and repair. Undercharge, stuck reversing valve, extremely hot return air temp.

**No**

The lock out relay is energized. The high temperature or high pressure switch had opened for one of the reasons listed above. Move the system switch from heat to off, then back to heat to reset the system. Find cause of lockout and repair.

263

INDOOR AND OUTDOOR UNITS
HIGH AND LOW VOLTAGE-DEFROST

| WALL THERMOSTAT SETTING | | |
|---|---|---|
| System Switch | Fan Switch | E.H. Light |
| HEAT | AUTO | OFF |

For defrost sequence of operation see "Defrost Control". This service tree assumes that the liquid refrigerant temperature entering the outdoor coil after the metering device is below 22°F.

To check out defrost control for proper operation see page 45.

The defrost relay's function is to shut down the outdoor fan and shift the reversing valve to the cooling position during defrost. It also activates relays to provide circuits to energize auxiliary heat and to keep the compressor operating until the defrost is complete.

Does the outdoor fan shut off?

**Yes**

**No**

Check the defrost relay for voltage across coil. If no voltage, replace defrost board. If voltage is present, then remove wires from defrost relay coil and check it for continuity. If no continuity, replace relay. If continuity, problem is in high voltage circuit through defrost relay.

Does the reversing valve shift?

**Yes**

**No**

Check the changeover relay for a bad coil or contacts stuck open. Check the reversing valve coil for an open circuit or for reversing valve being stuck. Replace defective component.

If the No. 1 outdoor thermostat is closed (if used) the auxiliary heat light on the indoor thermostat should be on. The first sequencer should be closed. Are the No. 1 and No. 2 heat strips on?

**Yes**

**No**

Check low voltage wiring from heat pump terminal board to the terminal board on the indoor unit. Check all fuses and circuit breakers on the indoor unit. Check the sequencers for proper operation. Check all heat strips for proper operation.

To check every component except the defrost board, jumper "common" on the defrost board to "out" on the defrost board. The unit should go into defrost and stay as long as the jumper is in place. When "common" and "out" are jumpered the defrost board is bypassed.

The defrost system is working properly. If two outdoor thermostats are used when the outdoor temperature is below the second balance point the second sequencer should be energized.

If no thermostats are used all heat strips should be on.

If the outdoor temperature is above the first balance point (outdoor thermostat No. 1) install a temporary jumper between S1 and W2 on the outdoor unit.

If the outdoor temperature is above the second balance point (outdoor thermostat No. 2,) install a temporary jumper between S1 and S2 on the outdoor unit terminal board.

With jumpers in place, the strip heaters should be on. If the strip heat does not work see Stage 2 heat low voltage and/or high voltage.

Does the unit only partially defrost?

**Yes**

**No**

There are 3 pins on the defrost board T1, T2, T3. Pin T1 will allow a defrost every 30 minutes. Pin T2 will allow a defrost every 60 minutes. Pin T3 will allow a defrost every 90 minutes. The amount of accumulated run time may be changed by moving the jumper from one pin to another.

If the coil is clear after a defrost cycle has ended, the unit is ok.

*Figure 8-104* (cont.)

| WALL THERMOSTAT SETTING | | |
|---|---|---|
| System Switch | Fan Switch | E.H. Light |
| HEAT | AUTO | OFF |

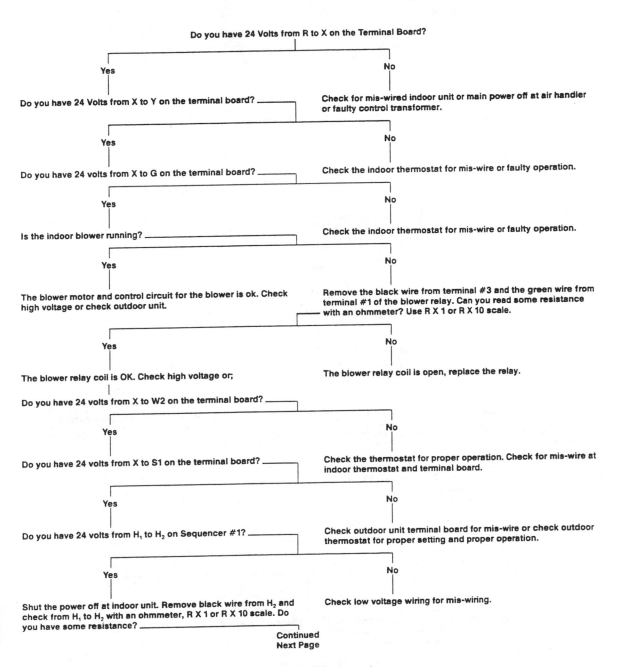

Do you have 24 Volts from R to X on the Terminal Board?

**Yes** — Do you have 24 Volts from X to Y on the terminal board?

**No** — Check for mis-wired indoor unit or main power off at air handler or faulty control transformer.

**Yes** — Do you have 24 volts from X to G on the terminal board?

**No** — Check the indoor thermostat for mis-wire or faulty operation.

**Yes** — Is the indoor blower running?

**No** — Check the indoor thermostat for mis-wire or faulty operation.

**Yes** — The blower motor and control circuit for the blower is ok. Check high voltage or check outdoor unit.

**No** — Remove the black wire from terminal #3 and the green wire from terminal #1 of the blower relay. Can you read some resistance with an ohmmeter? Use R X 1 or R X 10 scale.

**Yes** — The blower relay coil is OK. Check high voltage or;

Do you have 24 volts from X to W2 on the terminal board?

**No** — The blower relay coil is open, replace the relay.

**Yes** — Do you have 24 volts from X to S1 on the terminal board?

**No** — Check the thermostat for proper operation. Check for mis-wire at indoor thermostat and terminal board.

**Yes** — Do you have 24 volts from $H_1$ to $H_2$ on Sequencer #1?

**No** — Check outdoor unit terminal board for mis-wire or check outdoor thermostat for proper setting and proper operation.

**Yes** — Shut the power off at indoor unit. Remove black wire from $H_2$ and check from $H_1$ to $H_2$ with an ohmmeter, R X 1 or R X 10 scale. Do you have some resistance?

**No** — Check low voltage wiring for mis-wiring.

Continued
Next Page

*Figure 8-104*  (cont.)

265

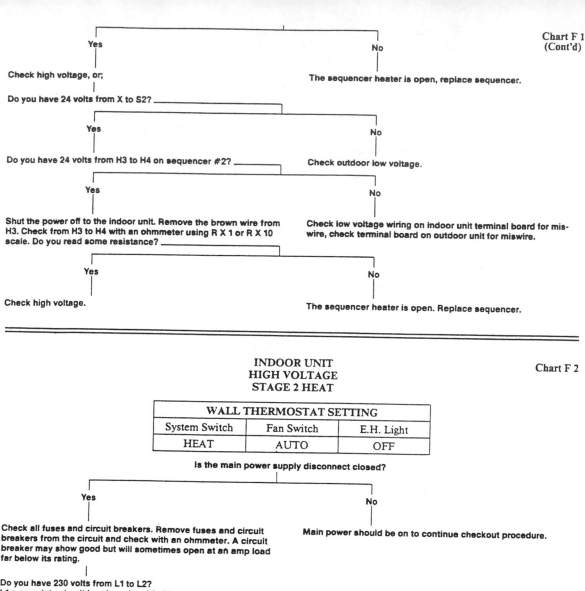

Yes

Check high voltage, or;

Do you have 24 volts from X to S2? _____

No

The sequencer heater is open, replace sequencer.

Yes

Do you have 24 volts from H3 to H4 on sequencer #2? _____

No

Check outdoor low voltage.

Yes

Shut the power off to the indoor unit. Remove the brown wire from H3. Check from H3 to H4 with an ohmmeter using R X 1 or R X 10 scale. Do you read some resistance? _____

No

Check low voltage wiring on indoor unit terminal board for mis-wire, check terminal board on outdoor unit for miswire.

Yes

Check high voltage.

No

The sequencer heater is open. Replace sequencer.

INDOOR UNIT
HIGH VOLTAGE
STAGE 2 HEAT

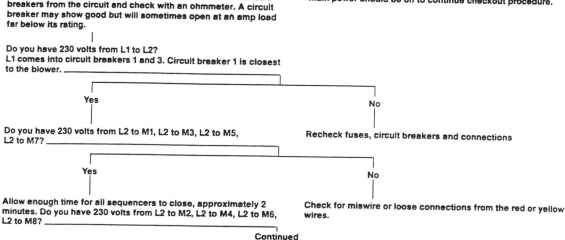

| WALL THERMOSTAT SETTING | | |
|---|---|---|
| System Switch | Fan Switch | E.H. Light |
| HEAT | AUTO | OFF |

Is the main power supply disconnect closed?

Yes

Check all fuses and circuit breakers. Remove fuses and circuit breakers from the circuit and check with an ohmmeter. A circuit breaker may show good but will sometimes open at an amp load far below its rating.

Do you have 230 volts from L1 to L2?
L1 comes into circuit breakers 1 and 3. Circuit breaker 1 is closest to the blower. _____

No

Main power should be on to continue checkout procedure.

Yes

Do you have 230 volts from L2 to M1, L2 to M3, L2 to M5, L2 to M7? _____

No

Recheck fuses, circuit breakers and connections

Yes

Allow enough time for all sequencers to close, approximately 2 minutes. Do you have 230 volts from L2 to M2, L2 to M4, L2 to M6, L2 to M8? _____

No

Check for miswire or loose connections from the red or yellow wires.

Continued
Next Page

266

*Figure 8-104* (cont.)

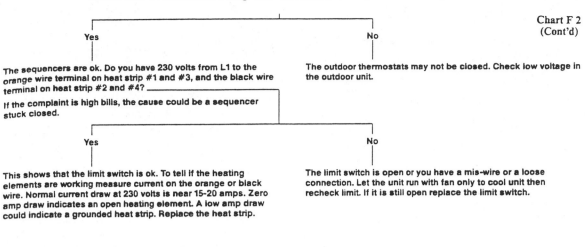

Chart F 2
(Cont'd)

Yes

No

The sequencers are ok. Do you have 230 volts from L1 to the orange wire terminal on heat strip #1 and #3, and the black wire terminal on heat strip #2 and #4? _____

If the complaint is high bills, the cause could be a sequencer stuck closed.

The outdoor thermostats may not be closed. Check low voltage in the outdoor unit.

Yes

No

This shows that the limit switch is ok. To tell if the heating elements are working measure current on the orange or black wire. Normal current draw at 230 volts is near 15-20 amps. Zero amp draw indicates an open heating element. A low amp draw could indicate a grounded heat strip. Replace the heat strip.

The limit switch is open or you have a mis-wire or a loose connection. Let the unit run with fan only to cool unit then recheck limit. If it is still open replace the limit switch.

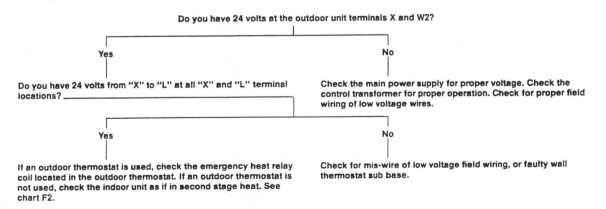

### INDOOR UNIT AND OUTDOOR UNIT
### EMERGENCY HEAT

Chart G 1

| WALL THERMOSTAT SETTINGS | | |
|---|---|---|
| System Switch | Fan Switch | E.H. Light |
| EM. HEAT | AUTO | ON/OFF |

Do you have 24 volts at the outdoor unit terminals X and W2?

Yes

No

Do you have 24 volts from "X" to "L" at all "X" and "L" terminal locations? _____

Check the main power supply for proper voltage. Check the control transformer for proper operation. Check for proper field wiring of low voltage wires.

Yes

No

If an outdoor thermostat is used, check the emergency heat relay coil located in the outdoor thermostat. If an outdoor thermostat is not used, check the indoor unit as if in second stage heat. See chart F2.

Check for mis-wire of low voltage field wiring, or faulty wall thermostat sub base.

*Figure 8-104* (cont.)

Note that each chart is keyed according to:

1. Wall thermostat settings
2. Indoor or outdoor unit
3. High voltage or low voltage
4. Stage 1 or stage 2

The service chart selection table can be used to determine the proper chart to start with when checking out any electrical service problem (Table 8-3).

To determine the proper chart to use for any electrical service problem, find the thermostat settings in the service chart selection table which match the actual thermostat settings when the electrical problem occurs.

Next determine if the problem appears to be in the outdoor unit or in the indoor unit. For

**TABLE 8-3**

Service Chart Selection Table

| Thermostat settings | | | | Service chart number | | | |
| | | | | Indoor unit | | Outdoor unit | |
| System switch | Fan switch | Mode | E.H. light | Low voltage | High voltage | Low voltage | High voltage |
|---|---|---|---|---|---|---|---|
| Cool | Auto | Cool | Off | A1 | A2 | A3 | A4 |
| Cool | Auto | Cool | On | | A2 | B1 | A4 |
| Heat | Auto | Stage 1 | Off | C1 | C2 | C3 | C4 |
| Heat | Auto | Stage 1 | On | | C2 | D1 | C4 |
| Heat | Auto | Defrost | Off | E1 | E1 | E1 | E1 |
| Heat | Auto | Stage 2 | Off | F1 | F2 | | |
| Emergency heat | Auto | Emergency heat | On/off | G1 | G1 | G1 | G1 |

Chart Index

| Chart no. | Page | Chart no. | Page | Chart no. | Page | Chart no. | Page |
|---|---|---|---|---|---|---|---|
| A1 | 25 | B1 | 29 | C4 | 32 | F2 | 36 |
| A2 | 25 | C1 | 29 | D1 | 33 | G1 | 37 |
| A3 | 26 | C2 | 30 | E1 | 34 | | |
| A4 | 27 | C3 | 31 | F1 | 35 | | |

example, if the compressor does not run, assume that the problem is in the outdoor unit. (If nothing runs, assume that the problem is in the indoor unit.)

Since it is usually more convenient, check high-voltage circuits first. Locate the high-voltage column under the unit where the problem is occurring.

The proper chart number will be found where the thermostat setting line intersects the high-voltage column.

# 9

# Heat Pump Startup and Standard Service Procedures

## INTRODUCTION

The job of starting up a heat pump system after the initial installation or after the unit has been shut-down for some time requires that certain steps be taken to prevent damage to the unit, especially to the compressor, and to make certain that the unit is operating as it was designed.

Also, there are standard service procedures that can be used when a problem is encountered with the unit. Each procedure is used to check out a particular problem and to determine if the component in question is at fault or if more checking is required to find the faulty component. There are also procedures listed for the replacement of certain components in a heat pump system.

Even though the procedures listed in this chapter are thorough there may be some specific procedures required for a particular unit brand or model. Also, the order in which

these procedures are performed is up to the particular individual performing them. However, these procedures should be used as a foundation for expansion related to each manufacturer's unit.

## *HEAT PUMP STARTUP PROCEDURES*

1. Check to be sure that electricity has been on to the outdoor unit for 24 hours.
2. Check the outdoor coil and clean if necessary.
3. Check all bearings; make necessary repairs.
4. Lubricate all bearings requiring service.
5. Check condition and tension of all belts. Replace or adjust as required (Figure 9-1).

270

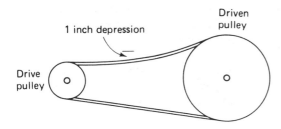

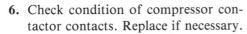

*Figure 9-1* Checking Belt Tension

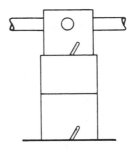

*Figure 9-3* Checking Temperature Drop or Rise on Indoor Coil

6. Check condition of compressor contactor contacts. Replace if necessary.

7. Check tightness of all electrical connections. Tighten loose connections.

8. Check and repair burnt or frayed wiring.

9. Set thermostat for cooling, and lower temperature setting below room temperature.

10. Check temperature rise on outdoor coil (Figure 9-2).

11. Check suction and discharge pressures.

12. Check refrigerant charge in system. If short, repair leak and add refrigerant.

13. Check the amperage draw to all motors.

14. Check temperature drop on indoor coil. If more than 24°F (13.33°C), check for dirty coil. If less than 18°F (10°C), check for inefficient compressor or refrigerant shortage on cooling. On heating, the temperature

rise should be approximately 30°F (16.67°C) (Figure 9–3).

15. Check thermostat calibration. Calibrate if necessary.

16. Clean or replace air filter.

17. Clean and vacuum indoor unit and area around unit.

18. Replace all covers.

## ELECTRIC HEATING STARTUP PROCEDURES _____

1. Turn off all electric power to unit.

2. Visually check for any burnt or bad wiring and replace as required.

3. Tighten all electrical connections.

4. Turn on electric power.

5. Set thermostat well above room temperature to ensure that all stages are demanding.

6. Check voltage drop on all elements after all relays are closed.

7. Check amperage to all elements. Replace any bad elements.

8. Check setting of heat anticipator. Adjust as required.

9. Set thermostat temperature selector below room temperature.

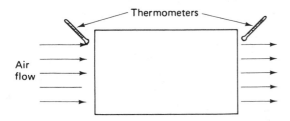

*Figure 9-2* Checking Temperature Rise on Outdoor Coil

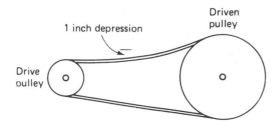

**Figure 9-4** Checking Belt Tension

10. After fan has stopped, turn off all electricity.
11. Check the condition of all bearings.
12. Lubricate all bearings requiring service.
13. Check condition and tension of belt. Replace or adjust as required (Figure 9–4).
14. Clean or replace air filter.
15. Clean and vacuum indoor unit and area around indoor unit.
16. Replace all covers.
17. Turn on electricity.

## STANDARD SERVICE PROCEDURES _____

*Procedure for Replacing Compressors:* Past experience has demonstrated that after a hermetic motor burnout has occurred, the refrigerant circuit must be cleaned to remove all contaminants. Without removal of these contaminants, a repeat burnout will occur. Failure to follow minimum cleaning recommendations as quickly as possible will result in excessive risk of a repeat burnout.

Cleaning of the circuit by flushing with refrigerant R–11 or similar refrigerants has been used in the past under certain conditions and with mixed degrees of success. This flushing method is seldom used today and is not recommended here.

The first step is to be certain that the compressor is burned. The following is the general procedure used.

1. When a compressor fails to start, it may appear to be burnt out. However, the fault can be in many other areas.
2. All other possibilities should be eliminated. Check for electrical and mechanical misapplications or malfunction.
3. Check the compressor motor for shorted, grounded, and open windings. If none is found, check the winding resistance with a precision ohmmeter to determine if turn-to-turn shorts exist.
4. Next, purge a small amount of refrigerant vapor from the compressor discharge. Smell the vapor in very small whiffs. A burned motor is usually indicated when a characteristic acrid odor is detected.

If the motor is found to be burned, the refrigerant should be discharged from the system, especially from small systems. The service technician should follow certain safety practices. In addition to the electrical hazards, the service technician should be aware of the dangers of receiving acid burns.

1. When discharging refrigerant (vapor or liquid) from the system in which the burnout occurred, avoid injury by not allowing the refrigerant to touch the eyes or skin. If the complete charge is to be purged, it should be discharged outside the building. The burnt refrigerant vapors will tarnish metal surfaces as well as contaminate the air used for breathing when released inside a building.
2. When it is necessary for the service technician to come in contact with the

oil or sludge in a burned-out compressor, rubber gloves should be worn to prevent possible acid burns.

When replacing the compressor in a system that has had a burnout, the system must be thoroughly cleaned to remove as many of the contaminants as possible. The following is a list of recommended procedures:

1. On systems of 5 hp and smaller, discharge the complete refrigerant charge to the outside atmosphere, preferably in the liquid form. On units larger than 5 hp, an attempt to save the refrigerant should be made by valving off the compressor (front-seating the service valves); then purge the refrigerant from the compressor.

2. Install an approved clean-up filter-drier in the suction line at the compressor. On units larger than 5 hp, this cannot be done until the refrigerant is pumped from the low side of the system (Figure 9–5). Be sure to install the suction line filter-drier between the reversing valve and the compressor suction connection.

3. Install an approved oversized liquid line cleanup drier in the liquid line (Figure 9–6). On units without pump down components, this cannot be done until the refrigerant is pumped from the high side of the system. Be sure to use a drier that will not deposit the contaminants back into the system when the refrigerant flow is reversed, or install the drier where the refrigerant will flow through it when the unit is operating in a given cycle (Figure 9–7). One is required for each cycle.

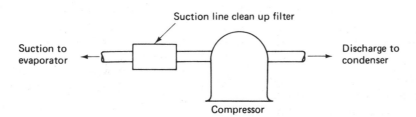

**Figure 9-5** Suction Line Cleanup Filter Drier

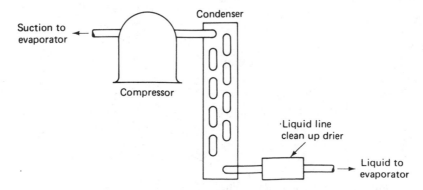

**Figure 9-6** Liquid Line Cleanup Drier

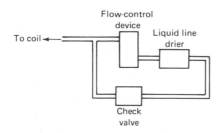

*Figure 9-7* Alternative Location for Liquid Line Driers on Heat Pump Systems

4. Check the refrigerant flow-control devices and clean them thoroughly or replace them.

5. Remove the inoperative compressor and install the proper replacement. Make all the connections leaktight.

6. Purge the system by blowing refrigerant through the system, forcing the refrigerant completely through all piping and coils.

7. Triple evacuate the system. The last evacuation should lower the system pressure to 1000 $\mu$ or lower, if time permits.

8. Recharge the system with refrigerant and put it back in operation. On systems larger than 5 hp, add refrigerant to complete the charge.

9. After the system has been in operation for approximately 24 hours, the acid content of the oil and the condition of the cleanup driers and filters should be checked. If acid is found in the oil, noted by discoloration, or the cleanup filters and driers show signs of stoppage, they should be replaced, the oil drained, if possible, and a new charge put into the system. This procedure should be repeated until a clean system is indicated.

*Procedure for Using the Gauge Manifold:* The gauge manifold is probably the most important tool in the service technician's toolbox. The various gauges can be used to check the system pressures, to charge refrigerant into the system, to evacuate the system, to add oil to the system, to purge noncondensables from the system, and for many other uses.

The gauge manifold set consists of a compound gauge, a pressure gauge, and a manifold that is equipped with hand valves to isolate the different connections or to allow their use in any combination as required (Figure 9–8). The ports to the gauge and the line connection are connected so that the gauges will indicate the pressure when connected to a pressure source. Flexible, leakproof hoses are used to make the connections from the gauge manifold to the system.

When connecting gauges to the system, a frequent task, care should be taken to prevent contaminants from entering the system. The hoses should be purged by allowing a small amount of refrigerant to escape from the fittings. The specific procedures for connecting the gauges to a system containing refrigerant are discussed next.

On systems where it is certain that both pressures are above 0 psig (0 kPa), use these procedures:

1. Front-seat the valves on the gauge manifold.

2. Back-seat the system service valves. This is to isolate the gauge ports from the rest of the system.

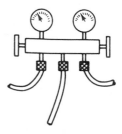

*Figure 9-8* Gauge Manifold Set

3. Make the hose connections to the system. Loosen one end of the center hose but do not disconnect the hose.

4. Crack the system service valves off the back seat. Do this slowly to prevent a sudden inrush of high-pressure vapor to the gauge.

5. Crack open one valve on the gauge manifold and allow a small amount of refrigerant vapor to escape out the center hose for a few seconds. Close the gauge manifold valve and repeat this process with the other valve.

6. The gauge manifold is now connected to the system and is ready for use (Figure 9–9).

On systems where the low-side pressure is below 0 psig (0 kPa), use these procedures:

1. Front-seat the valves on the gauge manifold.

2. Back-seat the system service valves. This is to isolate the gauge ports from the rest of the system.

3. Make the hose connections to the system. Tighten the hose connection to the discharge service valve, loosen the hose connection to the suction service valve, and plug the center hose connection to prevent the escape of refrigerant at this point.

4. Crack the system discharge service valve off the backseat. Do this slowly to prevent a sudden inrush of high-pressure vapor to the gauge.

5. Crack open the high-side gauge manifold valve and allow a small amount of refrigerant vapor to escape out the low-side hose connection at the system service valve. Tighten the hose connection after a few seconds. Close the gauge manifold valve.

6. Crack the system suction service valve off the backseat.

7. The gauge manifold is now connected to the system and is ready for use (Figure 9–9).

Gauges are delicate instruments and should be treated with care. Do not drop the gauges or subject them to pressures higher than the maximum pressure shown on the scale. Gauges should be kept in adjustment so that the proper pressures are indicated.

*Procedure for Purging Noncondensables from the System:* Air is considered a noncondensable under the pressures and temperatures normally encountered in an air-conditioning or refrigeration system. Air can enter the refrigerant circuit in several ways. The most common ways are a leak in the low side of the system or improper connection and use of the service gauges. In some cases it may not be practical or desirable to purge the complete refrigerant charge and evacuate the system. However, the air must be removed to prevent damage to the system due to chemical reactions and to help keep the system operating efficiently.

The air will normally be trapped in the top of the receiver and the condenser because of the liquid seal at the receiver or condenser outlet.

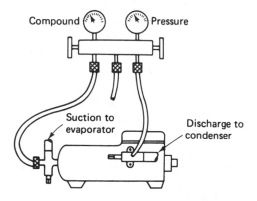

*Figure 9-9* Gauge Manifold Connected to System

Air in the system may be detected by a higher than normal condensing pressure caused by the trapped air. The amount of increase in pressure will be determined by the amount of trapped air.

To purge noncondensables from the system, use the following procedure.

1. Locate and remove the source of noncondensables.

2. Connect the gauges to the system (Figure 9–9).

3. If possible, pump the system down.

4. Stop the unit. Leave the condenser fan running on air-cooled units or block the water valve open on water-cooled units and leave the water pump running. Allow the unit to cool down for approximately 10 minutes. During this time the noncondensables will rise to the top of the condenser.

5. If purge valves are on the unit, use them for the purging process. If not, the gauge port on the compressor discharge may be used. To purge, slowly open the purge valve. Allow the vapor to bleed off very slowly for only a short period of time. Close the valve and let the unit sit idle. Be sure to purge slowly for only short periods of time to prevent the boiling off of excess refrigerant, the remixing of the air and refrigerant, and the purging of an excess of refrigerant. After the system has sat idle for a few minutes, repeat the purging process. Repeat this process three or four times.

6. Start the system and check the discharge pressure after a few minutes of operation. If the discharge pressure is still abnormally high, repeat the process starting with step 4. Repeat steps 4 through 6 until satisfactory operation is obtained.

*Procedure for Pumping a System Down:* Pumping a system down is the process generally used to put all the refrigerant into the condenser or receiver. Pumping down is used to save the refrigerant when work is required on components located from the receiver through the system to the suction valve in the compressor. Pumping a system down can be accomplished on systems equipped with service valves and a liquid line shutoff valve.

To pump a system down, use the following procedure:

1. Install the gauge manifold on the system (Figure 9–9).

2. Start the unit.

3. Front-seat the liquid line (King) valve.

4. Observe both the suction and discharge pressures. If there is a sharp increase in the discharge pressure, stop the compressor. Check to determine the reason for the increase in discharge pressure. If the condenser and receiver are full of refrigerant, the remaining charge must be removed from the system. When the suction pressure is reduced to about 1 or 2 psig (6.895 or 13.790 kPa), stop the compressor and observe the gauges. If the suction pressure increases to 10 or 15 psig (68.948 or 103.421 kPa), start the compressor and pump the system down to 1 or 2 psig again and stop the compressor. The suction pressure should remain at around this pressure. If not, repeat the pump-down process. If the pump down is repeated more than three times, the compressor discharge valves may be leaking. In this case the discharge service valve must be closed to prevent refrigerant bleeding into the low side of the system. *Caution:* Do not start the compressor when the discharge valve is closed.

5. On units equipped with a low-pressure control set at a pressure higher than 1 or 2 psig, it will be necessary to electrically bypass the control to keep the compressor running while pumping the system down.

6. Relieve any remaining pressure in the low side of the system by opening the low-side hand valve on the gauge manifold. Do not attempt to weld or solder on a system having refrigerant pressure inside.

7. The necessary repairs can now be made. It is desirable to install a new liquid line drier before recharging the system.

8. To put the system back in operation, open the compressor discharge service valve and open the liquid line (King) valve. Allow a small amount of refrigerant to escape through the gauge manifold; then close the gauge manifold low-side hand valve.

9. Start the unit and check the refrigerant charge. Add any required refrigerant to bring the system to full charge.

*Procedure for Pumping a System Out:* Pumping a system out is the process used, when repairs are to be made, to save the refrigerant in a system that does not have service valves. This procedure is also used when repairs are to be made to the high side of the system and a system pump down cannot be done. To accomplish this process, a portable condensing unit and a clean, dry refrigerant cylinder are required. The portable condensing unit is used to pump the refrigerant from the system and discharge it into the refrigerant cylinder, where it is stored while the repairs are being made. The refrigerant is then charged back into the system.

To pump a system out, use the following procedure:

1. Stop the unit.

2. Connect the gauge manifold to the system (Figure 9–9).

3. Connect the center hose on the gauge manifold to the suction service valve on the portable condensing unit.

4. Connect the liquid line connection on the portable condensing unit to the valve on the refrigerant cylinder. Leave this connection loose for purging air from the lines and from the portable condensing unit (Figure 9–10).

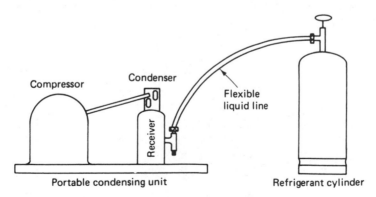

**Figure 9-10** Connections from Portable Condensing Unit to the Refrigerant Cylinder

5. Slowly crack the system service valves off the back seat until pressure is indicated on the gauges. Open the service valves on the portable condensing unit.

6. Open the hand valves on the gauge manifold and allow the refrigerant to blow out the connection on the refrigerant cylinder for a few seconds. Tighten the hose connection on the refrigerant cylinder.

7. Open the refrigerant cylinder valve.

8. Start the portable condensing unit and pump the refrigerant from the system and into the cylinder. To prevent overloading of the portable condensing unit, regulate the suction pressure by partially closing the hand valves on the gauge manifold. *Caution:* Be sure not to overcharge a cylinder with refrigerant. Use as many cylinders as required by weight.

9. Remove the refrigerant until only 1 or 2 psig of pressure (6.895 or 13.790 kPa) is left inside the system. Close off all valves to prevent the refrigerant escaping from the cylinder and the portable condensing unit.

10. Purge any remaining pressure from the refrigeration system and make the re-quired repairs. It is desirable to install a new liquid line drier before charging the system with refrigerant.

11. To put the system back in operation, a complete evacuation should be completed. Then charge the refrigerant from the cylinder and the portable condensing unit back into the system. It may be desirable to install a drier in the charging line to remove any contaminants from the system (Figure 9–11).

*Procedure for Leak Testing:* When the refrigerant has escaped from a system, the leak must be found and repaired or the refrigerant will escape again. Refrigeration systems must be vaportight for two reasons: (1) any leakage will result in a loss of refrigerant charge, and (2) leaks will allow air and moisture to enter the system when the pressure is reduced below 0 psig (0 kPa). Refrigerant leaks can and do occur at any time and at any point owing to low-quality workmanship, age of equipment, and vibration.

Because leak detection is a common service procedure, the service technician must be familiar with the various methods used. Each method has its advantages and disadvantages,

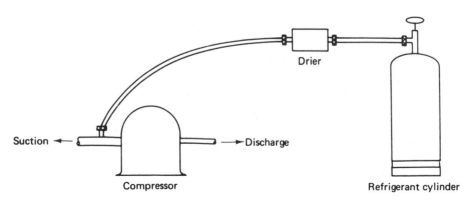

**Figure 9-11** Liquid Line Drier in Charging Line

depending on the circumstances surrounding the system. The three most popular methods are: electronic leak detection, halide-torch leak detection, and soap-bubble leak detection. However, the electronic method is not satisfactory in a heavy concentration of refrigerant because it is extremely sensitive; and the halide torch is not generally satisfactory when small leaks or blowing winds are encountered. Under most other conditions, these two tests are found most satisfactory. Because electronic leak detectors are very sensitive, the soap-bubble test is most satisfactory for finding small leaks in a suspected area. Sometimes more than one of these methods may be used to locate one leak.

*Procedure for Evacuating a System:* Evacuation is the process used to remove air and moisture from a refrigeration system. Evacuation is accomplished by the use of pumps specially designed for this purpose. A discarded refrigeration compressor is not suitable. Never run a motor compressor while the system is evacuated. To do so may result in serious damage to the motor winding. Evacuation is required any time a system has become contaminated or the compressor or system has been exposed to the atmosphere for long periods of time.

Purging a system will remove a good portion of the air, and driers will remove a part of any moisture from the system, but only up to the capacity of the drier. Therefore, there are still contaminants left in the system, and evacuation is the best means of being reasonably sure that the system is free of these contaminants.

There are basically two evacuation procedures: (1) simple evacuation and (2) triple evacuation. Simple evacuation is used on systems containing only a minimum of contaminants. Triple evacuation is used on systems containing a greater amount of contaminants.

For the simple evacuation method, use the following procedure:

1. Connect the gauge manifold to the system (Figure 9–9).
2. Purge all pressure from the refrigeration system by opening the system service valves and the gauge manifold hand valves.
3. Connect the center hose on the gauge manifold to the vacuum pump (Figure 9–12).
4. Start the vacuum pump and pump a vacuum of at least 1000 microns. A vacuum of 1500 microns is preferable.
5. Close off the gauge manifold hand valves.
6. Stop the vacuum pump. Do not stop the vacuum pump before closing the gauge manifold hand valves. This is to prevent air entering the system.
7. Disconnect the center hose of the gauge manifold from the vacuum pump and connect it to a cylinder containing the proper refrigerant (Figure 9–13).
8. Open the cylinder valve.
9. Loosen the center hose connection at the gauge manifold. Purge the hose for a few seconds; then tighten the connection.
10. Open the gauge manifold hand valves and admit refrigerant into the system.

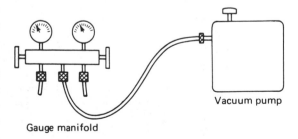

Vacuum pump

Gauge manifold

*Figure 9-12* Gauge Manifold Connected to Vacuum Pump

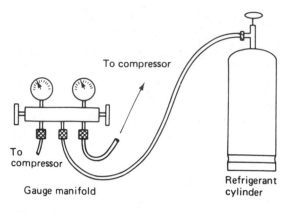

*Figure 9-13* Gauge Manifold Connected to Refrigerant Cylinder

11. Close the high-side hand valve on the gauge manifold.
12. Start the unit and add the proper charge of refrigerant.

For the triple-evacuation method, use the following procedure:

1. Connect the gauge manifold to the system (Figure 9–9).
2. Purge all pressure from the refrigeration system by opening the system service valves and the gauge manifold hand valves.
3. Connect the center hose on the gauge manifold to the vacuum pump (Figure 9–12).
4. Start the vacuum pump and pump a vacuum of approximately 1500 microns.
5. Close off the gauge manifold hand valves.
6. Stop the vacuum pump. Do not stop the vacuum pump before closing the gauge manifold hand valves. This is to prevent air entering the system.
7. Disconnect the center hose of the

gauge manifold from the vacuum pump and connect it to a cylinder containing the proper refrigerant (Figure 9–13).

8. Open the cylinder valve.
9. Loosen the center hose connection at the gauge manifold. Purge the hose for a few seconds; then tighten the connection.
10. Open the gauge manifold hand valves and admit refrigerant into the system until a pressure of about 5 psig is indicated on the gauges.
11. Close the refrigerant cylinder valve and the gauge manifold hand valves.
12. Disconnect the hose from the cylinder.
13. Open the gauge manifold hand valves and purge the pressure from the system.
14. Repeat steps 3 through 13.
15. Repeat steps 3 through 9 only. Pump a vacuum of 500 rather than 1500 microns.
16. Open the gauge manifold hand valves and admit refrigerant into the system until cylinder pressure is indicated on the gauges.
17. Close the high-side gauge manifold hand valve.
18. Start the unit and add the proper charge of refrigerant.

*Procedure for Charging Refrigerant into a System:* The performance of refrigeration and air-conditioning systems is highly dependent on the proper charge of refrigerant in the system. A system that is undercharged will operate with a starved evaporator. Low suction pressures, loss of system capacity, and possible compressor overheating are the results of an undercharged system. On the other hand, an overcharge of refrigerant will back up in the condenser. High discharge pressures and liquid refrigerant flooding of the compressor with possible compressor damage are the results of an overcharged system. Larger systems can tolerate a reasonable amount of overcharging or undercharging without severe effects, but some of the smaller systems have a critical charge. The system must be properly charged to obtain proper operation.

The amount of charge will depend on the size of the system in Btu, the length of the lines, the type of refrigerant, and the operating temperature. Therefore, each system must be considered separately. The unit nameplate will usually indicate what type of refrigerant is required and the approximate weight of the required refrigerant.

There are two methods of charging refrigerant into a system: (1) liquid charging and (2) vapor charging. Liquid charging is much faster than vapor charging and is, therefore, used extensively on large field built-up systems. Never charge liquid refrigerant into the compressor suction or discharge service valves. Liquid entering the compressor can damage the compressor valves. Vapor charging is normally done only when small amounts of refrigerant are to be added to a system. Vapor charging also allows the refrigerant to be charged into the compressor suction service valve.

For the liquid-charging method, use the following procedure:

1. Connect the gauge manifold to the system (Figure 9–9).

2. Open the system service valves and purge air from the lines.

3. Connect a charging line to the liquid line (King) valve. This line should include a liquid drier to prevent contaminants entering the system.

4. Connect the charging line to the liquid valve on a cylinder of the proper type of refrigerant (Figure 9–14).

5. Open the refrigerant cylinder liquid valve.

6. Loosen the charging line connection at the liquid line (King) valve and allow the refrigerant to escape for a few seconds.

7. If the correct charge weight is known, the refrigerant cylinder can be weighed to see when sufficient refrigerant has been charged into the system.

8. Close the liquid line (King) valve and start the compressor. Allow the liquid refrigerant to enter the liquid line until the correct weight has been charged into the system (see step 10).

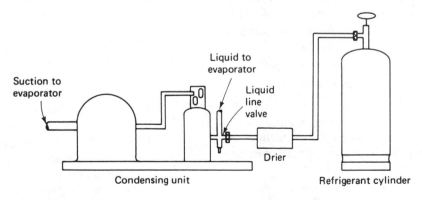

***Figure 9-14*** Connections for Liquid Charging a System

9. If the correct charge weight is not known, the liquid line valve must be opened periodically and the system operation observed. If more refrigerant is needed, close the liquid line valve again and charge more refrigerant into the system. Repeat this process until the proper charge is indicated (see step 10).

10. Closely watch the discharge pressure gauge. A sudden increase in discharge pressure indicates that the capacity of the condenser and receiver has been reached. Stop charging the unit immediately and open the liquid line valve. Any additional refrigerant must be charged into the system by the vapor method.

For the vapor-charging method, use the following procedure:

1. Connect the gauge manifold to the system (Figure 9–9).
2. Connect the center line on the gauge manifold to a cylinder of the proper refrigerant (Figure 9–13).
3. Open the refrigerant cylinder valve and the hand valves on the gauge manifold.
4. Loosen the hose connection at the system service valves and allow the refrigerant to escape for a few seconds. Retighten the connections.
5. Close the gauge manifold hand valves.
6. Crack the system service valves.
7. Start the unit.
8. Open the low-side gauge manifold hand valve and charge refrigerant into the system until the proper amount has been charged into the system.
9. Closely watch the discharge pressure gauge during the charging process to be certain that the system is not over-charged.

*Procedure for Determining the Proper Refrigerant Charge:* Determining the proper refrigerant charge is an important procedure. A system that does not contain the proper charge of refrigerant will not operate to maximum efficiency. There are several ways to determine if a system is properly charged: (1) weighing the charge, (2) using a sight glass, (3) using a liquid-level indicator, (4) using the liquid subcooling method, (5) using the superheat method, and (6) using the manufacturer's charging charts.

For the charge weight method, use the following procedure:

1. Connect the gauge manifold to the system service valves (Figure 9–9).
2. Purge all pressure from the system. Open both hand valves on the gauge manifold.
3. Connect the center line from the gauge manifold to a vacuum pump (Figure 9–12).
4. Start the vacuum pump and pump a deep vacuum on the system.
5. Close the gauge manifold hand valves.
6. Stop the vacuum pump. Do not stop the vacuum pump before closing the gauge manifold hand valves. This is to prevent air entering the system.
7. Disconnect the center line from the vacuum pump and connect it to a cylinder of the proper type of refrigerant (Figure 9–13). This cylinder may be a charging cylinder charged with the proper amount of refrigerant or a large cylinder placed on an accurate scale. Small systems require a charging cylinder, whereas large systems require a large cylinder.

8. Open the cylinder valve.

9. Loosen the center hose connection at the gauge manifold. Purge the hose for a few seconds; then tighten the connection.

10. Open both hand valves on the gauge manifold and charge the refrigerant into the system. This must be done suddenly so that all the refrigerant will enter the system before the vacuum vanishes.

11. Start the unit and check the operation.

For the sight glass method, use the following procedure:

1. Start the unit and allow it to operate for several minutes.

2. Check the flow of refrigerant through the sight glass. A flashlight may be needed to adequately see the flow (Figure 9–15).

3. A steady stream of bubbles indicates the system is low on refrigerant. If these bubbles are intermittent, allow the system to operate a while longer to see if they will disappear. If the bubbles remain, the system is low on refrigerant.

4. Connect the gauge manifold to the system service valves (Figure 9–9).

5. Connect the center line to a cylinder of the proper type of refrigerant (Figure 9–13).

6. Crack the system service valves and open the gauge manifold hand valves and loosen the connection on the refrigerant cylinder. Purge the lines for a few seconds. Tighten the connection.

7. Close the hand valves on the gauge manifold.

8. Open the refrigerant cylinder valve.

9. Open the low-side hand valve on the gauge manifold and admit refrigerant into the system while observing the sight glass and the discharge pressure gauge. When the bubbles disappear, close the low-side hand valve. Observe the sight glass. If the bubbles reappear, add more refrigerant into the system. Repeat this process until the bubbles do not reappear. A sudden increase in the discharge pressure indicates a system over-

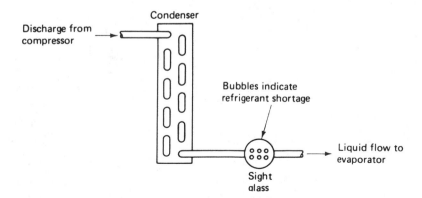

*Figure 9-15* Checking Refrigerant Charge Using a Sight Glass

charge. Stop charging the unit and remove some of the refrigerant.

For the liquid level indicator method, use the following procedure:

1. Start the unit and allow it to operate 10 to 15 minutes.

2. After the system has been in operation long enough for the pressures to stabilize, crack the liquid level test port. A continuous flow of liquid refrigerant from the port indicates sufficient charge. A continuous flow of vapor from the port indicates a shortage of refrigerant. The test port is usually located in the lower section of the condenser. Some units may have a liquid level indicator in the receiver tank (Figure 9–16).

3. To add refrigerant to the system, connect the gauge manifold to the system service valves. Close the gauge manifold hand valves. Connect the center charging line to a cylinder containing refrigerant of the proper type. Leave this connection loose (Figure 9–13).

4. Crack the system service valves open until pressure is indicated on the gauges.

5. Open the gauge manifold hand valves. Allow refrigerant to escape from the hose connection on the cylinder for a few seconds; then tighten the connection.

6. Close the high-side gauge manifold hand valve.

7. Open the refrigerant cylinder valve and charge refrigerant into the system.

8. Periodically open the liquid level test port and check for liquid refrigerant.

9. Continue to add refrigerant until a continuous stream of liquid refrigerant is indicated.

10. When liquid is indicated, close the low-

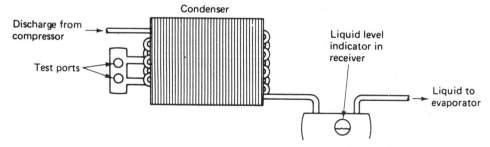

*Figure 9-16* Liquid-level Indicator Parts

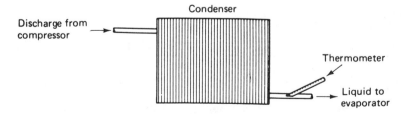

*Figure 9-17* Checking Liquid and Subcooling During Charging Process

side gauge manifold hand valve. Allow the unit to operate a few minutes and check for liquid again. If a continuous stream of liquid is not indicated, add more refrigerant and check again. Continue this process until the proper charge is indicated.

For the liquid subcooling method, use the following procedure:

1. Start the unit and allow it to operate for 10 to 15 minutes.

2. Strap a thermometer to the liquid line at the condenser outlet (Figure 9–17).

3. After the system has operated long enough for the pressures to stabilize, check the liquid temperature leaving the condenser with a thermometer.

4. Connect the gauge manifold to the system service valves. Close the gauge manifold hand valves (Figure 9–9).

5. Open the system service valves until pressure is indicated on the gauges.

6. Loosen the hose connections on the gauge manifold and allow the refrigerant vapor to escape for a few seconds; then retighten the connections.

7. Compare the liquid line temperature to the condensing temperature. The condensing temperature is determined by relating the discharge pressure to the corresponding temperature on a pressure-temperature chart. The liquid line temperature is indicated on the thermometer installed in step 2. The liquid line temperature should be approximately 5°F (2.35°C) below the condensing temperature. If less than 5°F (2.35°C), more refrigerant is needed in the system (Table 9-1).

8. To add refrigerant, connect the center charging hose on the gauge manifold

**Table 9-1**

Temperature-Pressure Chart

LIGHT FIGURES = VACUUM    BOLD FIGURES = PRESSURE

| °F | R-12 | R-13 | R-22 | R-500 | R-502 | R-717 Ammonia |
|---|---|---|---|---|---|---|
| −100 | 27.0 | 7.5 | 25.0 | 26.4 | 23.3 | 27.4 |
| −95 | 26.4 | 10.9 | 24.1 | 25.7 | 22.1 | 26.8 |
| −90 | 25.8 | 14.2 | 23.0 | 24.9 | 20.7 | 26.1 |
| −85 | 25.0 | 18.2 | 21.7 | 24.0 | 19.0 | 25.3 |
| −80 | 24.1 | 22.3 | 20.2 | 22.9 | 17.1 | 24.3 |
| −75 | 23.0 | 27.1 | 18.5 | 21.7 | 15.0 | 23.2 |
| −70 | 21.9 | 32.0 | 16.6 | 20.3 | 12.6 | 21.9 |
| −65 | 20.5 | 37.7 | 14.4 | 18.8 | 10.0 | 20.4 |
| −60 | 19.0 | 43.5 | 12 0 | 17.0 | 7.0 | 18.6 |
| −55 | 17.3 | 50.0 | 9.2 | 15.0 | 3.6 | 16.6 |
| −50 | 15.4 | 57.0 | 6.2 | 12.8 | 0.0 | 14.3 |
| −45 | 13.3 | 64.6 | 2.7 | 10.4 | 2.1 | 11.7 |
| −40 | 11.0 | 72.7 | 0.5 | 7.6 | 4.3 | 8.7 |
| −35 | 8.4 | 81.5 | 2.6 | 4.6 | 6.7 | 5.4 |
| −30 | 5.5 | 90.9 | 4.9 | 1.2 | 9.4 | 1.6 |
| −28 | 4.3 | 94.9 | 5.9 | 0.1 | 10.5 | 0.0 |
| −26 | 3.0 | 98.9 | 6.9 | 0.9 | 11.7 | 0.8 |
| −24 | 1.6 | 103.0 | 7.9 | 1.6 | 13.0 | 1.7 |
| −22 | 0.3 | 107.3 | 9.0 | 2.4 | 14.2 | 2.6 |
| −20 | 0.6 | 111.7 | 10.2 | 3.2 | 15.5 | 3.6 |
| −18 | 1.3 | 116.2 | 11.3 | 4.1 | 16.9 | 4.6 |
| −16 | 2.1 | 120.8 | 12.5 | 5.0 | 18.3 | 5.6 |
| −14 | 2.8 | 125.7 | 13.8 | 5.9 | 19.7 | 6.7 |
| −12 | 3.7 | 130.5 | 15.1 | 6.8 | 21.2 | 7.9 |
| −10 | 4.5 | 135.4 | 16.5 | 7.8 | 22.8 | 9.0 |
| −8 | 5.4 | 140.5 | 17.9 | 8.8 | 24.4 | 10.3 |
| −6 | 6.3 | 145.7 | 19.3 | 9.9 | 26.0 | 11.6 |
| −4 | 7.2 | 151.1 | 20.8 | 11.0 | 27.7 | 12.9 |
| −2 | 8.2 | 156.5 | 22.4 | 12.1 | 29.4 | 14.3 |
| 0 | 9.2 | 162.1 | 24.0 | 13.3 | 31.2 | 15.7 |
| 2 | 10.2 | 167.9 | 25.6 | 14.5 | 33.1 | 17.2 |
| 4 | 11.2 | 173.7 | 27.3 | 15.7 | 35.0 | 18.8 |
| 6 | 12.3 | 179.8 | 29.1 | 17.0 | 37.0 | 20.4 |
| 8 | 13.5 | 185.9 | 30.9 | 18.4 | 39.0 | 22.1 |
| 10 | 14.6 | 192.1 | 32.8 | 19.7 | 41.1 | 23.8 |
| 12 | 15.8 | 198.6 | 34.7 | 21.2 | 43.2 | 25.6 |
| 14 | 17.1 | 205.2 | 36.7 | 22.6 | 45.5 | 27.5 |
| 16 | 18.4 | 211.9 | 38.7 | 24.1 | 47.7 | 29.4 |
| 18 | 19.7 | 218.8 | 40.9 | 25.7 | 50.1 | 31.4 |
| 20 | 21.0 | 225.7 | 43.0 | 27.3 | 52.5 | 33.5 |
| 22 | 22.4 | 233.0 | 45.3 | 28.9 | 54.9 | 35.7 |
| 24 | 23.9 | 240.3 | 47.6 | 30.6 | 57.4 | 37.9 |
| 26 | 25.4 | 247.8 | 49.9 | 32.4 | 60.0 | 40.2 |
| 28 | 26.9 | 255.5 | 52.4 | 34.2 | 62.7 | 42.6 |
| 30 | 28.5 | 263.2 | 54.9 | 36.0 | 65.4 | 45.0 |
| 32 | 30.1 | 271.3 | 57.5 | 37.9 | 68.2 | 47.6 |
| 34 | 31.7 | 279.5 | 60.1 | 39.9 | 71.1 | 50.2 |
| 36 | 33.4 | 287.8 | 62.8 | 41.9 | 74.1 | 52.9 |
| 38 | 35.2 | 296.3 | 65.6 | 43.9 | 77.1 | 55.7 |
| 40 | 37.0 | 304.9 | 68.5 | 46.1 | 80.2 | 58.6 |
| 45 | 41.7 | 327.5 | 76.0 | 51.6 | 88.3 | 66.3 |
| 50 | 46.7 | 351.2 | 84.0 | 57.6 | 96.9 | 74.5 |
| 55 | 52.0 | 376.1 | 92.6 | 63.9 | 106.0 | 83.4 |
| 60 | 57.7 | 402.3 | 101.6 | 70.6 | 115.6 | 92.9 |
| 65 | 63.8 | 429.8 | 111.2 | 77.8 | 125.8 | 103.1 |
| 70 | 70.2 | 458.7 | 121.4 | 85.4 | 136.6 | 114.1 |
| 75 | 77.0 | 489.0 | 132.2 | 93.5 | 148.0 | 125.8 |
| 80 | 84.2 | 520.8 | 143.6 | 102.0 | 159.9 | 138.3 |
| 85 | 91.8 | — | 155.7 | 111.0 | 172.5 | 151.7 |
| 90 | 99.8 | — | 168.4 | 120.6 | 185.8 | 165.9 |
| 95 | 108.3 | — | 181.8 | 130.6 | 199.7 | 181.1 |
| 100 | 117.2 | — | 195.9 | 141.2 | 214.4 | 197.2 |
| 105 | 126.6 | — | 210.8 | 152.4 | 229.7 | 214.2 |
| 110 | 136.4 | — | 226.4 | 164.1 | 245.8 | 232.3 |
| 115 | 146.8 | — | 242.7 | 176.5 | 262.6 | 251.5 |
| 120 | 157.7 | — | 259.9 | 189.4 | 280.3 | 271.7 |
| 125 | 169.1 | — | 277.9 | 203.0 | 298.7 | 293.1 |
| 130 | 181.0 | — | 296.8 | 217.2 | 318.0 | 315.0 |
| 135 | 193.5 | — | 316.6 | 232.1 | 338.1 | 335.0 |
| 140 | 206.6 | — | 337.3 | 247.7 | 359.1 | 365.0 |
| 145 | 220.3 | — | 358.9 | 266.1 | 381.1 | 390.0 |
| 150 | 234.6 | — | 381.5 | 281.1 | 403.9 | 420.0 |
| 155 | 249.5 | — | 405.1 | 298.9 | 427.8 | 450.0 |
| 160 | 265.1 | — | 429.8 | 317.4 | 452.6 | 490.0 |

*Source:* Alco Controls Division, Emerson Electric Co.

to the valve on a cylinder containing the proper type of refrigerant. Do not tighten this connection (Figure 9–13).

9. Open the hand valves on the gauge manifold. Allow the refrigerant to escape a few seconds; then tighten the connection on the cylinder valve.

10. Close the high-side hand valve on the gauge manifold.

11. Open the cylinder valve and charge refrigerant into the system.

12. Observe the thermometer on the liquid line and the discharge temperature. When the desired subcooling is obtained, close the low-side hand valve on the gauge manifold to stop adding refrigerant to the system.

13. Allow the system to operate a few minutes to let the pressures stabilize, and check the amount of subcooling. If additional subcooling is required, add more refrigerant to the system. Continue this process until the 5°F (2.35°C) subcooling remains stable.

For the superheat method, use the following procedure:

1. Start the unit and allow it to operate for 10 to 15 minutes.

2. Strap a thermometer on the suction line about 6 in. from the compressor. Insulate the thermometer bulb so that accurate readings are indicated (Figure 9–18).

3. If a low-side pressure port is available, connect the low-side gauge to the system by connecting the low-side hose on the gauge manifold to the pressure port. If a pressure port is not available, strap a thermometer to a return bend about midway on the evaporating coil. Do not put the thermometer on a fin. Insulate the thermometer bulb so that accurate readings may be indicated (Figure 9–19).

4. If a pressure port is available, determine the difference between the suction line temperature and the saturation temperature equivalent to the suction pressure with the unit running. The saturation temperature is determined by relating the suction pressure to the corresponding temperature on a pressure-temperature chart. The temperature difference should be approximately 20° to 30°F (11.11 to 16.67°C); Table 9–1. If no pressure port is available, check the difference between the two thermometers. The difference in temperature should be approximately 20 to 30°F (11.11 to 16.67°C). When the unit is operating at a normal operating condition, a superheat of 20 to 30°F (−6.6 to −1.1°C) is satisfactory. A

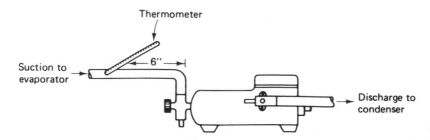

*Figure 9-18* Thermometer Strapped to Suction Line

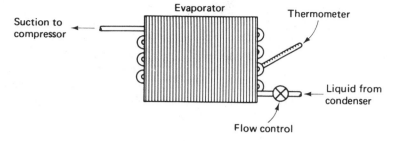

*Figure 9-19* Thermometer Strapped to Evaporating Coil
Return Bend

superheat lower than approximately 20°F (−6.6°C) indicates an overcharge of refrigerant. A superheat higher than approximately 30°F (−1.1°C) indicates an undercharge of refrigerant.

5. To add refrigerant, connect the low-side gauge on the gauge manifold to the pressure port on the system. If a low-side pressure port is not available, a saddle valve must be installed (Figure 9–20).

6. Connect the center charging hose on the gauge manifold to the valve on a cylinder of the proper type of refrigerant. Do not tighten this connection (Figure 9–13).

7. Open the gauge manifold hand valve and allow the refrigerant to escape for a few seconds. Tighten the connection on the cylinder valve.

8. Open the refrigerant cylinder valve and charge refrigerant into the system.

9. Observe the superheat by the method used in step 4. When the proper superheat is reached, stop charging refrigerant into the system.

10. Allow the unit to continue running until the pressures and temperatures have stabilized. If more refrigerant is needed, repeat the charging process until the desired superheat is stabilized.

For the manufacturer's charging chart method, use the following procedure:

1. Start the unit and allow it to operate for 10 to 15 minutes.

2. Connect the gauge manifold to the system service valves (Figure 9–17).

3. Connect the center charging hose on the

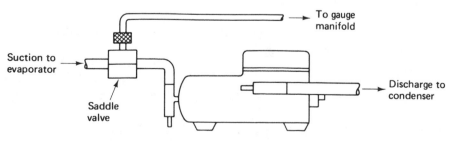

*Figure 9-20* Installation of Saddle Valve

gauge manifold to the valve on a cylinder of the proper type of refrigerant. Do not tighten this connection.

4. Open the hand valves on the gauge manifold and allow refrigerant to escape from the loose connection a few seconds. Tighten the connection on the cylinder valve.

5. Close the high-side hand valve on the gauge manifold.

6. Obtain a copy of the manufacturer's charging chart for that model unit (Figure 9–21).

7. Compare the pressures to those indicated on the chart.

8. If more refrigerant is needed, open the valve on the refrigerant cylinder and add refrigerant to the system.

9. Stop charging after a few minutes by closing the low-side hand valve on the gauge manifold and compare the pressure readings indicated on the chart. Repeat this procedure until the system pressures compare with those indicated on the chart.

*Procedure for Determining the Compressor Oil Level:* All refrigeration compressors require a specified amount of oil. This oil is required for lubrication of the moving parts and helps to make a refrigerant seal between the components. An abnormally low oil level would probably result in a loss of lubrication and compressor damage. An excess of lubricating oil will result in oil slugging, probable damage to the compressor valves, and lost system efficiency due to oil logging of the evaporator.

For the sight-glass method, use the following procedure:

1. Start the unit and allow it to operate for 10 to 15 minutes.

2. Check the oil level in the sight glass (Figure 9–22). A flashlight may be needed to accurately determine the level. The oil level should be at or slightly above the center of the sight glass while the unit is operating. If it is less than this, oil should be added. If more, the excess should be removed.

On sealed systems that do not have an oil sight glass, determining the amount of oil in the compressor is difficult. When a leak occurs and the amount of oil lost is small and can be reasonably calculated, add that amount to the system. If a large amount of oil has been lost, the compressor must be removed, the oil drained, and the correct amount added to the compressor before placing it back in operation.

*Procedure for Adding Oil to a Compressor:* Adding oil to a compressor is often required, and the service technician should be familiar with the three different procedures used, which depend on the type of system encountered and the type of tools at hand. These procedures are (1) the open-system method, (2) the closed-system method, and (3) the oil-pump method.

For the open-system method, use the following procedure:

1. Connect the gauge manifold to the system service valves (Figure 9–9).

2. Close the gauge manifold hand valves and open the compressor service valves.

3. Start the unit.

4. Front-seat the compressor suction service valve and run the unit until the low-side pressure is reduced to 1 or 2 psig (6.895 or 13.790 kPa). The low-pressure switch may need to be bypassed.

5. Stop the compressor.

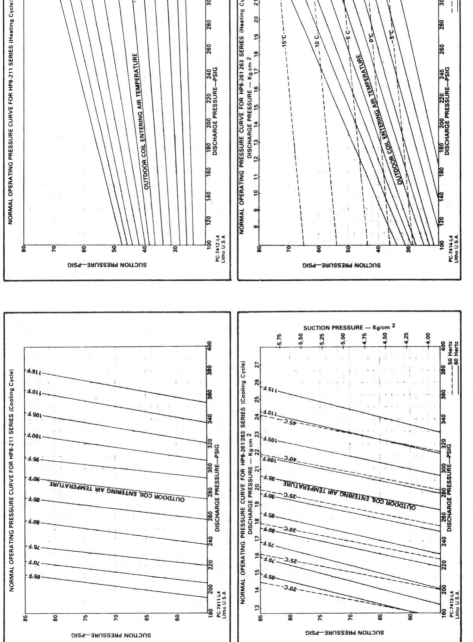

*Figure 9-21* Typical Manufacturer's Charging Chart (Courtesy of Lennox Industries, Inc.)

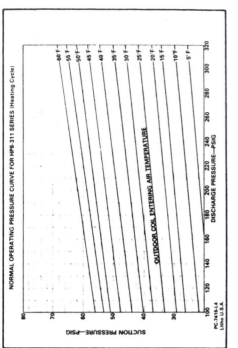

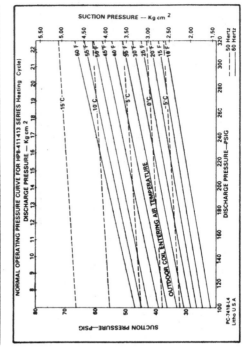

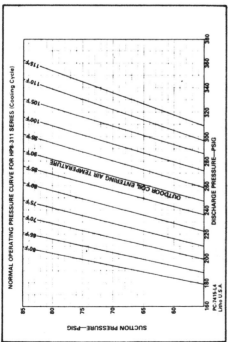

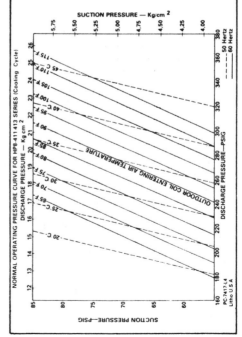

*Figure 9-21* (cont.)

290

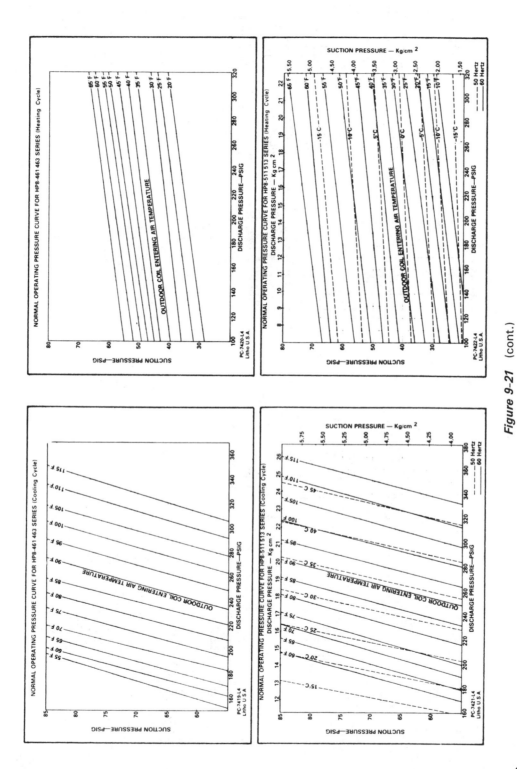

*Figure 9-21* (cont.)

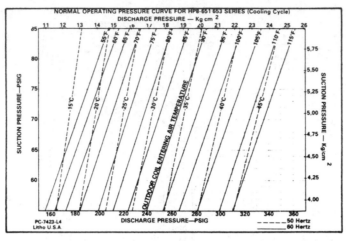

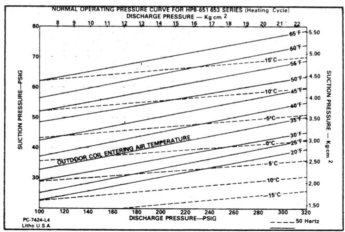

Figure 9-21 (cont.)

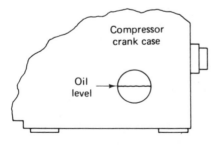

Figure 9-22 Checking Compressor Oil Level

6. Front-seat the compressor discharge service valve.

7. Open the low-side hand valve on the gauge manifold and relieve the slight pressure remaining in the system.

8. Remove the oil fill plug and pour the oil into the compressor crankcase until the proper level is reached. Use precaution to prevent contamination of the oil (Figure 9–23).

9. Close the low-side hand valve on the gauge manifold.

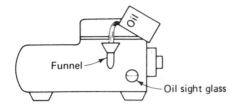

*Figure 9-23* Pouring Oil into a Compressor

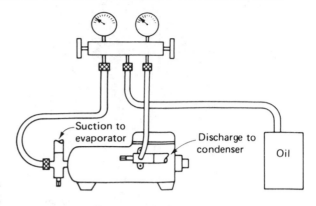

*Figure 9-24* Adding Oil to a Closed System

10. Slightly open the compressor suction service valve and allow a small amount of refrigerant to escape out of the oil fill hole.

11. Close off the compressor suction service valve.

12. Replace the oil fill plug and tighten.

13. Back-seat the compressor service valves.

14. Start the compressor and check the oil level in the compressor.

15. Remove the gauge manifold from the system.

For the closed-system method, use the following procedure:

1. Connect the gauge manifold to the system service valves (Figure 9-9).

2. Place the center charging line in a container of clean dry oil (Figure 9-24).

3. Open the system service valves until a pressure is indicated on the gauges.

4. Slightly open the low-side hand valve on the gauge manifold and purge a small amount of refrigerant through the lines and the oil.

5. Front-seat the system service valve.

6. Start the unit and draw a vacuum on the compressor crankcase.

7. Open the low-side hand valve on the gauge manifold and draw the oil into the compressor through the service valve. Be sure that the center line remains in the oil to prevent air entering the system.

8. When a sufficient amount of oil has been drawn into the compressor, close the hand valve on the gauge manifold.

9. Back-seat the system service valve, and place the unit in normal operation.

For the oil-pump method, use the following procedure:

1. Connect the low-side gauge of the gauge manifold to the system suction service valve.

2. Close the low-side hand valve on the gauge manifold.

3. Crack the system service valve until a pressure is indicated on the gauge.

4. Connect the center charging line to the oil pump. Do not tighten the connection.

5. Open the low-side gauge manifold hand valve and purge refrigerant through the loose connection for a few seconds; then tighten the connection.

6. Place the oil pump in a container of clean dry oil.

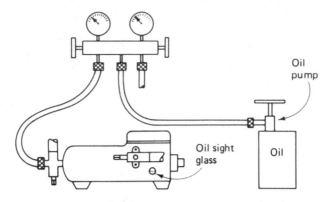

**Figure 9-25** Adding Oil to a System with an Oil Pump

7. Open the gauge manifold hand valve completely.

8. Move the system service valve to the midway position.

9. Pump oil into the system until the proper level is reached (Figure 9–25).

10. Back-seat the system service valve.

11. Close the low-side hand valve on the gauge manifold.

12. Remove the gauge manifold and place the system in normal operation.

*Procedure for Loading a Charging Cylinder:*
Charging cylinders are ideal for charging systems that use less than 5 lb (0.1417 kg) of refrigerant. Charging cylinders are calibrated in ounces for each type of refrigerant. Therefore, it is possible to charge the exact amount of refrigerant into the system. The service technician should be familiar with the steps involved in loading a charging cylinder.

To load a charging cylinder, use the following procedure:

1. Connect the low-pressure gauge of the gauge manifold to the charging cylinder (Figure 9–26).

2. Open the hand valve on the charging cylinder.

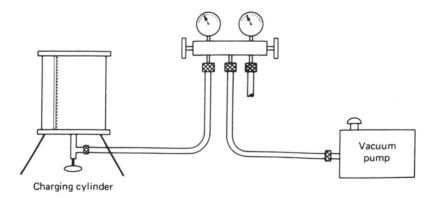

**Figure 9-26** Connections for Evacuating a Charging Cylinder

3. Open the low-side hand valve on the gauge manifold and purge all pressure from the charging cylinder.

4. Connect the center charging line on the gauge manifold to the vacuum pump.

5. Start the vacuum pump and draw as deep a vacuum as possible with the pump.

6. Close the low-side hand valve on the gauge manifold.

7. Remove the center charging line from the vacuum pump and connect it to the valve on a cylinder of the proper type of refrigerant. If there is a liquid valve, connect the line to it (Figure 9–27).

8. Open the refrigerant cylinder valve and loosen the center charging line connection on the gauge manifold. Allow the refrigerant to escape for a few seconds; then retighten the connection.

9. Invert the refrigerant cylinder so that liquid refrigerant will enter the charging line. If the line is connected to a liquid valve, the cylinder does not need to be inverted.

10. Open the low-side hand valve on the gauge manifold and allow liquid refrigerant to be drawn into the charging cylinder.

11. When the desired amount of refrigerant is drawn into the charging cylinder, close the refrigerant cylinder valve. If the refrigerant stops flowing before the desired amount is in the charging cylinder, the vent valve on top of the charging cylinder may be opened to permit the escape of vapor, thus allowing more liquid to enter (Figure 9–27).

12. Close the charging cylinder hand valve and disconnect the lines from the charging cylinder. Be sure that any liquid refrigerant in the lines does not come in contact with the skin or eyes.

### Procedure for Checking Compressor Electrical Systems

*Potential starting relay system with two-terminal overload.* Potential (voltage) starting relays are nonpositional. The contacts are normally closed. These relays are generally used on single-phase compressors of $\frac{1}{2}$ hp and larger.

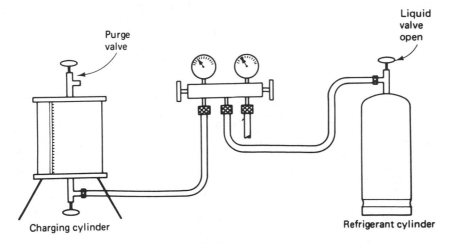

*Figure 9-27* Connections for Loading a Charging Cylinder

Use the following procedure to check this type of system. Refer to Figure 9–28.

1. Check to make certain that the proper voltage is being supplied to the unit. If proper voltage is supplied, continue to make the following checks. If the wrong voltage is being supplied, correct the problem with the power supply.

2. Disconnect the electrical power from the unit.

3. If a fan motor is used, disconnect one of the leads.

4. Make the following tests with an ohmmeter.

5. Check the continuity between points 1 and 2. Refer to Figure 9–28. If no continuity is found, the control con-

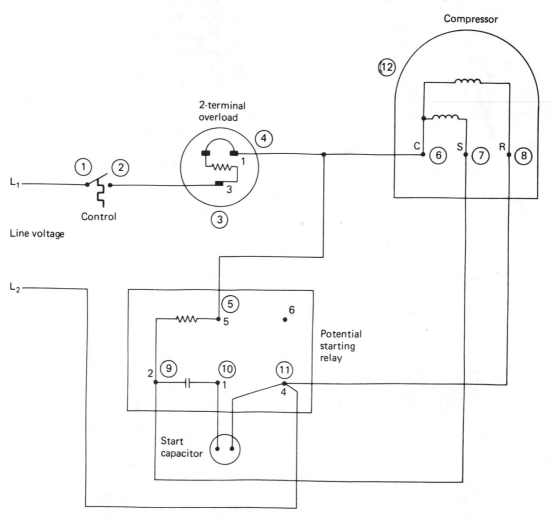

*Figure 9-28* Potential Starting Relay System with Two-terminal Overload

tacts are open. Close the contacts by setting the control to demand operation. If continuity is found, continue to make the following checks. If no continuity is found with the control on demand, replace the control.

6. Check the continuity between points 3 and 4. If continuity is found, continue to make the following checks. If no continuity is found, wait about 10 minutes and check the continuity again. The overload may be tripped. If there is no continuity now, replace the overload. Be sure to use an exact replacement.

7. Check the continuity between points 4 and 5. If continuity is found, continue to make the following checks. If no continuity is found, repair broken wire or loose connection.

8. Check the continuity between points 4 and 6. If continuity is found, continue to make the following checks. If no continuity is found, repair broken wire or loose connection.

9. Remove the wiring from the start terminal (point 7). Check the continuity between points 6 and 7. Compare this reading to that given by the compressor manufacturer for that particular compressor. If proper continuity is found, continue to make the following checks. If no or improper continuity is found, replace the compressor. The start winding is defective.

10. Remove the wiring from the run terminal (point 8). Check the continuity between points 6 and 8. Compare this reading to that given by the compressor manufacturer for that particular compressor. If proper continuity is found, continue to make

the following checks. If no or improper continuity is found, replace the compressor. The run winding is defective.

11. Check the continuity between points 5 and 9. If continuity is found, continue to make the following checks. If no continuity is found, the start relay coil is open. Replace the relay. Be sure to use an exact replacement.

12. Check the continuity between points 9 and 10. If continuity is found, continue to make the following checks. If no continuity is found, the start relay contacts are defective. Replace the relay. Be sure to use an exact replacement.

13. Set the ohmmeter on $R \times 100,000$. Check the continuity between points 10 and 11. If needle deflection is found, continue to make the following checks. If no needle deflection is found, the capacitor is open. Replace the capacitor. Be sure to use a proper replacement.

14. Set the ohmmeter on $R \times 1$. Check the continuity between points 10 and 11. If no continuity is found, continue to make the following checks. If continuity is found, the capacitor is shorted. Replace the capacitor. Be sure to use a proper replacement.

15. Check the continuity between points 6 and 12. If no continuity is found, continue to make the following checks. If continuity is found, the compressor motor is shorted. Replace the compressor.

16. Check the continuity between point 9 and the wire removed from point 7 (start terminal). If continuity is found, replace the wiring on the terminal and continue to make the following

checks. If no continuity is found, repair the wiring or loose connection.

17. Check the continuity between point 11 and the wire removed from point 8 (run terminal). If continuity is found, replace the wiring on the terminal and continue to make the following checks. If no continuity is found, repair the wiring or loose connection.

18. If the preceding steps do not indicate the trouble and the compressor will operate for a short period of time, disconnect the wire from point 10. Touch the wire to the same terminal and turn on the electricity. When the compressor starts, remove the wire from the terminal. A slight spark may occur. If the compressor continues to operate, replace the start relay. The contacts are not opening. Be sure to use an exact replacement. If the compressor does not continue to run when the wire is removed, reconnect the wire and proceed to the next step.

19. Check the amperage draw through the wire to point 6 (common terminal) while the compressor is trying to run. Compare the amperage draw to the locked rotor (LR) amperage of the motor. If the LR amperage is very close to that listed by the manufacturer, replace the compressor. The compressor has mechanical problems.

*Potential starting relay system with three-terminal overload.* Potential (voltage) starting relays are nonpositional. The contacts are normally closed. The relays are generally used on single-phase compressors of $\frac{1}{2}$ hp and larger. Use the following procedure to check this type of system (Figure 9-29).

1. Check to make sure that the proper voltage is being supplied to the unit.

If the proper voltage is being supplied, continue to make the following checks. If the wrong voltage is being supplied, correct the problem with the power supply.

2. Disconnect the electrical power from the unit.

3. If a fan is used, disconnect one of the leads.

4. Make the following checks with an ohmmeter.

5. Check the continuity between points 1 and 2. Refer to Figure 9-29. If continuity is found, continue to make the following checks. No continuity indicates that control contacts are open. Close the contacts by setting the control to demand operation. If there is no continuity with the control on demand, replace the control.

6. Check the continuity between points 2 and 3. If continuity is found, continue to make the following checks. If no continuity is found, repair the broken wire or loose connection.

7. Check the continuity between points 2 and 4. If continuity is found, continue to make the following checks. If no continuity is found, repair the broken wire or loose connection.

8. Remove the wiring from point 5 (start terminal). Check the continuity between points 4 and 5. Compare this reading to that given by the compressor manufacturer for that particular compressor. If proper continuity is found, continue to make the following checks. If no continuity or improper continuity is found, replace the compressor. The start winding is defective.

9. Remove the wiring from point 6 (run terminal). Check the continuity be-

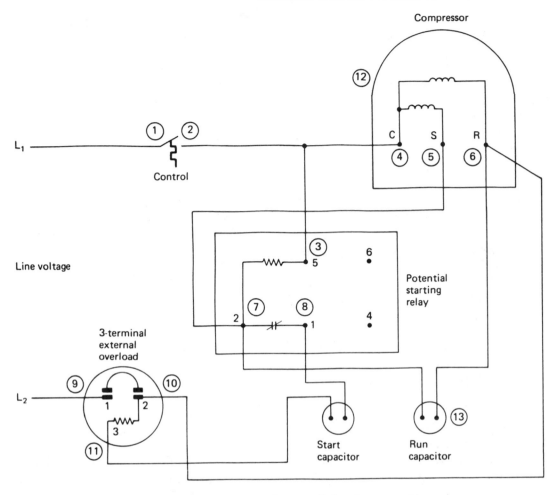

**Figure 9-29** Potential Starting Relay System with Three-Terminal Overload

tween points 4 and 6. Compare this reading to that given by the manufacturer for that particular compressor. If proper continuity is found, continue to make the following checks. If no or improper continuity is found, replace the compressor. The run winding is defective.

10. Check the continuity between points 3 and 7. If continuity is found, continue to make the following checks. If

no continuity is found, replace the relay. The coil is faulty. Be sure to use an exact replacement.

11. Check the continuity between points 7 and 8. If continuity is found, continue to make the following checks. If no continuity is found, replace the relay. The contacts are faulty. Be sure to use an exact replacement.

12. Set the ohmmeter on $R \times 100,000$. Check the continuity between points

8 and 11. If needle deflection is found, continue to make the following checks. If no needle deflection is found, the start capacitor is open. Replace the capacitor. Be sure to use an exact replacement.

13. Set the ohmmeter on R × 1. Check the continuity between points 9 and 11. If no continuity is found, continue to make the following checks. If continuity is found, the start capacitor is shorted. Replace the capacitor. Be sure to use a proper replacement.

14. Set the ohmmeter on R × 100,000. Check the continuity between point 7 and the wire removed from point 6 (run terminal). If needle deflection is found, continue to make the following checks. If no needle deflection is found, the run capacitor is open. Replace the capacitor. Be sure to use a proper replacement.

15. Set the ohmmeter on R × 1. Check the continuity between point 7 and the wire removed from point 6 (run terminal). If no continuity is found, continue to make the following checks. If continuity is found, the run capacitor is shorted. Replace the capacitor. Be sure to use a proper replacement.

16. Check the continuity between points 9 and 11. If continuity is found, continue to make the following checks. If no continuity is found, replace the overload. Be sure that the unit has been off for about 10 minutes to allow the overload to cool. Be sure to use an exact replacement.

17. Check the continuity between points 4 and 12. If no continuity is found, continue to make the following checks. If continuity is found, replace the compressor. The motor is grounded.

18. Check the continuity between point 7 and the wire removed from point 5 (start terminal). If continuity is found, continue to make the following checks. If no continuity is found, repair the wiring or loose connection. Reconnect the wiring to the terminal (point 5).

19. Check the continuity between point 13 and the wire removed from point 6 (run terminal). If continuity is found, continue to make the following checks. If no continuity is found, repair the wiring or loose connection. Reconnect this wire to the terminal (point 6).

20. Check the continuity between point 10 and the wire removed from point 6 (run terminal). If continuity is found, continue to make the following checks. If no continuity is found, repair the wiring or loose connection. Reconnect this wire to the terminal (point 6).

21. If the preceding steps do not indicate the trouble and the compressor will operate for a short period of time, disconnect the wire from point 8. Touch the wire to the same terminal and turn on the electricity. When the compressor starts, remove the wire from the terminal. A slight spark may occur. If the compressor continues to operate, replace the start relay. The contacts are not operating. Be sure to use an exact replacement. If the compressor does not continue to run when the wire is removed, reconnect the wire and proceed to the next step.

22. Check the amperage through the wire to point 4 (common terminal) while the compressor is trying to run. Compare the amperage draw to the locked rotor (LR) amperage of the motor. If

the LR amperage is very close to that listed by the manufacturer, replace the compressor. The compressor has mechanical problems.

*Current-type starting relay system with a two-terminal overload.* Current type starting relays are positional-type relays. They must be mounted in the position indicated by arrows on the relay. The contacts are normally open, and they are closed by an electromagnetic coil. They were opened by gravity, thus the purpose of mounting them in the proper position. These relays are generally used on fractional horse power motors up to about $\frac{1}{2}$ hp. Use the following procedure to check this type of system. Refer to Figure 9–30.

1. Check to make certain that the proper voltage is being supplied to the unit. If proper voltage is being supplied to the unit, continue to make the following checks. If the wrong voltage is being supplied, correct the problem with the power supply.

2. Disconnect the electrical power from the unit.

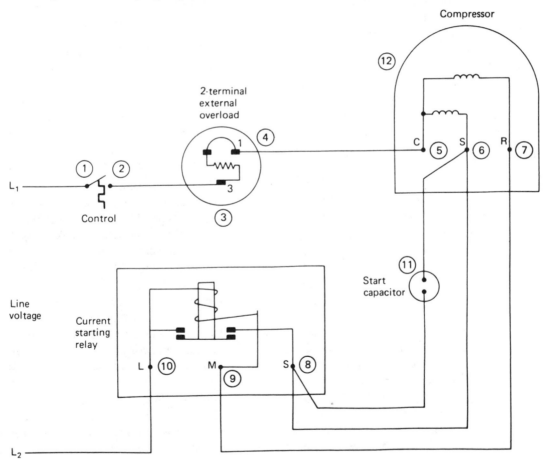

***Figure 9-30*** Current-type Starting Relay System with Two-terminal Overload

3. If a fan motor is being used, disconnect one of the leads.

4. Make the following checks with an ohmmeter.

5. Check the continuity between points 1 and 2. Refer to Figure 9–30. No continuity indicates that the control contacts are open. Close the contacts by setting the control to demand operation. If continuity is found, continue to make the following checks. If no continuity is found with the control on demand, replace the control.

6. Check the continuity between points 3 and 4. If continuity is found, continue to make the following checks. If no continuity is found, wait about 10 minutes and check the continuity again. The overload may be tripped. If there is no continuity now, replace the overload. Be sure to use an exact replacement.

7. Check the continuity between points 4 and 5. If continuity is found, continue to make the following checks. If no continuity is found, repair the wire or loose connections.

8. Remove the wire from point 6 (start terminal). Check the continuity between points 5 and 6. Compare this reading to that given by the manufacturer for that particular compressor. If proper continuity is found, continue to make the following checks. If no or improper continuity is found, replace the compressor. The start winding is defective.

9. Remove the wire from point 7 (run terminal). Check the continuity between points 5 and 7. Compare this reading to that given by the compressor manufacturer for that particular compressor. If the proper continuity is found,

continue to make the following checks. If no or improper continuity is found, replace the compressor. The run winding is defective.

10. Check the continuity between point 5 (common terminal) and point 12 (compressor housing). If no continuity is found, continue to make the following checks. If continuity is found, replace the compressor. The motor winding is grounded.

11. Check the continuity between points 10 and 8. If no continuity is found, continue to make the following checks. If continuity is found, replace the start relay. The contacts are closed and should be open. Be sure to use an exact replacement.

12. Check the continuity between points 10 and 9. If continuity is found, continue to make the following checks. If no continuity is found, replace the relay. The coil is open. Be sure to use an exact replacement.

13. Check the continuity between point 7 (run terminal) and point 9. If continuity is found, replace the wire on the terminal and continue to make the following checks. If no continuity is found, repair the wire or loose connections.

14. Check the continuity between point 6 (start terminal) and point 8. If continuity is found, replace the wire on the terminal and continue to make the following checks. If no continuity is found, repair the wire or loose connections.

15. If a starting capacitor is used, disconnect the wire from point 11. Set the ohmmeter on R × 100,000. Check the continuity between points 8 and 11. If needle deflection is found, continue to

make the following checks. If continuity is found, the capacitor is open. Replace the capacitor. Be sure to use a proper replacement.

16. Set the ohmmeter on R × 1. Check the continuity between points 8 and 11. If no continuity is found, continue to make the following checks. If continuity is found, the capacitor is shorted. Replace the capacitor. Be sure to use a proper replacement.

17. If the preceding steps do not indicate the trouble and the compressor hums but does not start, disconnect the wire from point 8 and touch it to point 10. Turn on the electricity. If the compressor starts, remove the wire from the terminal. A slight spark may occur. If the compressor continues to operate, replace the relay. Be sure to use an exact replacement. If the compressor does not start, reconnect the wire to point 8 and proceed to the next step.

18. Check the amperage draw in the wire to point 5 (common terminal) while the compressor is trying to run. Compare the amperage draw to the locked rotor (LR) amperage of the motor. If the LR amperage is very close to that listed by the manufacturer, replace the compressor. The compressor has mechanical problems.

*Permanent split capacitor (PSC) system with two-terminal overload.* PSC motors are low-starting-torque motors that have no starting relay or starting capacitor. This system is used on builders' models and room-type air-conditioning units. Use the following procedure to check this type of system. Refer to Figure 9–31.

1. Check to make certain that the proper voltage is being supplied to the unit. If proper voltage is being supplied, continue to make the following checks. If the wrong voltage is being supplied, correct the problem with the power supply.

2. Disconnect the electrical power from the unit.

3. If a fan motor is being used, disconnect one of the leads.

4. Make the following checks with an ohmmeter. Be sure to zero the ohmmeter.

5. Check the continuity between points 1 and 2. Refer to Figure 9–31. If continuity is found, continue to make the following checks. No continuity indicates that the control contacts are open. Close the contacts by setting the control to demand operation. If no continuity is found with the control on demand, replace the control.

6. Check the continuity between points 2 and 3. If continuity is found, continue to make the following checks. If no continuity is found, repair the wiring or loose connections.

7. Remove the wire from point 4 (start terminal). Check the continuity between points 3 and 4. Compare this reading to that given by the compressor manufacturer for that particular compressor. If proper continuity is found, continue to make the following checks. If no or improper continuity is found, replace the compressor. The start winding is defective.

8. Remove the wiring from point 5 (run terminal). Check the continuity between points 3 and 5. Compare this reading to that given by the compressor manufacturer for that particular compressor. If proper con-

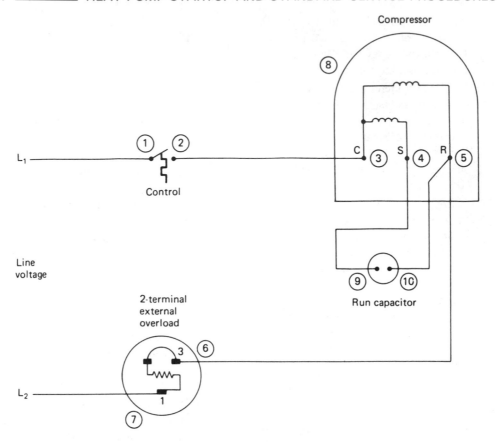

**Figure 9-31** Permanent Split Capacitor (PSC) System
with Two-terminal Overload

tinuity is found, continue to make the following checks. If no or improper continuity is found, replace the compressor. The run winding is defective.

9. Check the continuity between point 3 (common terminal) and point 8 (compressor housing). If no continuity is found, continue to make the following checks. If continuity is found, replace the compressor. The motor winding is grounded.

10. Set the ohmmeter on $R \times 100,000$. Check the continuity between points 9 and 10. If needle deflection is found, continue to make the following checks. If no needle deflection is

found, the capacitor is open. Replace the capacitor. Be sure to use a proper replacement.

11. Set the ohmmeter on $R \times 1$. Check the continuity between points 9 and 10. If no continuity is found, continue to make the following checks. If continuity is found, the capacitor is shorted. Replace the capacitor. Be sure to use a proper replacement.

12. Check the continuity between points 5 and 6. If continuity is found, continue to make the following checks. If no continuity is found, repair the wire or loose connections.

13. Check the continuity between points

6 and 7. If continuity is found, continue to make the following checks. If no continuity is found, wait about 10 minutes and check the continuity again. The overload may be tripped, If there is no continuity now, replace the overload. Be sure to use an exact replacement.

14. Reconnect all wiring. Connect a 12-in. piece of electrical wire to each terminal on a 130-$\mu$F × 370-V ac start capacitor. Touch the loose ends of one wire to point 4 (start terminal) and the other wire to point 5 (run terminal). Turn on the electricity. If the compressor starts immediately, remove the two wires from points 4 and 5. If the compressor continues to run, install a hard-start kit (Figure 9–32). If the compressor does not start or stops running when the wires are removed, proceed to the next step.

15. Check the amperage draw in the wire to point 3 (common terminal) while

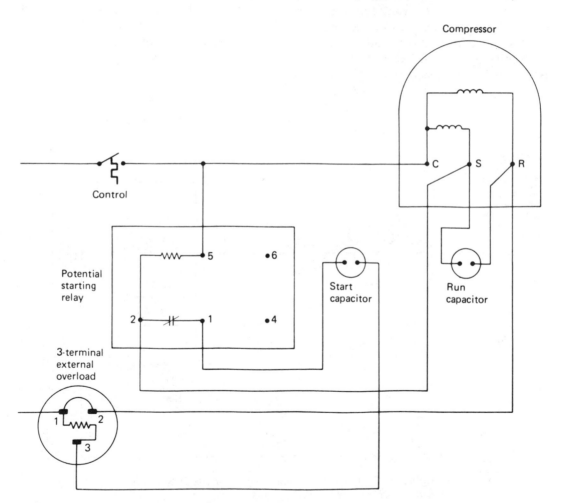

*Figure 9-32*  Connections Used for Installing a Hard-start Kit

the compressor is trying to run. Compare the amperage to the locked rotor (LR) amperage of the motor. If the LR amperage is very close to that listed by the manufacturer, replace the compressor. The compressor has mechanical problems.

*Permanent split capacitor (PSC) system with internal thermostat overload.* PSC motors are low-starting-torque motors that have no starting capacitor or starting relay. However, they do use a run capacitor. These systems have no external overload because a more sensitive internal overload is located inside the compressor housing. This system is used on builders' models and room air-condi-

tioning units. Use the following procedure to check this type of system. Refer to Figure 9-33.

1. Check to make certain that the proper voltage is being supplied to the unit. If proper voltage is being supplied, continue to make the following checks. If the wrong voltage is being supplied, correct the problem with the power supply.

2. Disconnect the electrical power from the unit.

3. If a fan motor is being used, disconnect one of the leads.

4. Make the following checks with an ohmmeter. Be sure to zero the ohmmeter.

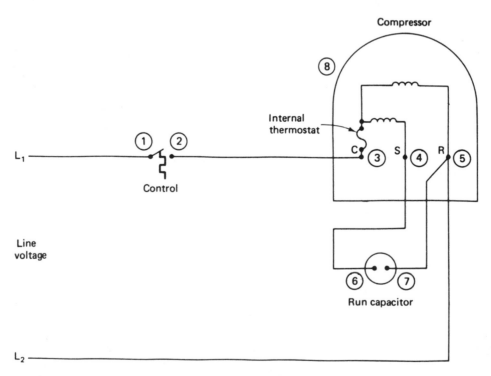

**Figure 9-33** Permanent Split Capacitor (PSC) System with Internal Thermostat

5. Check the continuity between points 1 and 2. Refer to Figure 9–33. If continuity is found, continue to make the following checks. No continuity indicates that the control contacts are open. Close the contacts by setting the control to demand operation. If no continuity is found with the control on demand, replace the control.

6. Check the continuity between points 2 and 3. If continuity is found, continue to make the following checks. If no continuity is found, repair the wiring or loose connections.

7. Remove the wire from point 4 (start terminal). Check the continuity between points 3 and 4. Compare this reading to that given by the compressor manufacturer for that particular compressor. If continuity is found, continue to make the following checks. If no or improper continuity is found, remove the wire from point 5 (run terminal). Check the continuity between points 4 and 5. If no continuity is found, it can be assumed that the start winding is open. Replace the compressor. Then, if no continuity is found, the internal overload may be tripped. Cool the compressor shell below 125 °F (51.78 °C) or until the hand may be held on the compressor without much discomfort. Again check the continuity between points 3 and 4. If no continuity is found, the internal overload is defective. Replace the compressor.

8. Remove the wire from point 5 (run terminal). Check the continuity between points 3 and 5. Compare this reading to that given by the compressor manufacturer for the particular compressor. If proper continuity is found, continue to make the following checks. If no or improper continuity is found, check the continuity between points 4 and 5. If continuity is found, it can be assumed that the internal overload is open. Cool the compressor shell below 125 °F (51.78 °C) or until the hand may be held on the compressor without much discomfort. Again check the continuity between points 3 and 5. If no continuity is found, the internal overload is defective. Replace the compressor.

9. Check the continuity between points 3 and 8. If no continuity is found, continue to make the following checks. If continuity is found, replace the compressor. The motor winding is grounded.

10. Set the ohmmeter on R × 100,000. Check the continuity between points 6 and 7. If needle deflection is found, continue to make the following checks. If no needle deflection is found, the capacitor is open. Replace the capacitor. Be sure to use a proper replacement.

11. Set the ohmmeter on R × 1. Check the continuity between points 6 and 7. If no continuity is found, continue to make the following checks. If continuity is found, the capacitor is shorted. Replace the capacitor. Be sure to use a proper replacement. If the capacitor is swollen, it should be replaced.

12. Reconnect all wiring. Connect a 12-in. piece of wire to each terminal on a 130 µF × 370 V ac start capacitor. Touch the loose end of one wire to point 4 (start terminal) and touch the loose end of the other wire to point 5 (run terminal). Turn on the electricity. If

the compressor starts immediately, remove the two wires from points 4 and 5. If the compressor continues to run, install a hard-start kit (Figure 9–32). If the compressor does not start or stops running when the wires are removed, proceed to the next step.

13. Check the amperage draw in the wire to point 3 (common terminal) while the compressor is trying to run. Compare the amperage draw to the locked rotor (LR) amperage of the motor. If the LR amperage is very close to that indicated by the manufacturer, replace the compressor. The compressor has mechanical problems.

*Permanent split capacitor (PSC) system with internal thermostat overload and external overload.* PSC motors are low-starting-torque motors that have no starting relay or starting capacitor. However, they do use a run capacitor. This particular compressor has both an internal thermostat and an external overload for added protection. The internal thermostat is temperature sensitive and interrupts the line voltage to the compressor motor common terminal. The external overload is both temperature sensitive and current sensitive. Its contacts interrupt the control circuit. Use the following procedure to check this type of system. Refer to Figure 9–34.

1. Check to make certain that the proper voltage is being supplied to the unit. If proper voltage is being supplied, continue to make the following checks. If the wrong voltage is being supplied, correct the problem with the power supply.

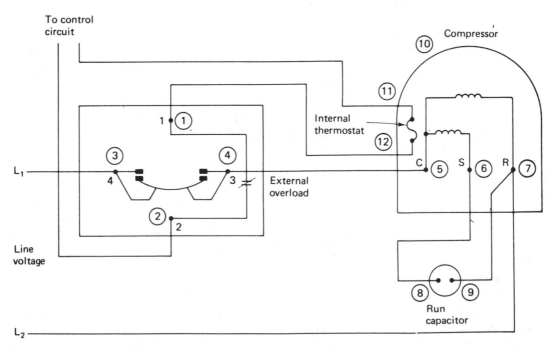

**Figure 9-34** Permanent Split Capacitor (PSC) System with Internal and External Overloads

2. Disconnect the electrical power from the unit.

3. If a fan motor is being used, disconnect one lead.

4. Make the following checks with an ohmmeter. Be sure to zero the ohmmeter.

5. Check the continuity between points 1 and 2. Refer to Figure 9–34. If continuity is found, continue to make the following checks. No continuity indicates that the overload is open. Wait about 10 minutes. Check the continuity again. If there is no continuity now, replace the external overload. Be sure to use an exact replacement.

6. Check the continuity between points 3 and 4. If continuity is found, continue to make the following checks. No continuity indicates that the external overload is defective. Replace the overload. Be sure to use an exact replacement.

7. Remove the wire from point 6 (start terminal). Check the continuity between pionts 5 and 6. Compare this reading to that given by the manufacturer for the particular compressor. If proper continuity is found, continue to make the following checks. If no or improper continuity is found, the start winding is defective. Replace the compressor.

8. Remove the wire from point 7 (run terminal). Check the continuity between points 5 and 7. Compare this reading to that given by the manufacturer for the particular compressor. If proper continuity is found, continue to make the following checks. If no or improper continuity is found, the run winding is defective. Replace the compressor.

9. Remove the wire from point 11. Check the continuity between points 11 and 12. If continuity is found, continue to make the following checks. If no or improper continuity is found, the internal thermostat is open. Cool the compressor below 125 °F (51.78 °C) or until the hand may be held on the compressor shell without much discomfort. Again check the continuity between points 11 and 12. If no continuity is found, replace the compressor. The internal thermostat is defective.

10. Check the continuity between points 5 and 10. If no continuity is found, continue to make the following checks. If continuity is found, replace the compressor. The motor winding is grounded.

11. Set the ohmmeter on R × 100,000. Check the continuity between points 8 and 9. If needle deflection is found, continue to make the following checks. If continuity is found, the capacitor is shorted. Replace the capacitor. Be sure to use a proper replacement.

12. Set the ohmmeter on R × 1. Check the continuity between points 8 and 9. If no continuity is found, continue to make the following checks. If continuity is found, the capacitor is shorted. Replace the capacitor. Be sure to use a proper replacement. If the capacitor is swollen, it should be replaced.

13. Reconnect all wiring. Connect a 12-in. piece of electrical wire to each terminal on a 130 $\mu$F × 370 V ac start capacitor. Touch the loose end of one wire to point 6 (start terminal) and the loose end of the other wire to point 7 (run

terminal). Turn on the electricity. If the compressor starts immediately, remove the two wires from points 6 and 7. If the compressor continues to run, install a hard-start kit (Figure 9–32). If the compressor does not start or stops running when the wires are removed, proceed to the next step.

14. Check the amperage draw in the wire to point 5 (common terminal) while the compressor is trying to run. Compare the amperage draw to the locked rotor (LR) amperage of the motor. If the LR amperage is very close to that indicated by the manufacturer, replace the compressor. The compressor has mechanical problems.

*Procedure for Checking Thermostatic Expansion Valves:* In checking complaints, if the expansion valve is suspected as the source of trouble, an orderly procedure for locating the exact difficulty and remedying it in the shortest time possible is desirable.

1. Check the suction pressure and discharge pressure. They act as the "pulse" of the refrigerating system and are the best guides to locating trouble.
2. Check the superheat. For this purpose, an accurate pocket thermometer may be taped to the suction line at the remote bulb location.

The following headings group the suggested possibilities of trouble indicated by the gauge and superheat readings.

*Low suction pressure, high superheat*

1. Expansion valve limiting flow:
   a. Inlet pressure too low from excessive vertical lift, too small liquid line, or excessively low condensing tempera-

ture. Resulting pressure difference across valve too small.
   b. Gas in liquid line due to pressure drop in the line or insufficient refrigerant charge. If there is no sight glass in the liquid line, a characteristic whistling noise may be observed at the expansion valve.
   c. Valve restricted by pressure drop through coil, requiring change to external equalizer (Figure 9–35).
   d. External equalizer line plugged or external equalizer connection capped without providing a new valve cage or body with internal equalizer.
   e. Moisture, wax, oil, or dirt plugging valve orifice. Ice formation or wax at valve seat may be indicated by sudden rise in suction pressure after shutdown and system has warmed up.
   f. Valve orifice too small.
   g. Superheat adjustment too high.
   h. Power assembly failure or partial loss of charge.
   i. Remote bulb of gas-charged thermostatic expansion valve has lost control owing to the remote bulb tubing or power head being colder than the remote bulb.
   j. Filter screen clogged.
   k. Wrong type of oil.
2. Restriction in system other than Thermostatic Valve. (Usually, but not necessarily, indicated by frost or lower than normal temperature at point of restriction.)
   a. Strainers clogged or too small.
   b. Solenoid valve failure or valve undersized.
   c. King valve at liquid receiver outlet too small or not fully opened.
   d. Plugged lines.

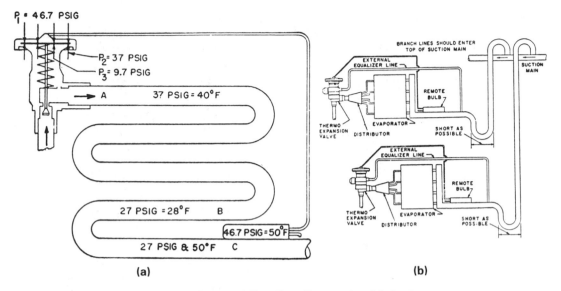

**Figure 9-35** External Equalizer Connection Methods: (a) Thermostatic Expansion Valve with Internal Equalizer on Evaporator with 10-psi Pressure Drop; (b) Schematic of Recommended Piping of Rising Suction Lines to a Common Suction Main

e. Hand valve stem failure or valve too small or not fully opened.

f. Liquid line too small.

g. Suction line too small.

h. Wrong type of oil in system, blocking liquid flow.

i. Discharge or suction service valve on compressor restricted or not fully opened.

*Low suction pressure, low superheat*

1. Poor distribution in evaporator causing liquid to short circuit through favored passes, throttling valve before all passes receive sufficient refrigerant.

2. Compressor oversized or running too fast owing to wrong pulley.

3. Uneven or inadequate coil loading, poor air distribution or brine flow.

4. Evaporator too small, sometimes indicated by excesive ice formation.

5. Evaporator oil logged.

*High suction pressure, high superheat*

1. Compressor undersized.

2. Evaporator too large.

3. Unbalanced system having an oversized evaporator, an undersized compressor, and a high load on the evaporator.

4. Compressor discharge valve leaking.

*High suction pressure, low superheat*

1. Compressor undersized.

2. Valve superheat setting too low.

3. Gas in liquid line with oversized thermo valve.

4. Compressor discharge valve leaking.

5. Pin and seat of expansion valve wire drawn, eroded, or held open by foreign material, which results in liquid flood-back.

6. Ruptured diaphragm or bellows in a constant pressure (automatic) expansion valve results in liquid flood-back.

7. External equalizer line plugged or external equalizer connection capped without providing a new valve cage or body with internal equalizer.

8. Moisture freezing valve in open position.

*Fluctuating suction pressure*

1. Poor control:
   a. Improper superheat adjustment.
   b. Trapped suction line (Figure 9–36).
   c. Improper remote bulb location or application (Figure 9–37).
   d. Flood-back of liquid refrigerant caused by poorly designed liquid distribution device or uneven coil circuit loading. Also improperly hung evaporator. Evaporator not plumb (See Figure 9–38).

   e. External equalizer tapped at common point on application with more than one valve on same evaporator.
   f. Faulty condensing water regulator causes change in pressure drop across valve.
   g. Evaporative condenser cycling, which causes radical change in pressure difference across expansion valve.
   h. Cycling of blowers or brine pumps.

*Fluctuating discharge pressure*

1. Faulty condensing water regulating valve.

2. Insufficient charge, usually accompanied by corresponding fluctuation in suction pressure.

3. Cycling of evaporative condenser.

4. Inadequate and fluctuating supply of cooling water to condenser.

*High discharge pressure*

1. Insufficient cooling water (inadequate supply or faulty water valve).

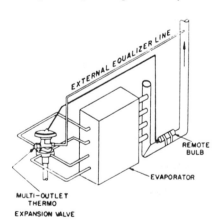

**INCORRECT**
Remote Bulb Location shown trapped.

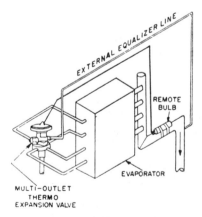

**CORRECT**
Remote Bulb Location shown free draining.

**Figure 9-36** Remote Bulb Installed on a Trapped Suction Line

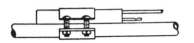

**EXTERNAL BULB ON SMALL SUCTION LINE**

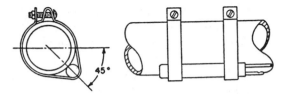

**EXTERNAL BULB ON LARGE SUCTION LINE**

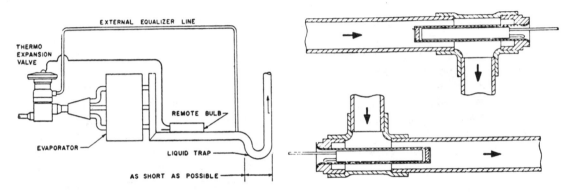

*Figure 9-37* Recommended Remote Bulb Location and Schematic Piping for Rising Suction Line

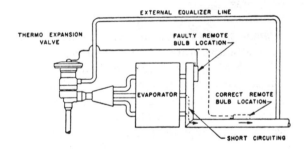

*Figure 9-38* Correct Remote Bulb Location on "Short Circuiting" Evaporator to Prevent "Flood-back"

2. Condenser or liquid receiver too small.
3. Cooling water above design temperature.
4. Air or noncondensable gases in condenser.

5. Overcharge of refrigerant.
6. Condenser dirty.
7. Air-type condenser improperly located to dispel hot discharge air.

*Expansion valve freeze-ups.* Expansion valve trouble may occur from the formation of ice crystals or the separation of wax out of the oil at the pin and seat, which will cause the refrigerant flow to be restricted or stopped entirely. Waxing generally occurs only at extreme low temperature; however, ice formation due to moisture in the system can occur whenever the evaporator is operating below freezing. Moisture is generally indicated by the starving of the evaporator. However, it may cause a valve to freeze open and cause floodback. When the compressor stops and the expansion valve and the system are allowed to warm up, the system will operate satisfactorily until the ice crystals form again. Another indication of moisture in the expansion valve can be noted if tapping the valve body will cause it to feed again for a short time. Under no circumstances should a torch be used on the valve body to melt ice accumulation at the pin and seat. If the ice must be melted to put the system in operation temporarily, apply hot rags. This is equally effective and will not damage the valve.

To remedy this, install a dryer of suitable size and operate the system above freezing for a few days to allow the moisture to be trapped in the dryer rather than to freeze at some location in the evaporator. If moisture persists, the refrigerant charge and oil should be dumped and the system dried by means of a vacuum and heat or by blowing dry nitrogen gas through the system.

Moisture can be admitted to the system as follows:

1. At the time of the original installation from moist air in the piping.
2. At the time of charging by not blowing the air and moisture out of the charging hose before tightening the fitting at the connection.
3. From improperly refilled refrigerant drums.
4. By allowing air to enter the system at the time of adding oil.
5. From a leaking or defective shaft seal.
6. By opening the system when in a vacuum.

In a methyl chloride system, regardless of the operating range, the presence of moisture creates an acid condition that attacks the system, creates sludge, and causes copper to plate upon the steel parts of the compressor as well as on the pin and seat of the expansion valve.

# PROCEDURE FOR TORCH BRAZING _____

*Article by Anthony Donofrio, Product Coordinator, Englehard Industries Division of Englehard Minerals and Chemicals Corp., Warwick R.I. Reprinted with permission.*

This article describes procedures that are important for making sound brazements of tubing in the air conditioning and refrigeration industry, using the phos-copper and the silver brazing filler metals.

The importance of brazing cannot be overlooked and/or overemphasized since it is a major part of the air-conditioning and refrigeration industry. Brazing is the process used in joining the major components of a refrigeration system into a closed circiut. Since the closed circuit contains a refrigerant, every brazed joint must be leak-free. If not, the refrigerant will escape, creating a severe inconvenience to the customer as well as a costly repair.

The purpose of this article is to provide the basic information about the correct method for torch brazing.

*Definitions:* Brazing is the application of heat above a temperature of 800 °F and below the melting points of the base metals to produce coalescence or bonding by surface adhesion forces between the molten filler metals and the surfaces of the base metals. The filler metal is distributed through the joint by using capillary action.

Brazing is not to be confused with soldering, even though the procedures are very similar. Soldering is the term used for metal-joining processes at temperatures below 800 °F.

In order for bonding and distribution by capillary attraction to occur, the filler metal must be able to "wet" the base metals. Wetting is the phenomenon in which the forces of attraction between the molecules of the molten filler metal and the molecules of the base metals are greater than the inward forces of attraction existing between the molecules of the filler metal.

The degree of wetting is a function of the compositions of the base metal and the filler metal and of the temperature. Good wetting can only occur on perfectly clean and oxide-free surfaces.

*Filler Metals:* The quality and strength of a brazement are more a function of the physical parameters of the joint and the brazing procedures used than which filler metal is applied to the joint. These parameters determine the selection of the best-suited and the most easily handled filler metal for a particular joint.

The phos-copper brazing alloys were specifically developed for joining copper and copper alloys. They are used for brazing copper, brass, bronze, or combinations of these.

When brazing with either brass or bronze, a flux must be used to prevent the formation of an "oxide coat" over these base metals. This coat would prevent wetting and the flow of the filler metal. However, when brazing copper to copper joints, these alloys are self-fluxing.

Because of the phosphorus embrittlement of the joint, phos-copper filler metals should not be used on ferrous metals or base metals containing more than 10% nickel. These alloys are also not recommended for use on aluminum bronze.

Unlike the phos-copper alloys, the silver brazing alloys contain no phosphorous. These filler metals are used to braze all ferrous, copper, and copper-based metals with the exception of aluminum and magnesium. They require a flux with all applications.

Extreme care should be taken when using the low-temperature cadmium-bearing silver alloy because of poisonous cadmium fumes being emitted.

Most refrigeration repair work can be ac-complished with a couple of filler metals. "Silvaloy 15" (15% silver content) is a commonly used phos-copper filler metal, and "Silvaloy 45" w/cd (45% silver, cadmium-bearing) is a commonly used silver brazing filler metal.

*Torch Brazing Procedures:*

*Copper-to-copper tube joints using the phos-copper brazing alloys*

I. Correct oxygen and fuel gas mixture [Figure 9-39]

*Excessive gas mixture*—a reducing type of flame denotes excessive fuel gas; a greater amount of fuel gas than oxygen. A slightly reducing flame heats and cleans the metal surface for quicker and better brazing [Figure 9-40].

*Balanced gas mixture*—the gas mixture contains an equal amount of oxygen and fuel gas. It produces a flame that heats the metal, but has no other effect [Figure 9-41].

*Excessive oxygen mixture*—the gas mixture contains an excessive amount of oxygen. It pro-

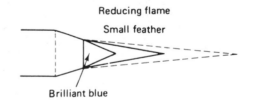

*Figure 9-39* Best Type of Flame for Brazing

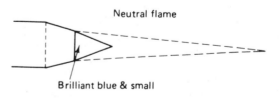

*Figure 9-40* Feather or Reducing Flame Disappeared

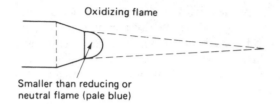

Oxidizing flame

Smaller than reducing or
neutral flame (pale blue)

**Figure 9-41** Worst Type of Brazing Flame

duces a flame that oxidizes the metal surface. A black oxide scale will form on the metal surface.

II. Cleanliness

General cleanliness is of prime importance to reliable brazing. All metal surfaces to be brazed must be cleaned of all dirt and foreign matter. When repair work is to be done, the metal surfaces to be joined must be either wire brushed and/or cleaned with sandpaper. A concerted effort must be made to keep oil, paint, dirt, grease, and aluminum off the surface of metals to be joined for these contaminants will

1. Keep brazing filler metal from flowing into the joint.
2. Prevent brazing filler metal from wetting or bonding to the metal surfaces.

III. Correct insertion and clearance of parts [Figure 9-42]

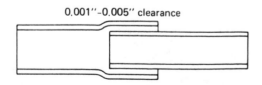

0.001"-0.005" clearance

**Figure 9-42** Minimum Insertion Distance is Equal to Diameter of Inner Tube; there Should be 0.001 to 0.005 in. of Clearance between the Walls of the Inner and Outer Tubes

IV. Correct amount and application of heat

Heat evenly over the entire joint circumference, and joint length. Use enough heat to get entire joint hot enough to melt the brazing filler metal without heating the filler metal directly with the flame [Figure 9-43]. Do not overheat the joint to the point that the metal to be joined begins to melt [Figure 9-44]. Always try to use the correct size torch tip, and a slightly reducing flame. Overheating enhances the base metal-filler metal interactions (e.g., formation of chemical compounds). In the long run, these interactions are detrimental to the life of the joint (Figure 9-45).

Male tube at brazing temperature, and female tube too cool. Result—filler metal ran away from the joint toward the heat source.

Filler metal and torch were applied at the same time. As illustrated, total joint area was not heated properly, and the heat never reached

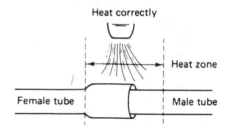

Heat correctly

Heat zone

Female tube    Male tube

**Figure 9-43** Heat Both Tubes at Joint and Distribute Heat Evenly

**Figure 9-44** Overheated Joint

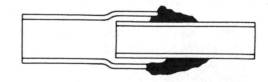

**Figure 9-45** Joint Too Cool for Brazing

male tube to allow capillary action of the filler metal [Figure 9–46].

Male tube inserted properly to diameter of female tube. Heat applied evenly to complete joint area. Filler metal was melted by heat of the tube [Figure 9–47].

V. Correct application of brazing rod [Figure 9–48]

Use the brazing rod as a temperature indicator. If the brazing rod melts when placed on the hot joint, the tubing is hot enough to begin brazing. For best results, preheat the rod slightly with the outer envelope of the flame, but the hot copper tubing (not the direct flame) should melt the brazing rod.

VI. Capillary attraction [Figure 9–49]

This is the phenomenon by which the brazing filler metal is drawn into the joint. It is caused by the attraction between the molecules of the brazing filler metal and the molecules of the metal surfaces to be joined. It can only work if: (1) the surface of the metal is clean, (2) the clearance between the metal surfaces is correct, (3) the metal at the joint area is hot enough to melt the brazing filler metal, and

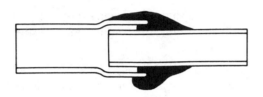

Figure 9-46 Filler Metal and Torch Applied at Same Time

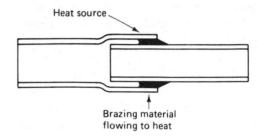

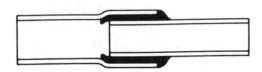

Figure 9-47 Heat Properly Applied to Good Fitting Joint

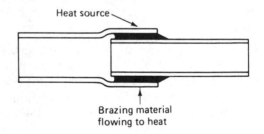

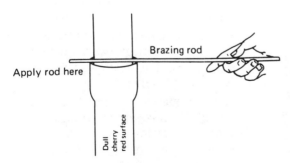

Figure 9-48 Correct Application of Brazing Rod

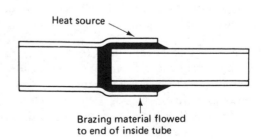

Figure 9-49 Capillary Attraction

(4) the brazing filler metal will flow toward the heat source as illustrated.

*Copper to brass tube joints using the phos-copper brazing alloys*

I. The above-mentioned procedures for copper to copper joints are strictly followed.

II. Before application of heat to the joint, a small amount of flux is applied to allow wetting of the brass by the filler metal.

III. Upon completion of the brazement, the flux residue is thoroughly cleaned off. Cleaning can be accomplished with hot water and mechanical brushing. Most fluxes are corrosive and must be completely removed from the joint.

*Steel to steel, copper, brass, or bronze using silver brazing alloys*

I. The above-mentioned procedures for joining copper to copper are strictly followed.

II. Before application of heat to the joint, flux is applied to allow wetting and flow to take place between the filler metal and the base metals.

III. Before application of the filler metal to the joint, a small amount of flux is applied to the rod. Rod is heated up and then dipped into the flux. This coats the filler metal with a thin layer of flux that provides for quick flow by preventing the formation of an oxide coat around the filler metal (zinc oxide).

IV Flux must be completely washed off upon completion of the joint.

*Fluxes:* A flux is analogous to a sponge absorbing water, only a flux is absorbing oxides. It can only absorb so much before it becomes useless as a flux.

When a flux becomes saturated with oxides, its viscosity increases. The ability of the flux to be displaced by the filler metal is therefore impeded. This leads to flux entrapment in the joint, which eventually causes corrosion and leaks.

One should use the least amount of flux that will get the job done. Then thoroughly clean the flux residue off after completing the brazement. Apply the flux along the surface of the joint and not into the joint itself. Allow the flux to flow into the joint ahead of the filler metal.

*Simplified Rules for Good Brazing*

1. Use a slightly reducing flame to get the maximum in heating and cleansing action.
2. Make sure metal surfaces are clean.
3. Examine parts for correct insertion and clearances.
4. Flux parts: (a) use minimum amount and (b) apply to outside of joint. No flux required for joining copper to copper with phos-copper filler metals.
5. Apply heat evenly around the joint to the proper brazing temperature.
6. Apply filler metal to joint. Make sure it is completely distributed throughout the joint by using the torch. Molten filler metals will follow the heat.
7. If flux is used, thoroughly clean off the residue.
8. A major part of good brazing is to complete the brazement as quickly as possible. Keep the heating cycle short. Avoid overheating.
9. Have adequate ventilation. Brazing may result in hazardous fumes from cadmium-bearing filler metals and fluoride fumes from fluxes.

# 10

# Representative Wiring
# Diagrams

## *General Electric*

### DEFROST CONTROL SYSTEMS

At present, the RANCO time temperature defrost method is being utilized on General Electric Weathertron® units. This method provides the dependability and simplification desirable for simplified service procedures. Formerly, the Dwyer/Klixon and Robertshaw methods have been extensively used for frost control. All three methods are reliable and when properly adjusted are relatively trouble free. All three defrost methods are covered in this booklet.

Before attempting adjustment of any system, four factors must be checked:

(1.) The refrigerant system must be properly charged.

(2.) Proper indoor airflow must be verified.

(3.) The outdoor coil must be clean with no lint, dirt, etc. creating abnormal restrictions.

(4.) Verify that the outdoor fan motor is wired to the proper voltage tap.

Several methods for verifying both airflow and system refrigerant charge are available. Detailed methods for charging and checking airflow are described in detail in Heat Pump Tune Up Publication #22-8063-2. Detailed charging procedures are included with the equipment and are located on access panels.

## SECTION I
## DWYER DEFROST SYSTEMS

Two requirements must be met before a defrost can be initiated. First the saturated vapor temperature of the refrigerant entering the outdoor coil out of the expansion device must be 32°F. or less. Second, the outdoor coil must be restricted 70-90% due to frosting. Dirty outdoor coils will cause premature defrost, and will result in higher operating cost.

Figure 1 shows the vacuum operated switch in the normal heating position. The O.D. coil does not have sufficient frosting to cause the vacuum effect of the O.D. fan to move the diaphragm.

Figure 2 shows the defrost termination switch in the normal heating position. The contacts are open and will not close unless the outdoor coil becomes 32°F. or less.

Figure 3 The defrost termination is now in the closed position. This occurs only when the outdoor coil refrigerant temp is 32° or below. The termination switch is a thermally operated, bi-metal switch that closes on temperature drop, (32°F.), and opens on temperature rise (55°F.).

Figure 4 The vacuum operated switch is now shown in the closed position. This occurs when the outdoor coil has frosted sufficiently causing the vacuum effect of the O.D. fan to move the diaphragm inward, closing the switch.

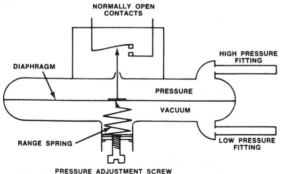

**FIGURE 1**

**COIL NOT FROSTED ENOUGH TO CREATE SUFFICIENT PRESSURE DROP ACROSS COIL TO INITIATE DEFROST.**

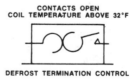

CONTACTS OPEN
COIL TEMPERATURE ABOVE 32°F

DEFROST TERMINATION CONTROL

**FIGURE 2**

**COIL FROSTED SUFFICIENTLY FOR THE VACUUM EFFECT OF THE O.D. FAN TO DRAW THE DIAPHRAGM INWARD, CLOSING THE SWITCH.**

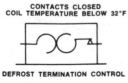

CONTACTS CLOSED
COIL TEMPERATURE BELOW 32°F

DEFROST TERMINATION CONTROL

**FIGURE 3**

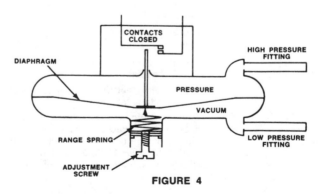

**FIGURE 4**

## TYPICAL WEATHERTON HEAT PUMP SCHEMATIC
## DWYER DEFROST SYSTEM

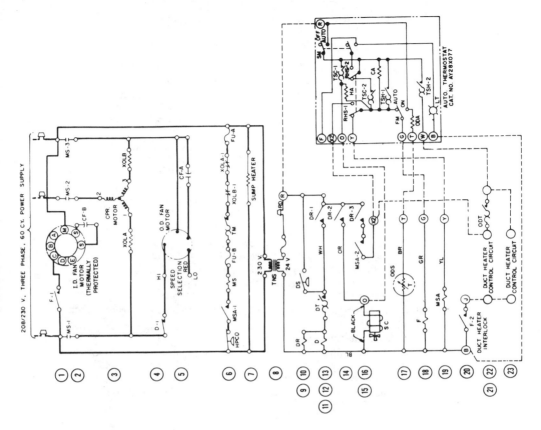

## DEFROST CYCLE

During the normal heating cycle the outdoor coil acts as an evaporator and under certain conditions (below 40°F.) will frost or ice up and require defrosting.

### Defrost Initiation

The first requirement for defrost initiation is that the defrost termination switch (12) close. This will occur when the outdoor coil is 32° or below and has the possibility of frost or ice build up on the coil. When the frost builds up on the coil to cause sufficient pressure differential, the defrost sensing switch (10) will close. When both switches close, a circuit is completed through DS (10) and DT (12) and energizes the D relay coil (11) and DR relay coil (9). D1 contacts (4) open, shutting off the O.D. fan motor. As the O.D. fan motor shuts off, the pressure differential is lost across the coil and the DS (10) returns to its normally open position. DR-1 contacts (13) lock in the D (11) and DR relay (9). DR-2 contacts (14) close completing the 24 volt circuit to the SC (15) switching the reversing valve (SC), DR-3 (16) completes a circuit through X2 to C and energizes the AH relay coil (21) provided the F relay coil (18) is energized to bring on electric heat to temper the I.D. air during defrost.

### Defrost Termination

When the outdoor coil liquid line temperature reaches 55°F. the defrost cycle will be completed. The DT switch (12) opens de-energizing the D relay (11), D1 contacts (4) close bringing on the O.D. fan. DR-2 (14) and DR-3 (16) contacts open and break the circuit to the reversing valve (SC) and supplementary heater control circuit.

321

## DEFROST ADJUSTMENT PROCEDURES

INITIATE PRESSURE ADJUSTMENT SCREW — Determines the amount of outdoor coil blockage required to cause the unit to go into defrost.

(A) Turning the screw clockwise will cause the unit to go into defrost with MORE coil blockage. If this screw is turned too far in, the coil may freeze completely before defrost, or may fail to go into defrost with 100% coil blockage.

(B) Turning the screw counterclockwise will cause the unit to go into defrost with LESS coil blockage. If this screw is backed out too far, the unit may short cycle defrost or go into defrost with little or no coil blockage.

If the outdoor temperature is above 40° the defrost termination switch will need to be temporarily bypassed.

1. Run the unit in the Heating cycle **with all panels and covers in place.**

2. Roll up overlapped sheets of newspapers or plastic material in a roll wide enough to cover the full width of the outdoor coil. (Fig. 1)

3. Place the roll at the base of the outdoor coil and unroll the papers slowly up the coil until the sensing switch trips. The switch must trip between 70 and 90% blockage.

4. If the switch does not trip within the limits, adjust the sensing switch as follows:

   Remove the side and control panel covers.

   If the switch tripped too soon turn the adjusting screw ½ turn **clockwise to raise** the trip point. If the switch tripped too late turn the adjusting screw ½ turn **counterclockwise to lower** the trip point. (Fig. 2)

   Replace the panels and repeat the previous check. Continue this procedure until the correct setting is reached.

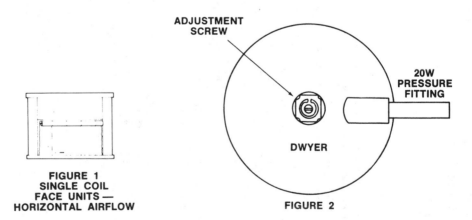

ADJUSTMENT
SCREW

20W
PRESSURE
FITTING

DWYER

**FIGURE 1
SINGLE COIL
FACE UNITS —
HORIZONTAL AIRFLOW**

**FIGURE 2**

## TROUBLE SHOOTING DWYER DEFROST SYSTEM

### Defrost Sensing Switch

General — The primary function of this component is to initiate a defrost cycle when the outdoor coil becomes restricted with frost or ice during heating operation. The switch, when operating properly, should initiate a defrost cycle when the outdoor coil is not more than 90% restricted or less than 70% restricted. The most common causes for the switch not to operate properly are:

Sensing tube(s) blocked by ice, dirt, insect, etc.

Loose or broken electrical connections.

Incorrect setting (wrong calibration).

Defective switch.

Leaking switch or diaphragm.

1.  To check sensing tube(s) — Disconnect the tube(s) and blow into it to see if tube(s) is clear.

2.  To check for broken or loose electrical connections — visually inspect the wires and connections in the switch circuit.

### Defrost Termination Switch

To check the operation of this switch:

IMPORTANT — Be sure the face of the switch has good contact with its mounting bracket. Ice or foreign matter between the switch and the bracket will cause improper operation.

Method A (OD Coil temperature below 32 F.). Operate the unit in the heating cycle, initiate defrost by blocking the outdoor coil. The unit should initiate defrost and terminate after the outdoor coil is clear but before the high pressure switch (HPCO) trips.

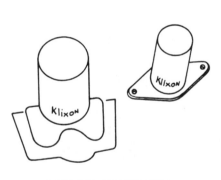

**DEFROST TERMINATION
SWITCH**

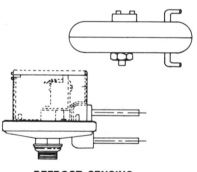

**DEFROST SENSING
SWITCH
(DWYER)**

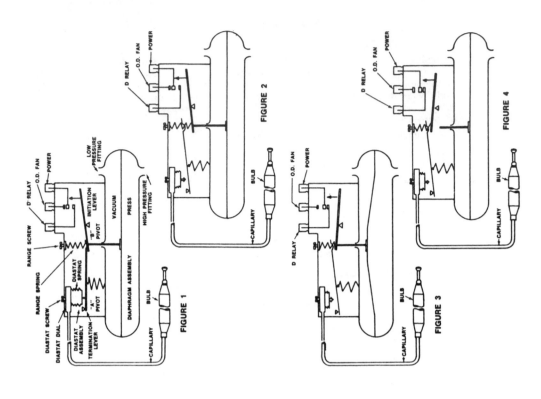

FIGURE 1

FIGURE 2

FIGURE 3

FIGURE 4

## SECTION II
## ROBERTSHAW DEFROST SYSTEMS

The Robertshaw combination air/temperature sensing method provided simplification of the electrical circuit by incorporating both the termination and initiation into one control.

## HOW THE SWITCH WORKS — MECHANICAL

Two requirements must be met before a defrost can be initiated. First, the saturated vapor temperature of the refrigerant entering the outdoor coil thru the expansion device must be 32°F. or less. Secondly, the outdoor coil must be restricted 70-90% due to frosting. Most areas have the greatest success at 75% blockage. Dirty outdoor coils will cause premature defrosts, even when system is properly adjusted. This condition will increase operating costs.

FIGURE 1 — Shows the switch in its normal heating operation position.

FIGURE 2 — Terminate lever is in the "up" position. This occurs only when the outdoor coil has 32°F., or less, refrigerant entering it. The terminate lever is moved up and down by contracting or expanding the alcohol charge stored in the sensing bulb, capillary, and Diastat assembly.

FIGURE 3 — Initiate lever in the down position. This will occur after the terminate lever has moved "up" and the outdoor coil has frosted sufficiently causing the vacuum effect at the outdoor fan hub to draw the diaphragm upwards.

FIGURE 4 — Defrost Termination — After the vacuum effect of the outdoor fan causes the diaphragm to initiate defrost it is turned "off" by the switch, this allows the diaphragm to return to its original position. The initiate lever is then held in the down position by the switch spring. When the liquid refrigerant temperature leaving the outdoor coil is increased to 55°-58° the pressure of the alcohol in the sensing bulb will re-set the terminate lever switching the switch back to Figure 1 position for normal heating operation.

324

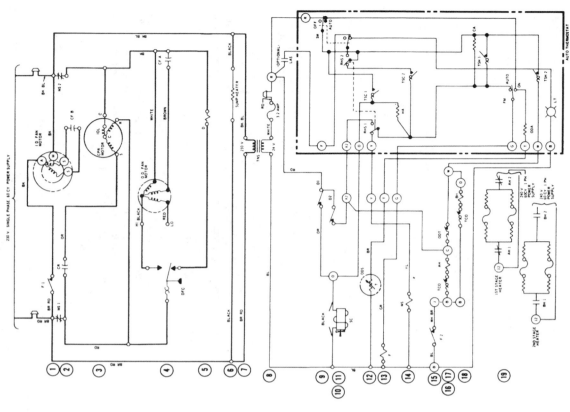

## DEFROST CYCLE

During the normal heating cycle operation the outdoor coil acts as an evaporator and under certain conditions (below 40°) will frost or ice up and require defrosting.

### Defrost Initiation:

When the defrost control, DFC (4), air switch senses a pressure difference of approximately .8″ water, and when the temperature of the liquid line supplying refrigerant to the outdoor coil is 32°F., or colder the unit will switch into defrost. Defrost control, DFC (4), switches energizing the D relay coil (5) while simultaneously turning off the OD fan. D-1 (9) & D-2 (10) contacts will close. Contacts D-1 (9) completes a 24 V. circuit to the SC (11) switching the S.O.V., D-2 (10) contacts complete a circuit through X2 to C and energizes the AH relay coil (16) provided the F relay coil (13) is energized to bring on some electric heat to temper the ID air during defrost.

### Defrost termination:

When the O.D. coil liquid line temperature reaches 55°F. the defrost cycle will be completed. The defrost control, DFC (4) will switch to normal operating position. D relay coil (5) is de-energized opening contacts D-1 (9) & D-2 (10) and breaks the 24 V. circuit to the S.O.V. and the supplementary heater control circuit.

## TYPICAL WEATHERTRON HEAT PUMP
## SCHEMATIC DIAGRAM
## ROBERTSHAW SYSTEM

325

## DEFROST SENSING SWITCH ADJUSTMENT FOR GENERAL ELECTRIC WEATHERTRON HEAT PUMPS USING ROBERTSHAW DEFROST CONTROL

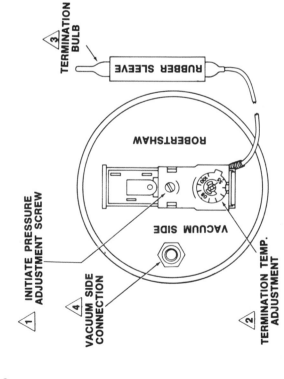

INITIATE PRESSURE ADJUSTMENT SCREW

VACUUM SIDE CONNECTION

TERMINATION TEMP. ADJUSTMENT

RUBBER SLEEVE

TERMINATION BULB

ROBERTSHAW

VACUUM SIDE

**ROBERTSHAW DEFROST CONTROL**

**FIGURE 1**

**1 INITIATE PRESSURE ADJUSTMENT SCREW** — Determines the amount of outdoor coil blockage required to cause the unit to go into defrost.

(A) Turning the screw clockwise will cause the unit to go into defrost with MORE coil blockage. If this screw is turned too far in, the coil may freeze completely before defrost, or may fail to go into defrost with 100% coil blockage.

(B) Turning the screw counterclockwise will cause the unit to go into defrost with LESS coil blockage. If this screw is backed out too far, the unit may short cycle defrost or go into defrost with little or no coil blockage, possibly stay in defrost without returning to normal heating function.

**2 DEFROST TERMINATION TEMPERATURE ADJUSTMENT** — This adjustment determines when the unit will complete the defrost cycle and return the unit to normal heating function.

(A) If the temperature setting is TOO HIGH, the unit may trip the high pressure cut out or compressor internal pressure relief valve before the defrost cycle is completed.

(B) If the temperature setting is TOO LOW, the outdoor coil may not defrost completely before the unit returns to the normal heating function.

(C) The temperature setting MUST be set at 55°-58° after defrost adjustments have been completed.

**3 TERMINATION BULB** — Measures liquid line temperature and determines when defrost is complete. Liquid line temperature at which the defrost will terminate depends upon the setting of the termination temperature dial.

(A) The rubber sleeve MUST be in place on the termination bulb for proper defrost operation.

(B) The termination bulb with rubber sleeve MUST be fully inserted in the liquid line receptacle for proper defrost operation.

(C) The capillary tube connecting the bulb to the control body should be routed to prevent rubbing against piping or components and be clear of the fan.

**4 VACUUM SIDE CONNECTION** — Connects to piping that senses pressure at the fan hub. Sensing piping must be clear of internal obstructions for proper defrost operation. Do not bend sensing piping from original factory routing since piping is routed to prevent accumulation of moisture inside the piping.

**5 PRESSURE SIDE CONNECTION** — (NOT SHOWN ON FIG. 1 DRAWING) — This connection is located on the opposite side of the control from the vacuum side connection. This connection may or may not be connected to piping, depending upon the particular model Weathertron to which the control is applied. If the unit is equipped with PRESSURE SIDE piping, the piping MUST be connected for proper defrost operation under normal operating conditions but MUST be disconnected while adjustments to the defrost control are being made.

### ADJUSTMENT PROCEDURE

1. Outdoor coil must be free of frost and dirt before adjusting defrost control. Disconnect O.D. fan and switch unit to cooling at thermostat to defrost O.D. coil. (Check liquid line temperature during this step). Do not exceed 60° or system will trip on high head pressure.

2. Disconnect pressure sensing piping from both sides of control. Blow through piping to be sure that it is not obstructed.

## ADJUSTMENT PROCEDURE, cont.

**CAUTION: Never apply pressure in excess of ½ P.S.I. to sensing piping while piping is connected to the control. Damage or rupture of the control diaphragm will result.**

3. Re-connect VACUUM SIDE piping. **DO NOT** re-connect PRESSURE SIDE piping until control adjustment is complete. Proper adjustment cannot be made with pressure side piping connected.

4. Operate the unit in the heating mode.

5. Turn INITIATE PRESSURE screw **CLOCKWISE ONE TURN. DO NOT** tighten screw, brass threads are easily damaged.

6. Turn TERMINATION TEMP. screw fully **COUNTERCLOCKWISE** against the stop. (FIGURE 2)

7. Block 75% of the outdoor coil surface. Top discharge airflow units **MUST** be blocked equally on all four sides. (FIGURE 3)

8. Permit unit to operate in heating mode for 5 minutes.

**TERMINATION TEMP.**

← INCR..

DIAL STOP

FIXED STOP & TEMP. INDEX

**SET TEMPERATURE DIAL FULLY COUNTERCLOCKWISE DURING CONTROL ADJUSTMENT.**

**FIGURE 2**

**◄SINGLE COIL FACE UNITS — HORIZONTAL AIRFLOW**

ROLLED NEWSPAPER

**◄FOUR COIL FACE UNITS — VERTICAL AIRFLOW**

**FIGURE 3**

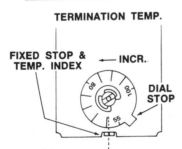

**TERMINATION TEMP.**

FIXED STOP & TEMP. INDEX

← INCR..

DIAL STOP

**TEMPERATURE DIAL MUST BE SET AT 55°-58° AFTER CONTROL ADJUSTMENT HAS BEEN COMPLETED**

**FIGURE 4**

9. **SLOWLY** turn INITIATE PRESSURE screw **COUNTERCLOCKWISE** until unit goes into defrost.

10. **IMMEDIATELY** return the TERMINATION TEMP. dial to 55°-58°. (FIGURE 4) (Unit will trip high pressure cut-out if permitted to operate in defrost for short period of time.)

**IMPORTANT: Vertical airflow models cannot be rechecked with panels and pressure sensing tube reconnected. Repeat steps 1 thru 10 to verify switch adjustment.**

11. DO NOT MAKE FURTHER ADJUSTMENTS TO DEFROST CONTROL.

12. **REPLACE** PRESSURE SIDE sensing piping.

13. Replace all panels and covers.

## SECTION III

## TIME TEMPERATURE DEFROST SYSTEM

### CONTROL DESCRIPTION (FIGURE 1)

The control is a combination time temperature device. The control requires no field adjustment and has a time override feature that will not permit a defrost cycle to last more than 10 minutes. A time selector permits field selection of 30, 45, or 90 minute defrost frequency. The factory setting is 90 minutes. A temperature sensing bulb will not permit defrost unless the refrigerant in the outdoor coil is below 26°F. The mechanism is driven by a 240 volt timer motor which is connected to the load side of the motor contactor so the timer operates only when the compressor operates. Some 440 volt equipment utilize a 24 volt timer motor.

### RANCO TYPE E-15 DEFROST CONTROL

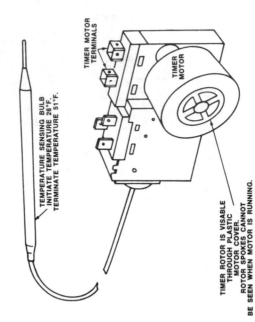

TEMPERATURE SENSING BULB
INITIATE TEMPERATURE 26°F.
TERMINATE TEMPERATURE 51°F.

TIMER MOTOR
TERMINALS

TIMER
MOTOR

TIMER ROTOR IS VISABLE
THROUGH PLASTIC
MOTOR COVER.
ROTOR SPOKES CANNOT
BE SEEN WHEN MOTOR IS RUNNING.

**FIGURE 1**

## RECHECK CONTROL OPERATION

IF PROPER OPERATION OF DEFROST CONTROL CANNOT BE ACCOMPLISHED — CHECK THE FOLLOWING LIST OF MUSTS BEFORE REPLACING THE CONTROL

1. Pressure sensing piping **MUST** be clear of obstructions.

2. PRESSURE SIDE sensing piping (if used) **MUST** be disconnected during adjustment procedure and **MUST** be re-connected after adjustment has been completed.

3. VACUUM SIDE piping **MUST** be connected both during adjustment and after adjustment has been completed.

4. TERMINATION TEMPERATURE dial **MUST** be set fully 55 degrees after adjustment has been completed. (Fig. 4)

5. Vertical airflow units **MUST** be blocked **EQUALLY** on all four sides during check-out and defrost control adjustments.

6. System refrigerant charge **MUST** be near the **CORRECT** system charge for proper defrost operation. (Check head and suction pressures during normal heating function and compare measured pressures to pressure charts attached to unit panel.

   A. If outdoor temperature is BELOW 65 degrees F., use proper heating charging method located on control panel of outdoor unit. (Performance Chart or Hot Gas line method)

   B. If outdoor temperature is ABOVE 65 degrees F., operate unit in cooling mode, use superheat charging chart for charging. (This method applies to capillary flow control systems only.)

   C. Use performance charts on systems using TEV in heating or cooling mode.

7. Replacement defrost controls require adjustment after installation in most cases. When a defrost control is replaced, the unit **MUST** be checked for proper defrost operation and adjustment when necessary.

## DEFROST INITIATION (FIGURE 2)

Defrost is initiated when the timer motor cam slot reaches the selected defrost time; 30, 45 or 90 minutes **and** the temperature sensing bulb is below 26°F. Both conditions must be met before defrost will initiate.

If the timer cam is in the defrost position and the temperature sensing bulb is above 26°F., the control will skip defrost and will not attempt another defrost until the selected time period has again elapsed.

When defrost is initiated the control switch stops the outdoor fan and energizes a defrost relay which reverses the refrigerant cycle and energizes the electric heater circuits to temper the indoor supply air during the defrost cycle.

## DEFROST TERMINATION

Defrost duration depends upon the amount of frost buildup on the outdoor coil. Normal duration is between 2 and 5 minutes; however, in extremely humid weather when the temperature is in the low 30 degree range, longer defrost cycles may be experienced.

Defrost cycles are terminated by the temperature sensing bulb. The bulb is sensing the liquid line temperature at the outdoor coil.

As liquid temperature increases, head pressure is also increasing, therefore the sensing bulb must end the defrost cycle before head pressure limiting devices cut out. This increase is rapid since the outdoor fan has been shut down by the defrost circuit.

The E-15 sensing bulb terminates defrost when it reaches 51°F.

## TIMER CAM OPERATION (FIGURE 3)

The time cam slot that permits the defrost switch to operate is cut for a 10 minute time period. The leading edge of the slot has a detent cut in the slot bottom. This detent permits the switch arm to drop lower than the slot. The switch arm must be in the detent section of the slot to initiate defrost. The switch arm stays in the detent for approximately 1 minute. If the temperature sensing bulb is **above** 26°F. during this minute, the control will skip to the next time cycle. If the temperature sensing bulb is **below** 26°F. defrost will initiate and continue until the bulb reaches 51°F. or 10 minutes elapse, whichever occurs first.

## RANCO E-15 TIME CAM FUNCTIONS

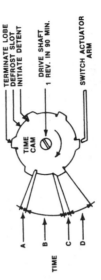

TERMINATE LOBE
DEFROST SLOT
INITIATE DETENT

DRIVE SHAFT
1 REV. IN 90 MIN.

TIME CAM

SWITCH ACTUATOR ARM

A
B
C
D

TIME

TIME A — 1 MINUTE — INITIATE DETENT — SWITCH ACTUATOR MUST BE IN
   DETENT SLOT TO INITIATE DEFROST.

TIME B — 9 MINUTES — DEFROST SLOT — DEFROST CAN CONTINUE BUT
   CANNOT INITIATE.

TIME C — 2 MINUTES — TERMINATING AREA — TIME CAM WILL TERMINATE
   DEFROST IS NOT PREVIOUSLY TERMINATED BY TEMPERATURE BULB.

TIME D — 3 MINUTES — TERMINATED — DEFROST CANNOT OCCUR REGARDLESS
   OF TEMPERATURE BULB ACTION.

FIGURE 3

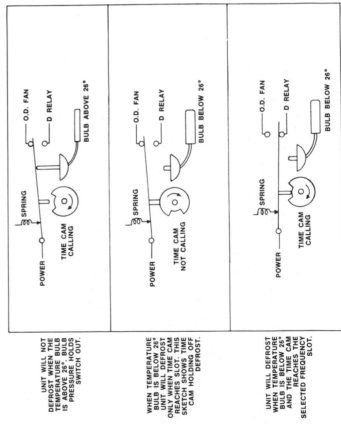

## SIMPLIFIED OPERATION
## RANCO TYPE E-15 DEFROST CONTROL

POWER — SPRING — TIME CAM CALLING — O.D. FAN — D RELAY — BULB ABOVE 26°

UNIT WILL NOT DEFROST WHEN THE TEMPERATURE BULB IS ABOVE 26°. BULB PRESSURE HOLDS SWITCH OUT.

POWER — SPRING — TIME CAM NOT CALLING — O.D. FAN — D RELAY — BULB BELOW 26°

WHEN TEMPERATURE BULB IS BELOW 26° UNIT WILL DEFROST ONLY WHEN TIME CAM REACHES SLOT. THIS SKETCH SHOWS TIME CAM HOLDING OFF DEFROST.

POWER — SPRING — TIME CAM CALLING — O.D. FAN — D RELAY — BULB BELOW 26°

UNIT WILL DEFROST WHEN TEMPERATURE BULB IS BELOW 26° AND THE TIME CAM REACHES THE SELECTED FREQUENCY SLOT.

FIGURE 2

## DEFROST FREQUENCY SELECTION (FIGURE 4)

Defrost frequency will be set for 90 minutes at the factory. **DO NOT** arbitrarily change the frequency. Some climatic areas may require a 45 minute cycle. These areas are where high humidity conditions exist at temperatures in the low 30° range for a substantial number of winter heating hours. **Always** select the longest possible time cycle for maximum operating efficiency.

To set the frequency, simply turn the knurled drive shaft knob until **either** end of the drive shaft screw slot aligns with the selected number stamped on the knob.

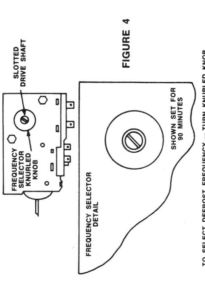

SLOTTED DRIVE SHAFT

FREQUENCY SELECTOR KNURLED KNOB

FREQUENCY SELECTOR DETAIL

SHOWN SET FOR 90 MINUTES

**FIGURE 4**

TO SELECT DEFROST FREQUENCY — TURN KNURLED KNOB UNTIL EITHER END OF THE SCREWDRIVER SLOT ALIGNS WITH THE FREQUENCY DESIRED

## MANUAL DEFROST OPERATION (FIGURE 4)

To manually operate the defrost, insert a screwdriver in the drive shaft slot and rotate the shaft clockwise. Rotate the shaft **slowly** since the entire drive mechanism is engaged. The screw slot is cut for clockwise rotation only. NEVER rotate the shaft counterclockwise with other tools or the mechanism will be destroyed.

As the shaft is rotated, several "clicks" will be heard depending upon the time selector setting. Continue rotation for at least a full revolution until the control initiates.

THE TEMPERATURE BULB **MUST** BE BELOW 26°F. FOR DEFROST INITIATION.

If a full revolution of the drive shaft is made and the unit will not manually defrost, disable the outdoor fan and operate the unit in the heating mode until the temperature bulb is below 26°F. An indication of bulb temperature can be determined by suction pressure. Suction pressure must be lower than 50 PSIG (26° saturation) before the bulb will be below 26°F. When the outdoor ambient temperature is below 40°F., the temperature bulb will normally be below 26°F.

## DEFROST CYCLE ELECTRICAL

### DEFROST INITIATION

When the temperature sensing bulb is 26°F. or colder and the timer motor (7) reaches the selected time slot (30, 45 or 90 minutes), defrost will be initiated. Defrost control DFT-1 (5) switches energizing the D relay coil (6) while simultaneously turning off the outdoor fan motor (4). Contacts D1 (9) and D2 (10) close. Contact D1 (9) completes a circuit to the SC (8) switching the reversing valve. Contact D2 (10) completes a circuit through X2 and energizes the AH (17) relay coil, provided the F relay coil (14) is energized, to bring on some electric heat to temper the I.D. air during defrost.

### DEFROST TERMINATION

When the O.D. coil liquid line temperature reaches 51°F. the defrost cycle will be completed. If control malfunction or system malfunction prevents the temperature sensing bulb from terminating, the time override cam will end the defrost cycle 10 minutes after initiation. The DFT-1 (5) will switch to normal heating position completing the circuit to the outdoor fan motor (4). Defrost relay (6) is de-energized, opening contacts D1 (9) and D2 (10) which breaks the 24 Volt circuit to the S.O.V. and supplementary heater control circuit.

### TROUBLESHOOTING THE TIMER

The rotor of the timer is visible through the plastic timer cover. If a malfunction occurs in the control, observe the timer rotor for rotation. If the rotor is not rotating, check power at timer terminal block. If power is present and the timer is not rotating, the entire control must be replaced. The motor and timer are not stocked as separate replacement parts.

## LOSS OF CHARGE IN THE TERMINATION BULB

### HEATING CYCLE

Regardless of outdoor coil temperature, defrost will be initiated when the time motor reaches the selected time slot. The defrost cycle can terminate only on the 10 minute override feature.

### COOLING CYCLE

When the cam slot reaches the selected time, the unit will attempt defrost. The outdoor fan motor is de-energized and the electric heat is energized. The unit will remain in this position until the 10 minute override returns the unit to the normal cooling cycle. During this 10 minute cycle, the compressor internal pressure relief valve will normally open causing the compressor to cycle on the internal overload.

## TYPICAL WEATHERTON HEAT PUMP SCHEMATIC
## RANCO TIME TEMPERATURE SYSTEM

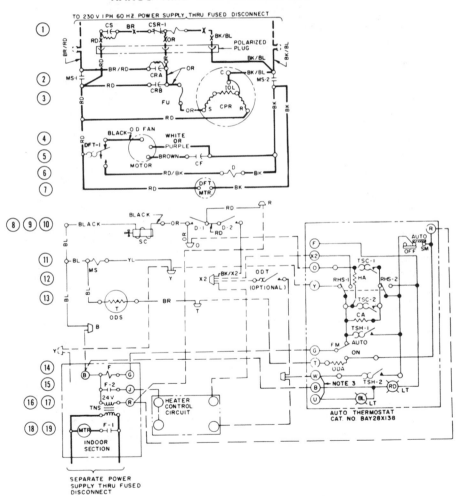

## TYPICAL WEATHERTRON® HEAT PUMP SCHEMATIC DIAGRAM

### COOLING CYCLES

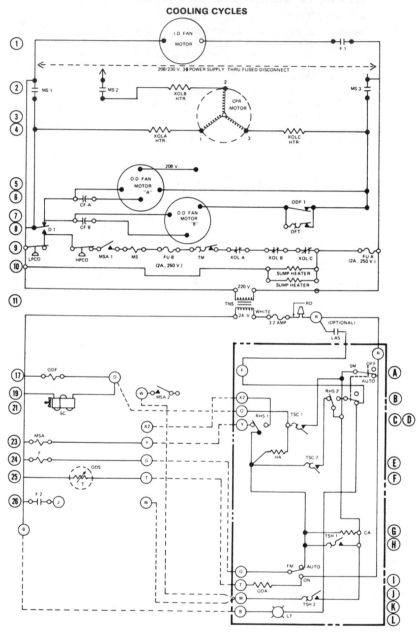

## TYPICAL WEATHERTRON® HEAT PUMP SCHEMATIC DIAGRAM

### HEATING CYCLES

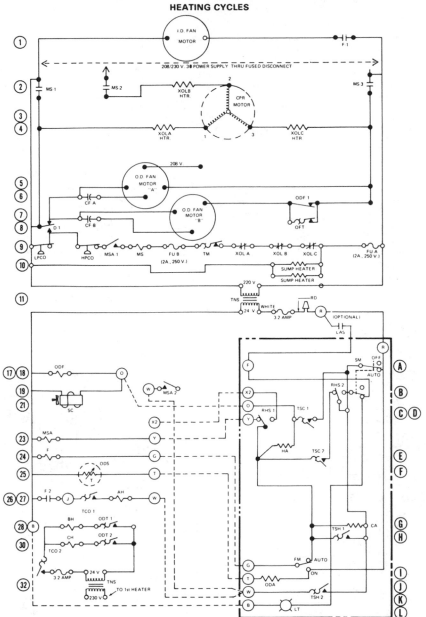

## TYPICAL WEATHERTRON® HEAT PUMP SCHEMATIC DIAGRAM

### DEFROST CYCLE

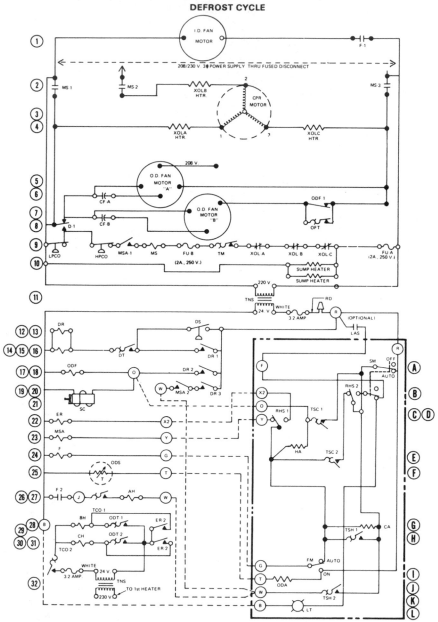

# TYPICAL WEATHERTRON® HEAT PUMP SCHEMATIC DIAGRAM

## ELECTRIC HEAT CYCLE

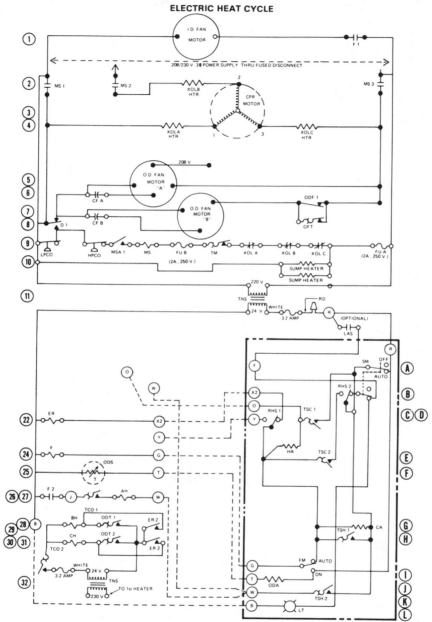

## TYPICAL WEATHERTRON® HEAT PUMP SCHEMATIC DIAGRAM
### COOLING — HEATING — DEFROST — ELECTRIC HEAT CYCLES

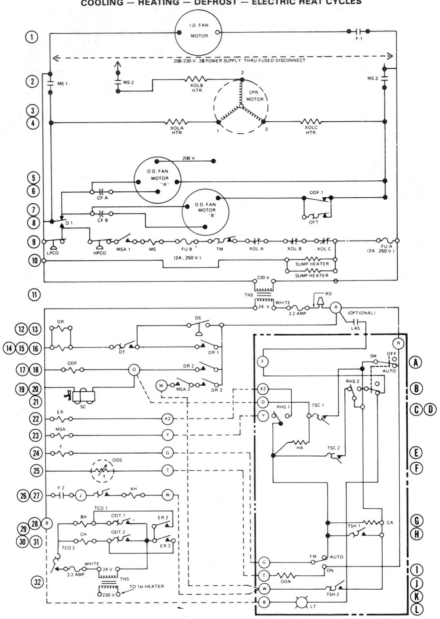

# WA, WC090 & 120 MODELS ELECTRICAL COMPONENTS AND CIRCUITS

## COOLING OPERATION
### ELECTRICAL COMPONENTS & SEQUENCE

**STEP 1: Thermostat system switch in "OFF" position. (Disconnect turned "ON").**

This position supplies line voltage to the sump heaters (10) and transformer (11) for control voltage. The ODS (25) — ODA (J) outdoor anticipator circuit is energized thru the thermostat at all times.

**STEP 2: First stage cooling — (system switch (A) switched to "Auto").**

Thermostat first stage calls. TSC-1 (D) closes completing the 24V. circuit thru O energizing the switch-over valve solenoid coil (21) and ODF (17) relay, nothing else occurs. (TSC-1's only function is to electrically make ready the switch-over valve operation and to open the normally closed ODF-1 (6) contacts allowing the OFT (7) to cycle the O.D. fan motor "B" as the outdoor temperature indicates).

**STEP 3: Second stage cooling — (After .7° to 1.5° Temperature Rise).**

Thermostat second stage indicates cooling is needed. TSC-2 (F) closes completing the 24V. circuits thru Y to the MSA (23) relay coil and thru G to the F (24) relay coil,

— simultaneously closing —

(1) MSA-1 (9) and MSA-2 (19) contacts. (MSA-1 (9) completes line voltage circuit to MS (9) relay.)
(2) MS-1, MS-2, & MS-3 (2) contacts.
(3) F-1 (1) & F-2 (26) contacts.

Compressor and O.D. fans (5 & 7) start. I.D. fan starts and safety interlock in Electric Heat Circuit closes. Normally TSC-1 remains energized during the cooling season and TSC-2 controls the system.

## HEATING OPERATION

**STEP 1: Thermostat system switch in "OFF" position. (Disconnect turned "ON").**

This position supplies voltage to the sump heaters (10) and transformer (11) for control voltage. The ODS (25) — ODA (J) outdoor anticipator circuit is energized thru the thermostat at all times.

**STEP 2: First stage heating — (system switch (A) switched to "Auto").**

Thermostat calls for heat. TSH-1 (H) closes completing the 24V. circuits thru Y to the MSA (23) relay and G to the F (24) relay,

— simultaneously closing —

(1) MSA-1 (9) and MSA-2 (19) contacts — MSA-1 (9) completes line voltage circuit to MS (9).
(2) MS-1, MS-2, and MS-3 (2) contacts.
(3) F-1 (1) and F-2 (26) contacts

Compressor and O.D. fans (5 & 7) start. I.D. fan starts and safety interlock in electric heat circuit closes.

The refrigerant system is now operating in the heating mode, note that the S.O.V. (21) is "de-energized". A circuit from TSH-1 (H) to HA (E) in series with SC (21) is also completed. The fixed heat anticipator (E) prevents SC (21) from energizing because HA = 3,000 OHMS and the SC = 5 OHMS. This circuit works jointly with the ODS-ODA heat anticipation circuit to minimize the swing range of the thermostat caused by extended run cycles. ODF (17) will not energize because of HA (E) ODF-1 (6) are normally closed contacts.

**STEP 3: Second stage heating — (After a .7° to 1.5° temperature drop).**

Thermostat indicates additional heating is required. TSH-2 (K) closes completing a 24V. circuit thru W to the AH (27) relay coil. AH heater supplies 230V. to supplementary heater control transformer (32). ODT-1 (28) and ODT-2 (30) provide control to BH (28) and CH (30) to stage heat as needed. TSH-2 will control the electric heat and TSH-1 will operate the compressor until the thermostat is satisfied.

## DEFROST CYCLE

During the normal heating cycle operation the outdoor coil functions as an evaporator and under certain conditions (below 40° F.) will frost or ice over and defrosting will be necessary. Two conditions are required to initiate defrost; (1) Refrigerant temperature entering the O.D. coil must be 32° or less. (2) 75% of the O.D. coil surface area covered with frost or ice.

**Defrost Initiation:**

When the defrost system requirements are met, the following electrical functions occur:

(1) DT (15) closes at 32°.
(2) DS (13) closes at 75% coil restriction.
(3) DR (12) and D (14) relays are energized.
(4) O.D. fan motors "A" and "B" are de-energized by D-1 (8) contacts.
(5) DR-1 (16) contacts lock-in D (14) and DR (12) relays thru the DT (15).
(6) DR-2 (18) contacts energize ODF (17 and SC (21) switching the system to cooling operation.
(7) DR-3 (20) contacts energizes the W circuit and brings on tempering heat to counteract the cooling effect of the indoor coil during defrost. Electric heat will stage as determined by ODT-1 (28) and ODT-2 (30).

System defrost may at times be interrupted by the thermostat. This is a normal function. When the thermostat is satisfied, TSH-1 (H) will open the control circuit to Y and de-energize the MSA (23), MSA-1 (9) and MSA-2 (19) will open to de-energize the MS (9) which controls the compressor, and open the control circuit to W external to the thermostat to turn-off the tempering heat. The next thermostat heat call will complete defrost as needed.

**Defrost Termination:**

When the O.D. coil liquid line temperature is increased to 55° F. the defrost cycle will be completed. The DT (15) will open de-energizing the D (14) and DR (12) relays returning the system to the normal heating mode operation.

## RESISTANCE HEAT OPERATION

In the event of a refrigerant system failure, or other mechanical failure, the Weathertron® thermostat provides an emergency electric heat switch which electrically isolates the Y circuit in the thermostat by mechanical means. This is done to prevent further damage to the refrigerant system components. To activate the emergency heat switch, remove the thermostat cover. The switch is located behind the fan/system black support bar, push the switch to the right. A signal light, located in the upper right hand corner of the thermostat will be energized, and will burn continuously with the switch in the "ON" position. The light is a reminder to the customer that the system is on electric heat operation only.

Resistance heat switch RHS-1 (C) and RHS-2 (B) switched. RHS-1 (C) contacts move from Y terminal to X2 terminal internally in the thermostat. This electrically isolates the MSA (23) relay coil and prevents the refrigerant circuit from operating on a heat call, and energizes the ER (22) relay which by-passes the ODT's in the heater circuits. RHS-2 (B) provides the control circuit thru TSH-2 (K) which energizes the heaters. The heating control point will be .7° to 1.5° F. below the setting indicated on the thermostat selector dial.

## WEATHERTRON®
## CONDENSED VERSION — ELECTRICAL SEQUENCE

**Thermostat set for cooling season — disconnect turned on.**

| | |
|---|---|
| (G ) | Cooling Anticipator energized. |
| (D ) | TSC-1 makes 1.5° before thermostat calls. |
| (21 ) | S.O.V. coil energized. |
| (17 ) | ODF coil energized |
| (6 ) | ODF-1 contact opens. |
| (F ) | TSC-2 makes when thermostat calls for cooling. |
| (C ) | Y circuit is completed. |
| (1 ) | G circuit is completed. |
| (23 ) | MSA coil energizes. |
| (24 ) | F coil energizes. |
| (9 ) | MSA-1 energizes MS coil which closes (2) MS-1, MS-2 and MS-3. |
| (1 ) | F-1 (1) and F-2 (26) close. |

### Thermostat Set for Heating Season

#### First Stage Heating

| | |
|---|---|
| (H ) | TSH-1 makes. |
| (23 ) | MSA coil energizes. |
| (24 ) | F coil energizes. |
| (9 ) | MSA-1 (9) energizes MS (9) closing MS-1, MS-2 and MS-3 (2). |
| (1 ) | F-1 contact closes. |
| (26 ) | F-2 contact closes. |

#### Second Stage Heating

| | |
|---|---|
| (K ) | TSH-2 closes. |
| (27 ) | AH coil makes. |
| (32 ) | Power supplied to heater control transformer ODT's stage heat. |

#### Defrost

| | |
|---|---|
| (15 ) | DT closes. |
| (13 ) | DS closes. |
| (12 ) | DR coil energizes. |
| (14 ) | D coil energizes. |
| (5/7) | OD fans stop. |
| (16 ) | DR-1 contacts close locking in circuit. |
| (18 ) | DR-2 contacts close. |
| (21 ) | S.O.V. coil energizes. |
| (20 ) | DR-3 contacts close. |
| (27 ) | AH coil energizes. |

#### Defrost Termination

| | |
|---|---|
| (15 ) | DT opens. |
| (14 ) | D relay de-energizes. |
| (5/7) | OD fan motors start. |
| (12 ) | DR relay de-energizes. |
| (16 ) | DR-1 contact opens. |
| (18 ) | DR-2 contact opens. |
| (21 ) | S.O.V. coil de-energized. |
| (20 ) | DR-3 contact opens. |
| (27 ) | AH relay de-energized. |

### Electric Heat
*Manually switch to Electric Heat Position*

| | |
|---|---|
| (C ) | RHS-1 isolates Y, makes X2. |
| (22 ) | ER relay by-passes ODT's. |
| (B ) | RHS-2 provide control thru TSH-2 (K) and makes circuit (L). |

## GENERAL ELECTRIC DEFROST SYSTEM
## COMPONENT DESCRIPTION

The G.E. Morrison defrost control system is a time/temperature control, consisting of two motor driven rotating timer cams and a remote mounted refrigerant temperature sensor.

The timer cams are driven by a 240-volt motor which is energized by the motor starter on a call for heating or cooling. Features of the control include: selection of defrost cycle frequency — 45 or 90 minutes, a 10-minute (approximate) time override to terminate defrost if not previously terminated by the temperature sensor, manual advance for ease of check-out and motion indicator for drive gear check-out.

## COMPONENT DESCRIPTION

The defrost relay is a 3-pole double throw type which utilizes one of the normally closed contacts to operate the O.D. fan motor. When the relay is energized, the O.D. fan motor circuit is opened and a holding circuit for the defrost relay is completed. The remaining contacts energize the S.O.V. solenoid and auxiliary heater relays.

## DEFROST RELAY

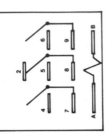

## DEFROST TERMINATOR

The temperature sensing element is a normally open S.P.S.T. switch which senses refrigerant temperature at the O.D. coil, and is calibrated to close when the refrigerant temperature reaches 26°, making ready the defrost sequence. The switch re-opens at 51°, terminating the defrost cycle.

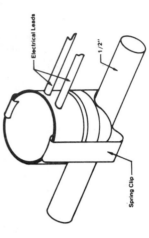

## DEFROST TIMER CONTROL

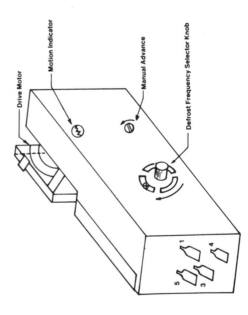

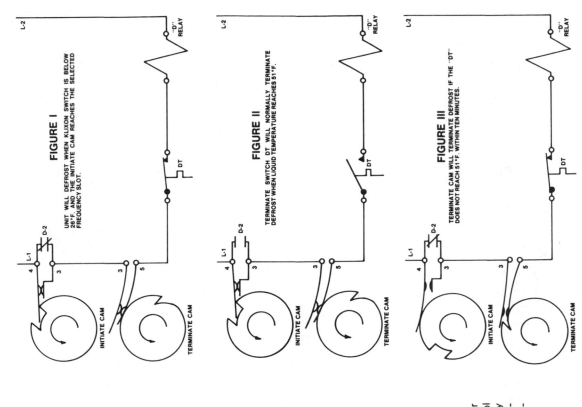

**FIGURE I**

UNIT WILL DEFROST WHEN KLIXON SWITCH IS BELOW 26°F. AND THE INITIATE CAM REACHES THE SELECTED FREQUENCY SLOT.

**FIGURE II**

TERMINATE SWITCH DT WILL NORMALLY TERMINATE DEFROST WHEN LIQUID TEMPERATURE REACHES 51°F.

**FIGURE III**

TERMINATE CAM WILL TERMINATE DEFROST IF THE "DT" DOES NOT REACH 51°F. WITHIN TEN MINUTES.

## SIMPLIFIED INITIATION AND TERMINATION SEQUENCE

### INITIATION:

Figure I: The defrost cycle is initiated when the initiate cam reaches the selected frequency of 45 or 90 minutes, and the temperature sensing element (referred to as defrost terminator) "DT" is closed, having reached 26°F. refrigerant temperature entering the O.D. coil. This completes a circuit to the defrost relay "D", closing relay contact D-2 which provides a lock-in circuit for the defrost relay.

### NORMAL TERMINATION:

Figure II: Termination of the defrost is accomplished one of two ways. Under normal operating conditions, the defrost terminator switch "DT" will open when the liquid refrigerant temperature leaving the O.D. coil reaches 51°F., de-energizing the defrost relay terminating the defrost cycle.

### TIME OVERRIDE TERMINATION:

Figure III: If 51°F. liquid temperature is not attained, or other malfunction occurs, the termination cam of the defrost control will open the circuit to the defrost relay by opening a normally closed contact which is in series with the terminator, de-energizing the defrost relay circuit, after a delay of approximately ten (10) minutes.

341

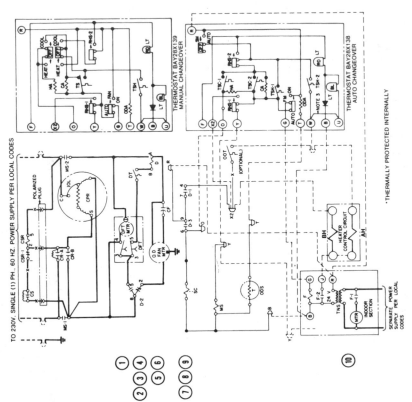

TO 230V, SINGLE (1) PH., 60 HZ  POWER SUPPLY PER LOCAL CODES

THERMOSTAT BAY28X139
MANUAL CHANGEOVER

THERMOSTAT BAY28X138
AUTO CHANGEOVER

*THERMALLY PROTECTED INTERNALLY

NOTES

1. FOR USE WITH SUPPLEMENTARY HEATERS. SEE WIRING DIAGRAM ON HEATER CONTROL BOX.

2. LOW VOLTAGE (24 V.) FIELD WIRING MUST BE 18 AWG MIN.

3. FACTORY ADDED JUMPER.

4. THE 12TH DIGIT OF THE MODEL NUMBER ON THE RATING NAMEPLATE MAY CHANGE WITHOUT AFFECTING THE ELECTRICAL CIRCUIT.

5. USE COPPER CONDUCTORS. IF ALUMINUM OR COPPER-CLAD ALUMINUM POWER WIRING IS USED, CONNECTORS WHICH MEET ALL APPLICABLE CODES AND ARE ACCEPTABLE TO THE INSPECTION AUTHORITY HAVING JURISDICTION SHALL BE USED.

6. START CONTROL ACCESSORY BAY41X189 MUST BE PROVIDED ON BWB942 WHENEVER USED WITH BLOWERLESS INDOOR SECTION AND CONTROL CENTER BAY24X039.

① ④
② ③
⑤ ⑥
⑦ ⑧ ⑨
⑩

## DEFROST CYCLE
## ELECTRICAL SEQUENCE OF OPERATION

When the heat pump is operating at temperatures below 40°F. outdoors, the coil will accumulate frost or ice as a result of low coil operating temperatures.

In order to initiate a defrost cycle to clear the coil of frost or ice, two conditions must be met. First, the refrigerant temperature entering the O.D. coil must be at or below 26°F. as detected by the defrost terminator switch DT (4). The second condition is that timer must have completed the time requirement of 45 or 90 minutes run time. If these conditions are met, the initiate cam switch (1) will close completing a circuit thru the normally closed override terminate switch (3) and DT (4) to the defrost relay D (6), closing relay contacts D-1 (9), D-2 (2) and D-3 (8). Contacts D-1 (9) and D-3 (8) complete circuits to the auxiliary heater relays (10), and to the S.O.V. solenoid coil (7). Contact D-2 (2) opens the circuit to the O.D. fan motor (5), and provides a holding circuit for the D relay (6).

Termination of the defrost cycle is normally accomplished by the defrost terminator switch DT (4), opening when the liquid temperature leaving the O.D. coil reaches 51°F. If termination does not occur within ten (10) minutes, the override switch (3) will open de-energizing the D relay (6), terminating the cycle.

342

# CIRCLED REFERENCE NOMENCLATURE

(1) — F-1 Contacts — I.D. Fan Motor
(2) — MS-1, MS-2, MS-3 Contacts, XOLB
(3) — Compressor
(4) — XOLA, XOLC
(5) — O.D. Fan Motor "A"
(6) — ODF-1 Contacts
(7) — O.D. Fan Motor "B", OFT
(8) — D-1 Contacts
(9) — Pilot Circuit
(10) — Sump Heaters
(11) — Transformer (system)
(12) — DR Relay Coil
(13) — DS
(14) — D Relay Coil
(15) — DT
(16) — DR-1 Contacts

(17) — ODF Relay Coil
(18) — DR-2 Contacts
(19) — MSA-2 Contacts
(20) — DR-3 Contacts
(21) — S.O.V. Coil
(22) — ER Relay Coil
(23) — MSA Relay Coil
(24) — F Relay Coil
(25) — ODS
(26) — F-2 Contacts
(27) — AH Relay Coil
(28) — BH Relay Coil
(29) — ER-2 Contacts
(30) — CH Relay Coil
(31) — ER-2 Contacts
(32) — Heater Control Transformer

(A) = Thermostat Switch
(B) = RHS-2
(C) = RSH-1
(D) = TSC-1
(E) = HA
(F) = TSC-2

(G) = CA
(H) = TSH-1
(I) = FM
(J) = ODA
(K) = TSH-2
(L) = Light

## COMPONENT NOMENCLATURE

### GLOSSARY

Amp. or AMP. — Amperes
C. — Centigrate
CCW — Counter Clockwise
CFM — Cubic Feet per Minute
Cu. In. — Cubic Inches
DPDT — Double Pole Double Throw
DPST — Double Pole Single Throw
FL, F/L — Full Load
FLA — Full Load Amperes
F. — Fahrenheit
°F. — Degrees Fahrenheit
H₂O — Water
HP — Horsepower

Hz. — Hertz
ID — Indoor
In. — Inches
Kg-cm² — Kilograms/Square Centimeter
Lb. — Pound
LR, L/R — Locked Rotor
m³ — Cubic Millimeter
Meg. — Million
MFD — Micro Farad
Min. — Minimum
mm — Millimeter
OD — Outdoor
Oz. — Ounce

△ P — Increment of Pressure
PSI — Pounds per Square Inch
PSIG — Pounds per Square Inch Gauge
3PST — Three Pole Single Throw
RPM — Revolution Per Minute
SAE — Society of Automotive Engineers
S/O — Switchover
SPDT — Single Pole Double Throw
SPST — Single Pole Single Throw
Temp. — Temperature
V. — Volt or Volts
VA — Volt/amperes
WB or W.B. — Wet Bulb

### COLOR CODES

COLOR OF WIRE

BK/BL    Black Wire with Blue Marker

COLOR OF MARKER

| | | | |
|---|---|---|---|
| BK | Black | RD | Red |
| BL | Blue | WH | White |
| BR | Brown | YL | Yellow |
| OR | Orange | GR | Green |

## SYMBOLS

|  |  |
|---|---|
| —— | Factory Wiring — 24 V. |
| —— | Factory Wiring — Line V. |
| - - - | Field Wiring — 24 V. |
| - - - | Field Wiring — Line V. |
| | Relay Contact — SPST |
| | Relay Contact — SPDT |
| o—⊢o | Contact Normally Open |
| o—⊬o | Contact Normally Closed |
| | Temperature Sensing Switch |
| | Pressure Sensing Switch |
| o—⊣⊢o | Capacitor |
| | Wire Nut or Connector |
| | Fuse |
| | Magnetic Coil |
| | Motor Winding |
| | Resistor or Heating Element |
| | Variable Resistor |
| | Thermistor |
| | Light |
| | Defrost Control Switch |
| | Switchover Valve Solenoid |
| | Internal Overload Protector Three Phase |
| | Timed — Thermal Switch |
| | Fusible Link |
| | Internal Overload Protector |
| | Contacts — SPDT |
| | Defrost Control Temperature — Time Actuated |
| | Compressor |
| ⏚ | Ground |
| | Indicating Light *B - Blue *R - Red |
| | Junction of Conductors |
| | 1 Ph. Motor |
| | 3 Ph. Motor |
| | Polarized Connector → Male Plug >— Female Plug |
| | Pressure Sensing Switch Opens on Falling Pressure |
| | Pressure Sensing Switch Opens on Rising Pressure |
| | Relay Contact — SPDT |
| | Switch, Time Delay |
| | Switchover Valve Solenoid |
| | Temperature Sensing Switch |
| | Temperature Sensing Switch Opens on Falling Temperature |
| | Transformer |
| | Wire Nut or Connector |
| | Terminal Board |

## LEGEND

| | | | |
|---|---|---|---|
| AH | Supplementary Heater Contactor | LT | Light |
| BH | Supplementary Heater Contactor | LVTB | Low Voltage Terminal Board |
| CA | Cooling Anticipator | MS | Compressor Motor Contactor |
| CB | Circuit Breaker | MSA | Compressor Motor Contactor — Auxiliary |
| CF | Fan Capacitor | MTR | Motor |
| CN | Wire Connector | | |
| CR | Run Capacitor | ODA | Outdoor Temperature Anticipato |
| CPR | Compressor | ODF | Outdoor Fan Relay |
| CS | Start Capacitors | ODS | Outdoor Temperature Sensor |
| CSR | Capacitor Switching Relay | ODT | Outdoor Thermostat |
| D | Defrost Relay | RH | Emergency Heat Relay |
| D & DR | Defrost Relay | RHS | Resistance Heat Switch |
| DFC | Defrost Control | SC | Switchover Valve Solenoid |
| DFT-I | Defrost Timer | SM | System Switch (Room Thermostat) |
| DS | Diaphragm Switch (Defrost Sensing) | T | Thermistor |
| DT | Defrost Termination Switch | TCO | Temperature Limit Switch |
| F | Indoor Fan Relay | TDL | Discharge Line Thermostat |
| FM | Manual Fan Switch | TDR | Time Delay Relay |
| FU | Fuse | TM | Compressor Motor Thermostat |
| HA | Heating Anticipator | TNS | Transformer |
| HPCO | High Pressure Cut-Out | TS | Heating & Cooling Thermostat |
| HTR | Heater | TSC | Cooling Thermostat |
| HVTB | High Voltage Terminal Board | TSH | Heating Thermostat |
| IOL | Internal Overload Protector | UA | Universal Anticipator |
| LAS | Low Airflow Sensing | XOL | External Overload Protector |
| LPCO | Low Pressure Cut-Out | XOLS | External Overload Protector (Supplementary) |

# Amana Refrigeration

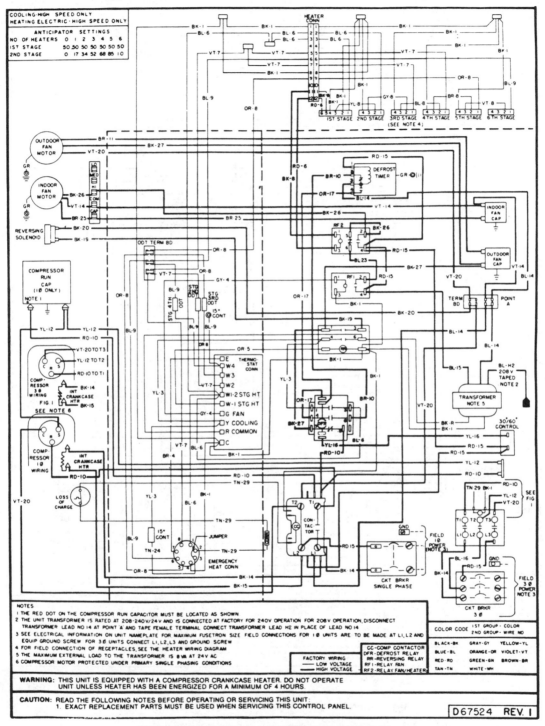

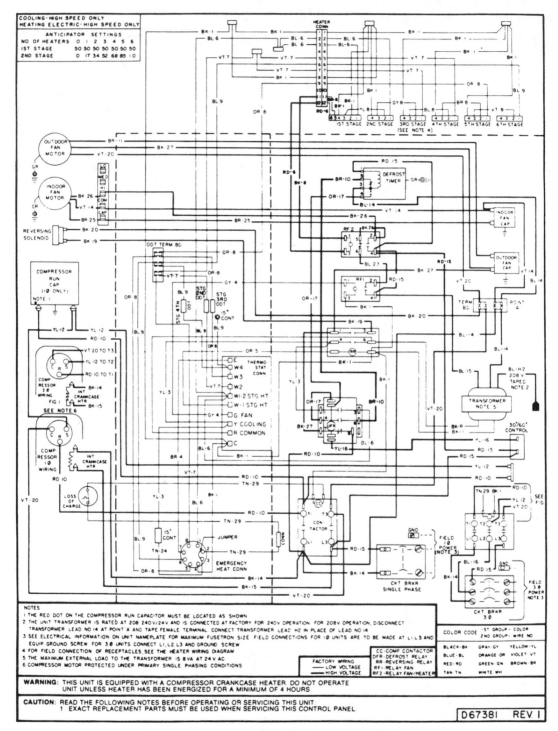

UNIT WIRING DIAGRAM 60M BTUH UNIT, 1∅ AND 3∅

346

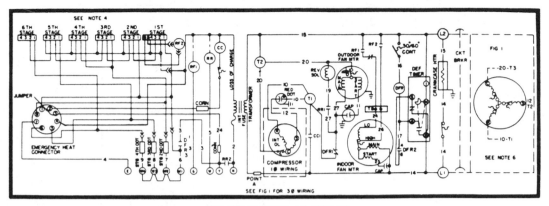

**42M BTUH UNIT, 1∅ AND 3∅**

(Reference D67380)

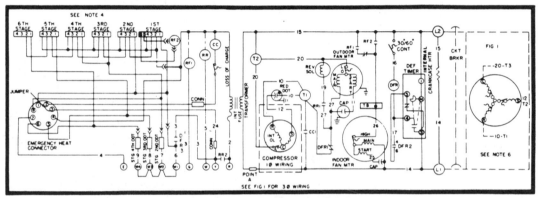

**48M BTUH UNIT, 1∅ AND 3∅**

(Reference D67524-1)

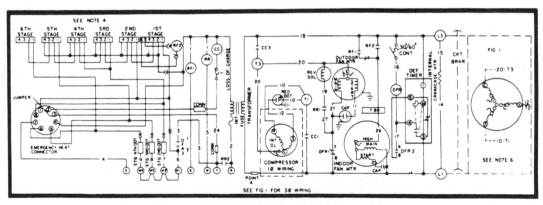

**60M BTUH UNIT, 1∅ AND 3∅**

(Reference D67381)

**NOTES:**

4. For field connection of receptacles, see heater wiring diagram.

6. Compressor motor protected under primary single phasing conditions.

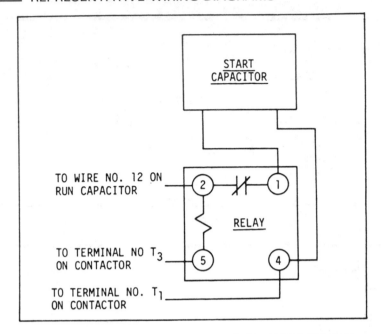

FIGURE 23-START CAPACITOR AND RELAY WIRING DIAGRAM

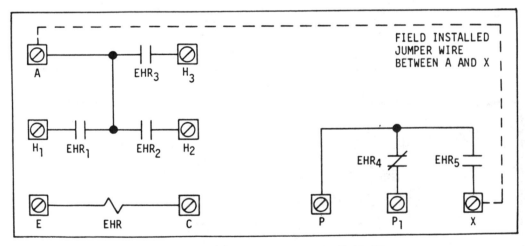

TYPICAL SCHEMATIC-EMERGENCY HEAT RELAY

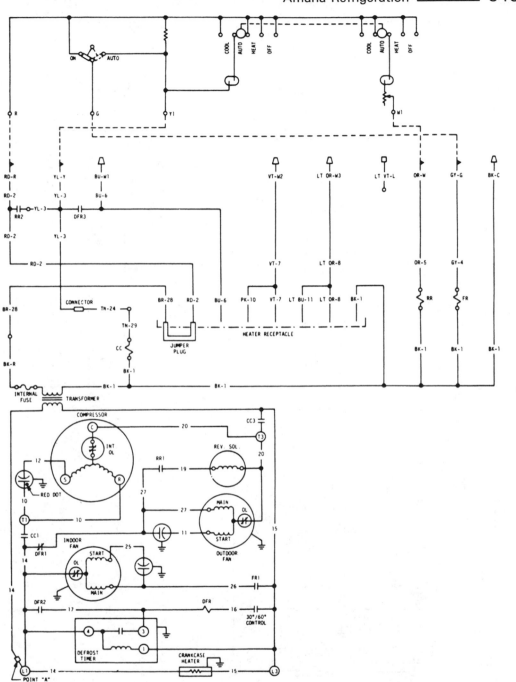

TYPICAL WIRING SCHEMATIC- PH2-1E, PH2.5-1E, PH3-1E

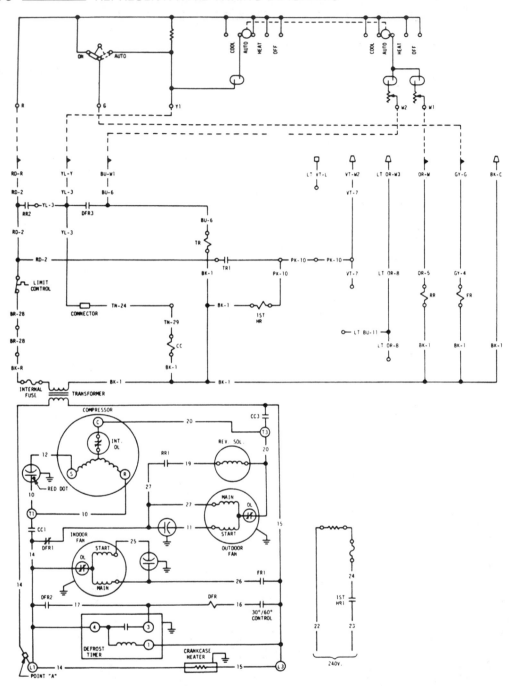

TYPICAL WIRING SCHEMATIC-D54444-1-PH2-1E, PH2.5-1E, PH3-1E

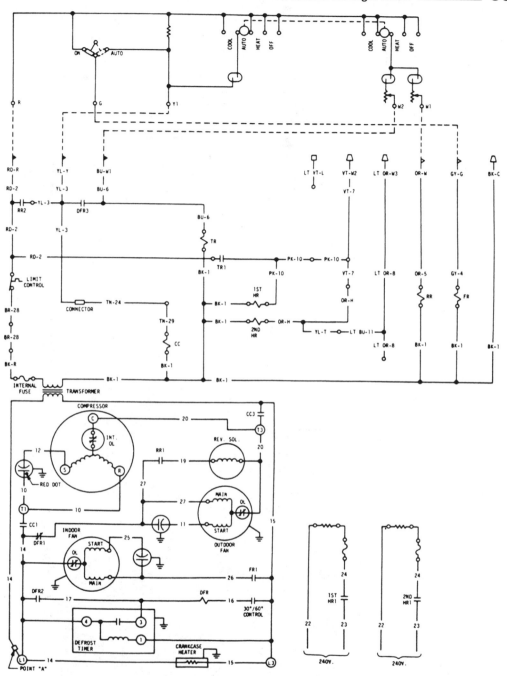

TYPICAL WIRING SCHEMATIC-(1)D54444-1, (1)D54444-2-PH2-1E, PH2.5-1E, PH3-1E

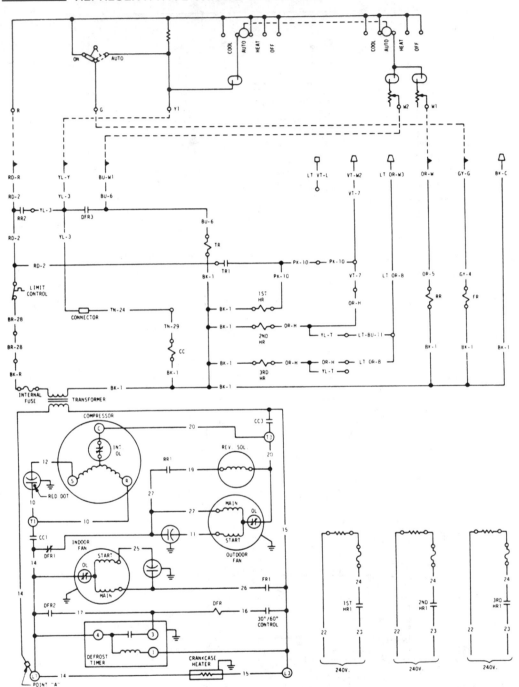

TYPICAL WIRING SCHEMATIC - (1) D54444-1, (2) D54444-2 - PH3-1E

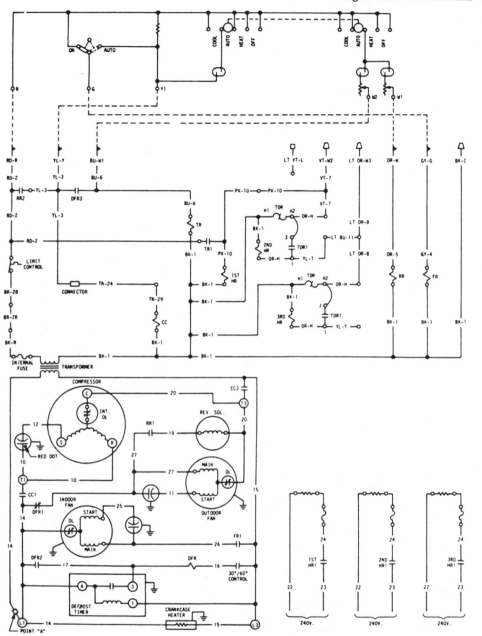

TYPICAL WIRING SCHEMATIC- (1) D54444-1, (2) D54444-2, (2) A46850-1 - PH3-1E
NOTE:  WIRING FOR PH2-1E AND PH2.5-1E IS IDENTICAL EXCEPT ONLY 2 HEATERS
AND ONE TIME DELAY RELAY MAY BE INSTALLED.

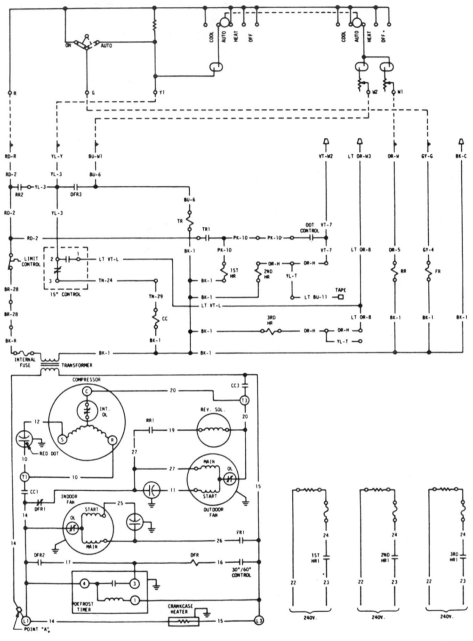

TYPICAL WIRING SCHEMATIC-      (1) D54444-1,      (2) D54444-2,      (1) C48224-2,
(1) B47760-2--PH3-1E
NOTE: WIRING FOR PH2-1E AND PH2.5-1E IS IDENTICAL EXCEPT ONLY TWO HEATERS MAY BE
      USED--THEREFORE EITHER THE 15° CONTROL OR OUTDOOR TEMPERATURE CONTROL MAY
      BE WIRED WITH THE 2ND STAGE ELECTRIC HEATER.

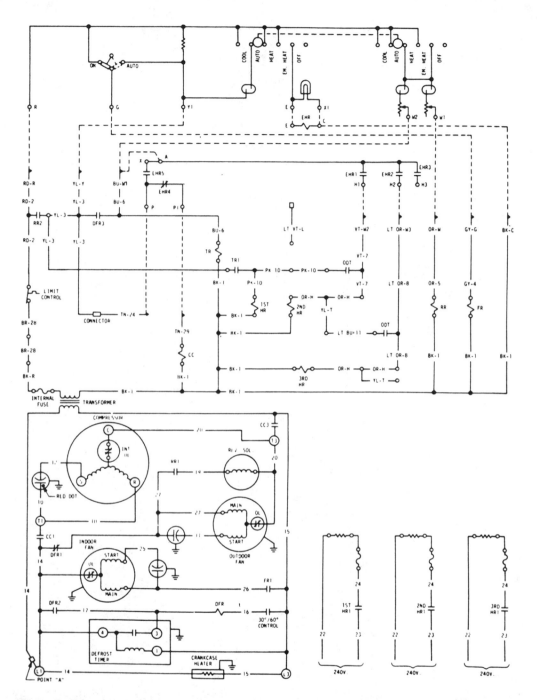

TYPICAL WIRING SCHEMATIC- (1) D54444-1, (2) D54444-2, (1) A46202-2,
(2) C48224-2-- PH3-1E
NOTE: WIRING FOR PH2-1E AND PH2.5-1E IS IDENTICAL EXCEPT ONLY TWO HEATERS
MAY BE USED.

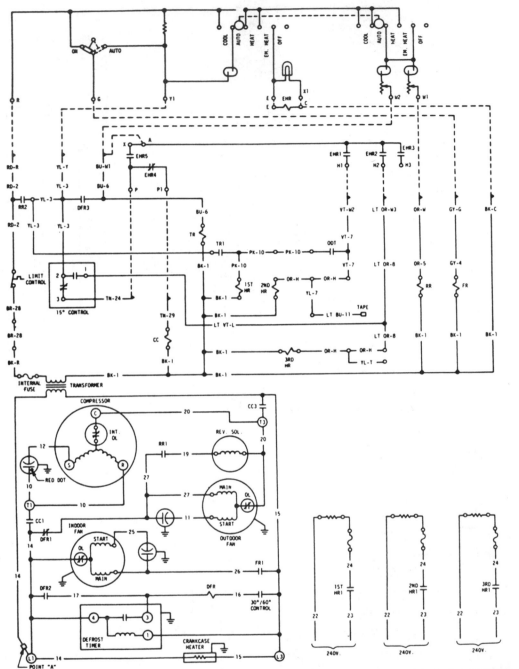

TYPICAL WIRING SCHEMATIC- (1) D54444-1, (2) D54444-2, A46202-2, C48224-2, B47760-2-- PH3-1E

NOTE: WIRING FOR PH2-1E AND PH2.5-1E IS IDENTICAL EXCEPT ONLY TWO HEATERS MAY BE USED--THEREFORE EITHER THE 15° CONTROL OR OUTDOOR TEMPERATURE CONTROL MAY BE WIRED WITH THE 2ND STAGE ELECTRIC HEATER

# *BDP Company* ——————————

**Model:** 541A/SHP
**Size:** 030 & 036
**Series:** A
**Voltage:** 208-230-60-1

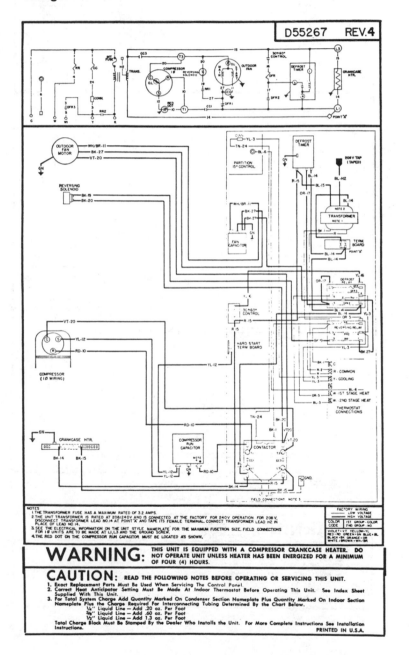

D55267    REV.4

NOTES
1. THE TRANSFORMER FUSE HAS A MAXIMUM RATED OF 3.2 AMPS.
2. THE UNIT TRANSFORMER IS RATED AT 208/240V AND IS CONNECTED AT THE FACTORY FOR 240V OPERATION FOR 208V DISCONNECT TRANSFORMER LEAD NO.14 AT POINT "A" AND TAPE ITS FEMALE TERMINAL. CONNECT TRANSFORMER LEAD H2 IN PLACE OF LEAD NO.14.
3. SEE THE ELECTRICAL INFORMATION ON THE UNIT STYLE NAMEPLATE FOR THE MAXIMUM FUSETRON SIZE. FIELD CONNECTIONS FOR 1Ø UNITS ARE TO BE MADE AT L1,L3 AND THE GROUND SCREW.
4. THE RED DOT ON THE COMPRESSOR RUN CAPACITOR MUST BE LOCATED AS SHOWN.

FACTORY WIRING
————— LOW VOLTAGE
————— HIGH VOLTAGE

| COLOR | 1ST GROUP - COLOR |
| CODE | 2ND GROUP - NO. |

VIOLET=VT  YELLOW=YL
RED=RD  GREEN=GN  BLUE=BL
BLACK=BK  ORANGE=OR
WHITE=WH  BROWN=BR

**Model:** 541A/SHP
**Size:** 048
**Series:** B
**Voltage:** 208-230-60-1
208-240-60-3

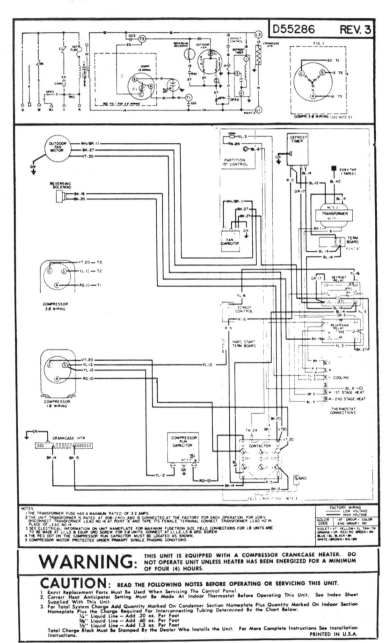

**WARNING:** THIS UNIT IS EQUIPPED WITH A COMPRESSOR CRANKCASE HEATER. DO NOT OPERATE UNIT UNLESS HEATER HAS BEEN ENERGIZED FOR A MINIMUM OF FOUR (4) HOURS.

**CAUTION:** READ THE FOLLOWING NOTES BEFORE OPERATING OR SERVICING THIS UNIT.

1. Exact Replacement Parts Must Be Used When Servicing The Control Panel.
2. Correct Heat Anticipator Setting Must Be Made At Indoor Thermostat Before Operating This Unit. See Index Sheet Supplied With This Unit.
3. For Total System Charge Add Quantity Marked On Condenser Section Nameplate Plus Quantity Marked On Indoor Section Nameplate Plus the Charge Required For Interconnecting Tubing Determined By the Chart Below.
   1/4" Liquid Line — Add .20 oz. Per Foot
   3/8" Liquid Line — Add .60 oz. Per Foot
   1/2" Liquid Line — Add 1.3 oz. Per Foot
Total Charge Block Must Be Stamped By the Dealer Who Installs the Unit. For More Complete Instructions See Installation Instructions.

PRINTED IN U.S.A.

**Model:** 541A/SHP
**Size:** 048 & 060
**Series:** A
**Voltage:** 208-230-60-1
208-240-60-3

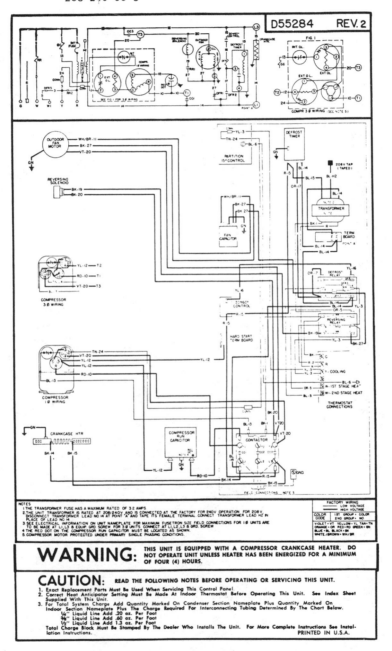

D55284    REV. 2

NOTES
1. THE TRANSFORMER FUSE HAS A MAXIMUM RATED OF 3.2 AMPS.
2. THE UNIT TRANSFORMER IS RATED AT 208/240V AND IS CONNECTED AT THE FACTORY FOR 240V OPERATION. FOR 208V. DISCONNECT TRANSFORMER LEAD NO.4 AT POINT "A" AND TAPE. ITS FEMALE TERMINAL CONNECT TRANSFORMER LEAD NO.2 IN PLACE OF LEAD NO.4
3. SEE ELECTRICAL INFORMATION ON UNIT NAMEPLATE. FOR MAXIMUM FUSETRON SIZE FIELD CONNECTIONS FOR 1Ø UNITS ARE TO BE MADE AT L1,L2,L3 & EQUIP. GRD SCREW. FOR 3Ø UNITS CONNECT AT L1,L2,L3 & GRD. SCREW.
4. THE RED DOT ON THE COMPRESSOR RUN CAPACITOR MUST BE LOCATED AS SHOWN.
5. COMPRESSOR MOTOR PROTECTED UNDER PRIMARY SINGLE PHASING CONDITIONS.

FACTORY WIRING
———— LOW VOLTAGE
———— HIGH VOLTAGE

| COLOR CODE | 1ST GROUP • COLOR |
| | 2ND GROUP • NO |

VIOLET • VT   YELLOW • YL   TAN • TN
ORANGE • OR   RED • RD   GREEN • GN
BLUE • BL   BLACK • BK
WHITE /BROWN • WH /BR

**WARNING:** THIS UNIT IS EQUIPPED WITH A COMPRESSOR CRANKCASE HEATER. DO NOT OPERATE UNIT UNLESS HEATER HAS BEEN ENERGIZED FOR A MINIMUM OF FOUR (4) HOURS.

**CAUTION:** READ THE FOLLOWING NOTES BEFORE OPERATING OR SERVICING THIS UNIT.
1. Exact Replacement Parts Must Be Used When Servicing This Control Panel.
2. Correct Heat Anticipator Setting Must Be Made At Indoor Thermostat Before Operating This Unit. See Index Sheet Supplied With This Unit.
3. For Total System Charge Add Quantity Marked On Condenser Section Nameplate Plus Quantity Marked On Indoor Section Nameplate Plus The Charge Required For Interconnecting Tubing Determined By The Chart Below.
   ¼" Liquid Line Add .20 oz. Per Foot
   ⅜" Liquid Line Add .60 oz. Per Foot
   ½" Liquid Line Add 1.3 oz. Per Foot
Total Charge Block Must Be Stamped By The Dealer Who Installs The Unit. For More Complete Instructions See Installation Instructions.
PRINTED IN U.S.A.

**Model:** 541B/SHP
**Size:** 057
**Series:** A
**Voltage:** 208/230-60-1

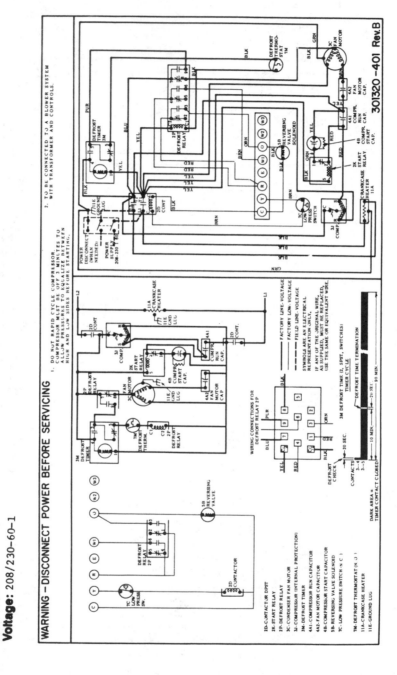

NOTE: Rev. A has note, single pole contactor is used.

**Model: 541B/SHP**

**Size:** 041, 047, & 057

**Series:** A

**Voltage:** 208/230-60-3

## WARNING – DISCONNECT POWER BEFORE SERVICING

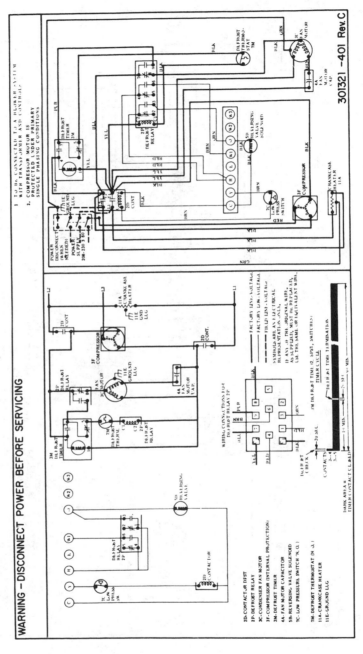

1. TO BE CONNECTED TO A BLOWER SYSTEM WITH TRANSFORMER AND CONTROLS.
2. COMPRESSOR MOTOR IS PROTECTED UNDER PRIMARY SINGLE PHASING CONDITIONS.

301321-401 Rev.C

2D-CONTACTOR DFST
2P-DEFROST RELAY
3C-CONDENSER FAN MOTOR
3F-COMPRESSOR (INTERNAL PROTECTION)
3M-DEFROST TIMER
4A-FAN MOTOR CAPACITOR
5B-REVERSING VALVE SOLENOID
7C-LOW PRESSURE SWITCH (N.O.)
7M-DEFROST THERMOSTAT (N.O.)
11A-CRANKCASE HEATER
11E-GROUND LUG

**NOTE:** Rev. B has note, single pole contactor is used.

Model: 541B/SHP
Size: 021, 028, & 034
Series: A
Voltage: 208-230-60-1

362

WARNING – DISCONNECT POWER BEFORE SERVICING

301314-401 Rev.A

1. SINGLE POLE CONTACTOR IS USED.
2. DO NOT RAPID CYCLE COMPRESSOR. COMPRESSOR MUST BE OFF 3 MINUTES TO ALLOW PRESSURES TO EQUALIZE BETWEEN HIGH AND LOW SIDES BEFORE STARTING.
3. TO BE CONNECTED TO A BLOWER SYSTEM WITH TRANSFORMER AND CONTROLS.

2A—CONTACTOR SPST (N. O.)
2P—DEFROST RELAY TPDT
3C—OUTDOOR FAN MOTOR
3J—COMPRESSOR
3M—DEFROST TIMER: 2 SPST SWITCHES
4G—DUAL RUN CAPACITOR, COMPRESSOR AND FAN MOTOR
4K—PTC START THERMISTOR
5B—REVERSING VALVE SOLENOID, ENERGIZED IN COOLING
7C—LOW PRESSURE SWITCH SPST (N. C.)
7M—DEFROST THERMOSTAT SPST (N. O.)
11A—CRANKCASE HEATER
11E—CHASSIS GROUND LUG

FACTORY LINE-VOLTAGE
FACTORY LOW-VOLTAGE
FIELD LINE-VOLTAGE

SYMBOLS ARE AN ELECTRICAL REPRESENTATION ONLY.
IF ANY OF THE ORIGINAL WIRE, AS SUPPLIED, MUST BE REPLACED, USE THE SAME (OR EQUIVALENT) WIRE.

WIRING CONNECTIONS FOR DEFROST RELAY 2P

3M DEFROST TIMER (2, SPST, SWITCHES)
TIMER CYCLE
DEFROST TIME TERMINATION

DEFROST CHECK

DARK AREA = TIMER CONTACT CLOSED

CONTACTS
3 - 4
3 - 5

20 SEC.
10 MIN.
20 SEC.
90 MIN.

DEFROST THERMOSTAT 7M
POWER DISCONNECT (WHEN NEEDED)
POWER SUPPLY 208-230
DEFROST TIMER 3M
DEFROST RELAY 2P
REVERSING VALVE SOLENOID 5B
LOW PRESS SWITCH 7C
COMPRESSOR 3J
CRANKCASE HEATER 11A
DUAL RUN CAPACITOR 4G
FAN MOTOR 3C
GROUND LUG
PTC 4K

CRANKCASE HEATER 11A
CONTACTOR 2A
DEFROST RELAY 2P
FAN MOTOR 3C
DEFROST THERM. 7M
DEFROST TIMER 3M
GROUND LUG 11E
DUAL RUN CAPACITOR 4G
COMPRESSOR 3J

REVERSING VALVE 5B
DEFROST RELAY 2P
CONTACTOR 2A
LOW PRESS SW. 7C

**Model:** 541B/SHP

**Size:** 047

**Series:** A

**Voltage:** 208/230-60-1

WARNING – DISCONNECT POWER BEFORE SERVICING

1. SINGLE POLE CONTACTOR IS USED.
2. DO NOT RAPID CYCLE COMPRESSOR. COMPRESSOR MUST BE OFF 3 MINUTES TO ALLOW PRESSURES TO EQUALIZE BETWEEN HIGH AND LOW SIDES BEFORE STARTING.
3. TO BE CONNECTED TO A BLOWER SYSTEM WITH TRANSFORMER AND CONTROLS.

1A-CONTACTOR SPST
2K-START RELAY (N.C.)
1P-DEFROST RELAY
1J-CONDENSER FAN MOTOR
1J-COMPRESSOR (INTERNAL PROTECTION)
3M-DEFROST TIMER
4A1-COMPRESSOR RUN CAPACITOR
4A2-FAN MOTOR CAPACITOR
4B-COMPRESSOR START CAPACITOR
5B-REVERSING VALVE SOLENOID
7C-LOW PRESSURE SWITCH (N.C.)
7M-DEFROST THERMOSTAT (N.O.)
11A-CRANKCASE HEATER
11E-GROUND LUG

———— FACTORY LINE-VOLTAGE
———— FACTORY LOW-VOLTAGE
———— FIELD LINE-VOLTAGE

SYMBOLS ARE AN ELECTRICAL REPRESENTATION ONLY.

IF ANY OF THE ORIGINAL WIRE, AS SUPPLIED, MUST BE REPLACED, USE THE SAME OR EQUIVALENT WIRE.

DARK AREA = TIMER CONTACT CLOSED

30I319 -401 Rev. A

**Model:** 541A/SHP
**Size:** 018 & 024
**Series:** A
**Voltage:** 208-230-60-1

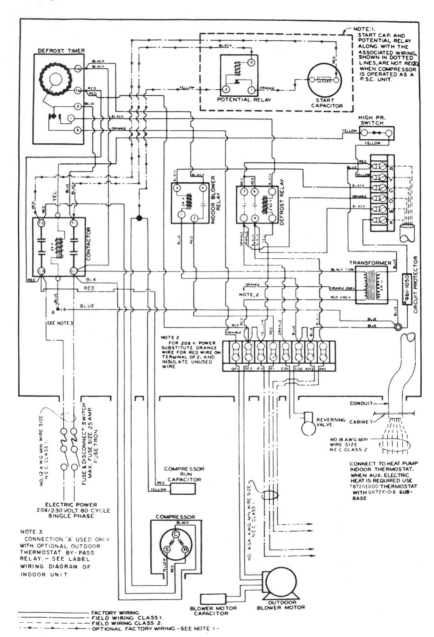

# York Division Of Borg-Warner Corp.

## TYPICAL FIELD WIRING CONNECTIONS

### SPLIT-SYSTEM HEAT PUMP
#### MODELS CHP018 THRU CHP060
#### 208/230-1-60 AND 208/230-3-60

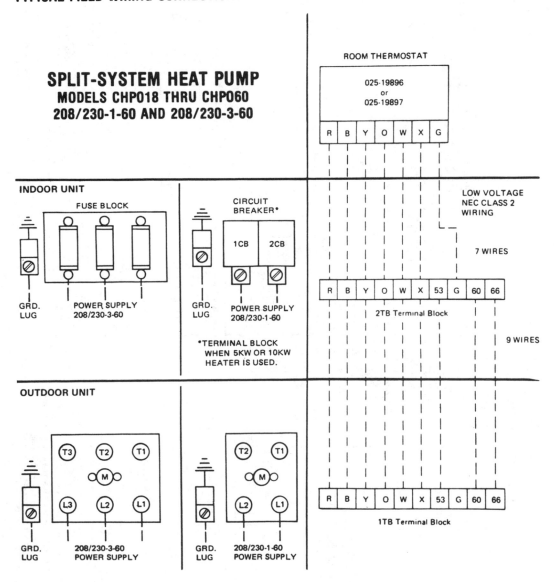

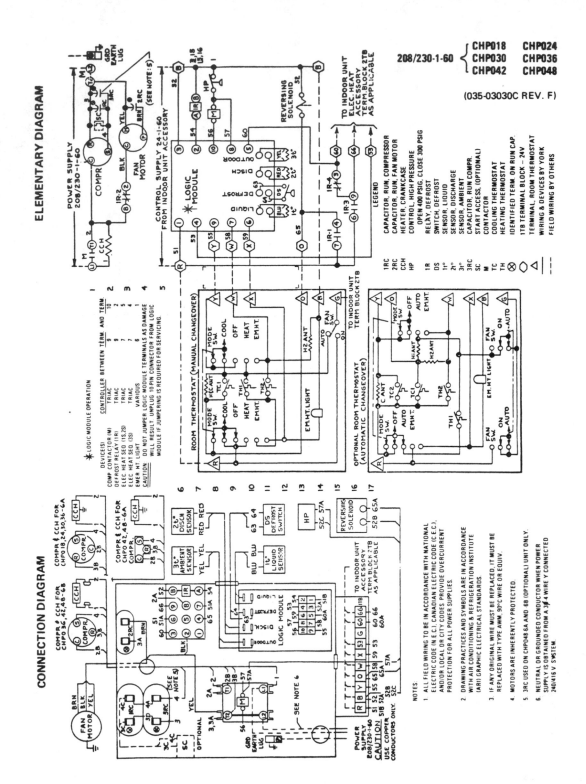

ELEMENTARY DIAGRAM

208/230-1-60 { CHP018 CHP024 / CHP030 CHP036 / CHP042 CHP048

(035-03030C REV. F)

LEGEND

| 1RC | CAPACITOR, RUN, COMPRESSOR |
| 2RC | CAPACITOR, RUN, FAN MOTOR |
| CCH | HEATER, CRANKCASE |
| HP | CONTROL, HIGH PRESSURE OPEN 400 PSIG, CLOSE 300 PSIG |
| 1R | RELAY, DEFROST |
| DS | SWITCH, DEFROST |
| 1° | SENSOR, LIQUID |
| 2° | SENSOR, DISCHARGE |
| 3° | SENSOR, AMBIENT |
| 3RC | CAPACITOR, RUN COMPR. |
| SC | START ACCESS. (OPTIONAL) |
| M | CONTACTOR |
| TC | COOLING THERMOSTAT |
| TH | HEATING THERMOSTAT |
| 1TB | IDENTIFIED TERM. ON RUN CAP. |
| | TERMINAL BLOCK – 24V |
| | TERMINAL, ROOM THERMOSTAT |
| | WIRING & DEVICES BY YORK |
| | FIELD WIRING BY OTHERS |

ROOM THERMOSTAT (MANUAL CHANGEOVER)

OPTIONAL ROOM THERMOSTAT (AUTOMATIC CHANGEOVER)

LOGIC MODULE OPERATION

| DEVICE(S) | CONTROLLER BETWEEN TERM. AND TERM. | | |
|---|---|---|---|
| COMP CONTACTOR (M) | TRIAC | 9 | 10 |
| DEFROST RELAY (1R) | TRIAC | 9 | 5 |
| ELEC HEAT SEQ (1S,2S) | TRIAC | 7 | 4 |
| ELEC HEAT SEQ (3S) | TRIAC | 6 | 1 |
| EMER HT LIGHT | VARIOUS | | |

CAUTION: DO NOT JUMPER LOGIC MODULE TERMINALS AS DAMAGE WILL RESULT. UNPLUG 10-PIN CONNECTOR FROM LOGIC MODULE IF JUMPERING IS REQUIRED FOR SERVICING.

CONNECTION DIAGRAM

CAUTION: USE COPPER CONDUCTORS ONLY.

NOTES

1. ALL FIELD WIRING TO BE IN ACCORDANCE WITH NATIONAL ELECTRIC CODE (N.E.C.), CANADIAN ELECTRIC CODE (C.E.C.) AND/OR LOCAL OR CITY CODES PROVIDE OVERCURRENT PROTECTION FOR ALL POWER SUPPLIES.

2. DRAWING PRACTICES AND SYMBOLS ARE IN ACCORDANCE WITH AIR CONDITIONING & REFRIGERATION INSTITUTE (ARI) GRAPHIC ELECTRICAL STANDARDS.

3. IF ANY ORIGINAL WIRE MUST BE REPLACED, IT MUST BE REPLACED WITH TYPE AWM, 90°C WIRE OR EQUIV.

4. MOTORS ARE INHERENTLY PROTECTED.

5. 3RC USED ON CHP048 6A AND 6B (OPTIONAL) UNIT ONLY.

6. NEUTRAL OR GROUNDED CONDUCTOR WHEN POWER SUPPLY IS OBTAINED FROM A 3 OR 4 WIRE Y CONNECTED 240/416 V SYSTEM.

366

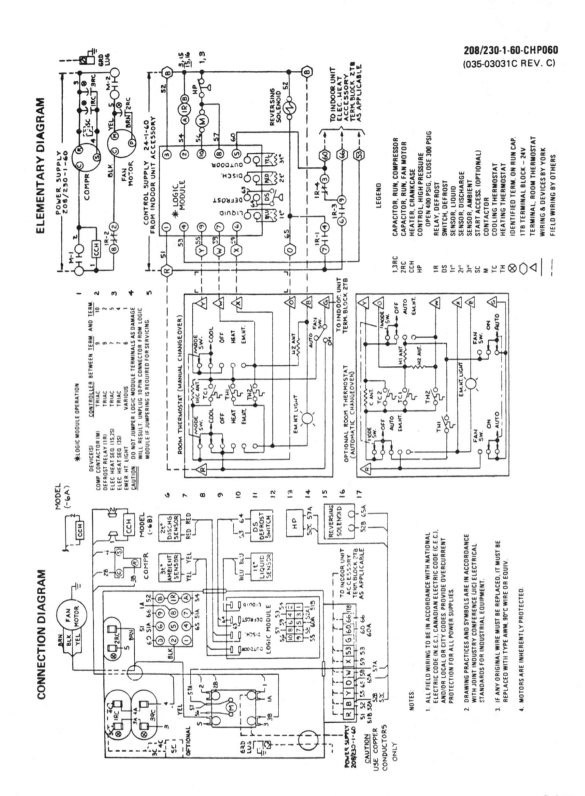

208/230-1-60-CHP060
(035-03031C REV. C)

367

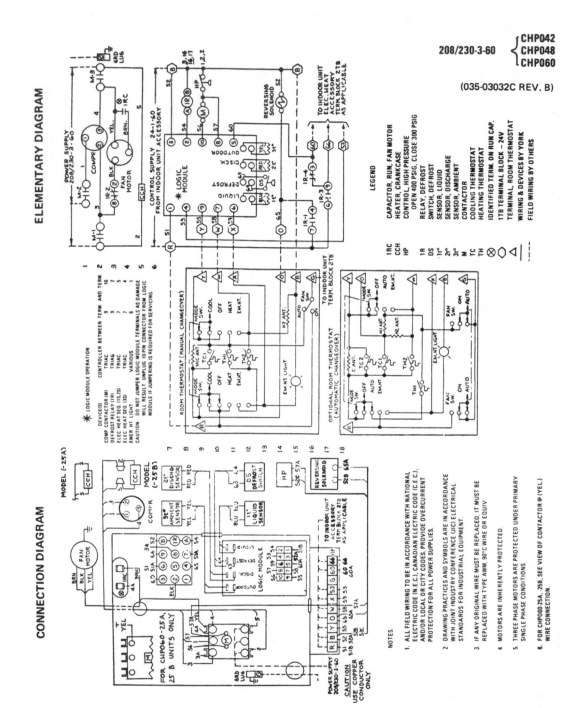

## ELEMENTARY DIAGRAM

208/230-3-60

CHP042
CHP048
CHP060

(035-03032C REV. B)

### LEGEND

| | |
|---|---|
| 1RC | CAPACITOR, RUN, FAN MOTOR |
| CCH | HEATER, CRANKCASE |
| HP | CONTROL, HIGH PRESSURE |
| | OPEN 400 PSIG, CLOSE 300 PSIG |
| 1R | RELAY, DEFROST |
| DS | SWITCH, DEFROST |
| 1° | SENSOR, LIQUID |
| 2° | SENSOR, DISCHARGE |
| 3° | SENSOR, AMBIENT |
| M | CONTACTOR |
| TC | COOLING THERMOSTAT |
| TH | HEATING THERMOSTAT |
| | IDENTIFIED TERM. ON RUN CAP. |
| | 1TB TERMINAL BLOCK – 24V |
| | TERMINAL, ROOM THERMOSTAT |
| | WIRING & DEVICES BY YORK |
| | FIELD WIRING BY OTHERS |

## CONNECTION DIAGRAM

### NOTES:

1. ALL FIELD WIRING TO BE IN ACCORDANCE WITH NATIONAL ELECTRIC CODE (N.E.C.), CANADIAN ELECTRIC CODE (C.E.C.), AND/OR LOCAL OR CITY CODES. PROVIDE OVERCURRENT PROTECTION FOR ALL POWER SUPPLIES.

2. DRAWING PRACTICES AND SYMBOLS ARE IN ACCORDANCE WITH JOINT INDUSTRY CONFERENCE (JIC) ELECTRICAL STANDARDS FOR INDUSTRIAL EQUIPMENT.

3. IF ANY ORIGINAL WIRE MUST BE REPLACED, IT MUST BE REPLACED WITH TYPE AWM, 90°C WIRE OR EQUIV.

4. MOTORS ARE INHERENTLY PROTECTED.

5. THREE PHASE MOTORS ARE PROTECTED UNDER PRIMARY SINGLE PHASE CONDITIONS.

6. FOR CHP060-25A, 25B, SEE VIEW OF CONTACTOR @ (YEL) WIRE CONNECTION.

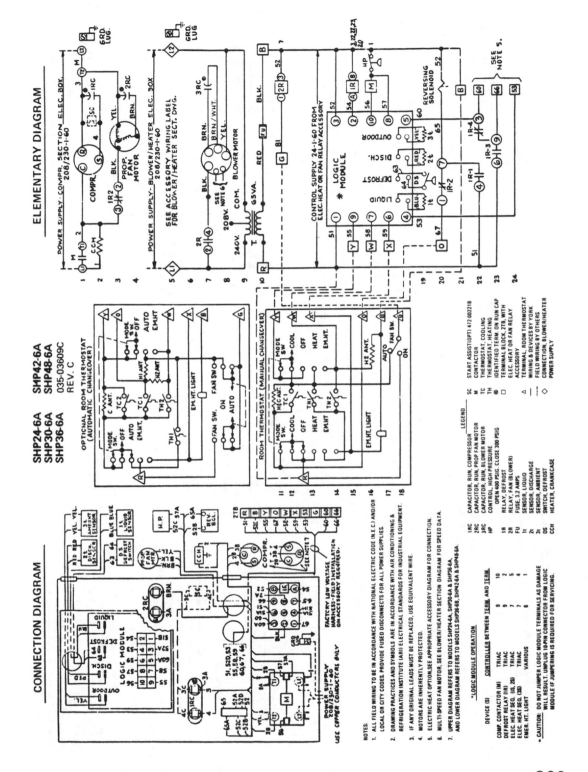

369

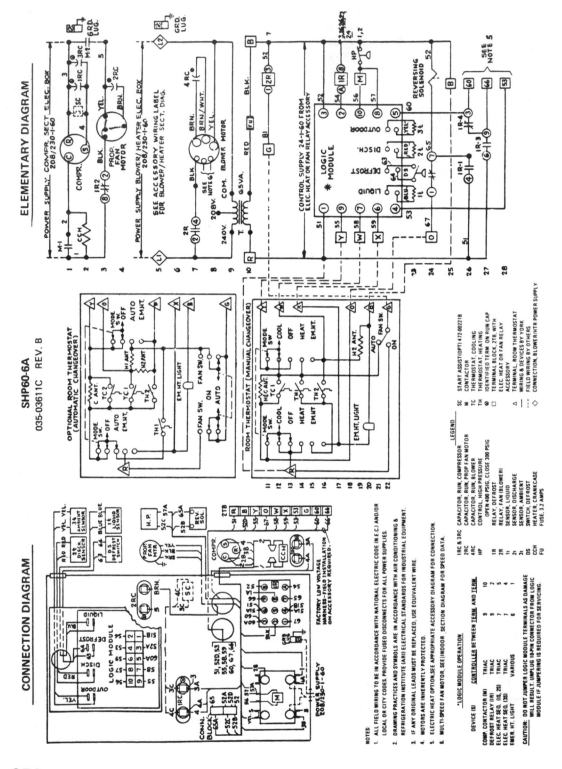

## ELEMENTARY DIAGRAM

SHP60-6A
035-03611C  REV. B

## CONNECTION DIAGRAM

NOTES:

1. ALL FIELD WIRING TO BE BE IN ACCORDANCE WITH NATIONAL ELECTRIC CODE (N.E.C.) AND/OR LOCAL OR CITY CODES. PROVIDE FUSED DISCONNECTS FOR ALL POWER SUPPLIES.

2. DRAWING PRACTICES AND SYMBOLS ARE IN ACCORDANCE WITH AIR CONDITIONING & REFRIGERATION INSTITUTE (ARI) ELECTRICAL STANDARDS FOR INDUSTRIAL EQUIPMENT.

3. IF ANY ORIGINAL LEADS MUST BE REPLACED, USE EQUIVALENT WIRE.

4. MOTORS ARE INHERENTLY PROTECTED.

5. ELECTRIC HEAT OPTION, SEE APPROPRIATE ACCESSORY DIAGRAM FOR CONNECTION.

6. MULTI-SPEED FAN MOTOR, SEE INDOOR SECTION DIAGRAM FOR SPEED DATA.

CAUTION: DO NOT JUMPER LOGIC MODULE TERMINALS AS DAMAGE WILL RESULT. UNPLUG 18-PIN CONNECTOR FROM LOGIC MODULE IF JUMPERING IS REQUIRED FOR SERVICING.

### LEGEND

| | |
|---|---|
| SC | START ASSIST (OPT) 471-00221B |
| M | CONTACTOR |
| TC | THERMOSTAT, COOLING |
| TH | THERMOSTAT, HEATING |
| □ | IDENTIFIED TERM ON RUN CAP |
|  | TERMINAL BLOCK, 2TB, WITH |
|  | ELEC. HEAT OR FAN RELAY ACCESSORY |
| △ | TERMINAL, ROOM THERMOSTAT |
| --- | FIELD WIRING BY OTHERS |
| ◇ | CONNECTION, BLOWER/HTR POWER SUPPLY |
| 1RC & 3RC | CAPACITOR, RUN, COMPRESSOR |
| 2RC | CAPACITOR, RUN, PROP-FAN MOTOR |
| 4RC | CAPACITOR, RUN, BLOWER |
| HP | CONTROL, HIGH PRESSURE |
|  | OPEN 400 PSIG, CLOSE 300 PSIG |
| 1R | RELAY, DEFROST |
| 2R | RELAY, FAN (BLOWER) |
| 1L | SENSOR, LIQUID |
| 2L | SENSOR, DISCHARGE |
| 3L | SENSOR, AMBIENT |
| DS | SWITCH, DEFROST |
| CCH | HEATER, CRANKCASE |
| FU | FUSE, 3.2 AMPS |

### *LOGIC MODULE OPERATION

| | CONTROLLER BETWEEN TERM. AND TERM. | | |
|---|---|---|---|
| DEVICE (S) | | | |
| COMP. CONTACTOR (1M) | TRIAC | 9 | 10 |
| DEFROST RELAY (1R) | TRIAC | 9 | 2 |
| ELEC. HEAT SEQ. (1S, 2S) | TRIAC | 7 | 4 |
| ELEC. HEAT SEQ. (3S) | TRIAC | 7 | 1 |
| EMER. HT. LIGHT | VARIOUS | 6 | 1 |

370

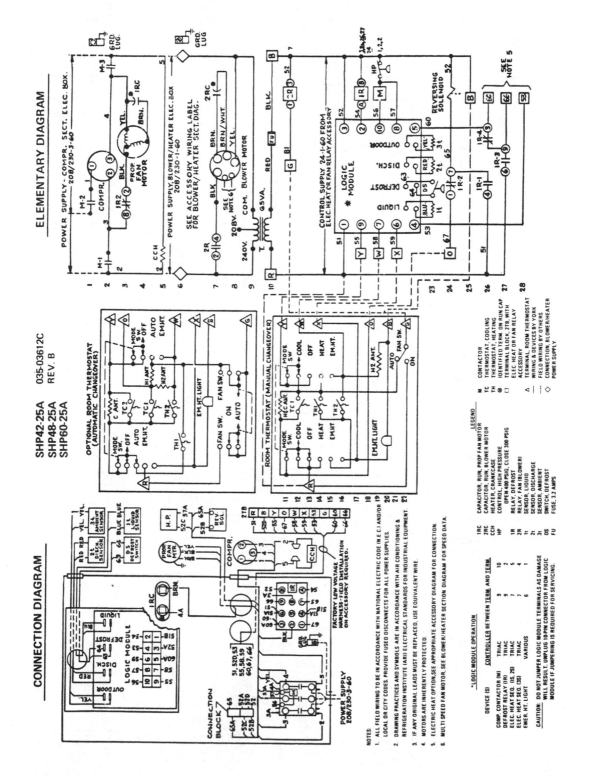

371

# HEAT PUMP MODELS
## CHPI60 AND SHP60
## WITH HEAT ACCESSORY HPC25-6A

---

### ELEMENTARY DIAGRAM

---

035-03044C REV. C

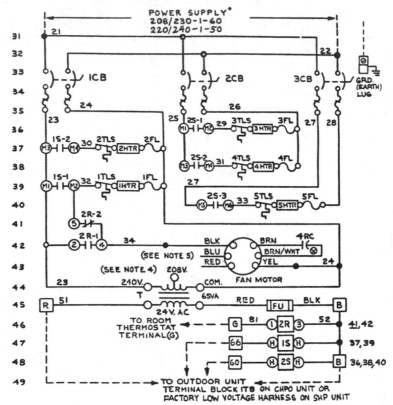

NOTES:

1. ALL FIELD WIRING TO BE IN ACCORDANCE WITH NATIONAL ELECTRIC CODE (N.E.C.), CANADIAN ELECTRIC CODE (C.E.C.), AND/OR LOCAL OR CITY CODES. PROVIDE FUSED DISCONNECTS OR CIRCUIT BREAKERS FOR ALL POWER SUPPLIES.

2. DRAWING PRACTICES AND SYMBOLS ARE IN ACCORDANCE WITH AIR CONDITIONING AND REFRIGERATION INSTITUTE (ARI) GRAPHIC ELECTRICAL STANDARDS.

3. IF ANY ORIGINAL WIRE MUST BE REPLACED, IT MUST BE REPLACED WITH TYPE AWN, 105°C WIRE OR EQUIV.

4. DIAGRAM SHOWN FOR 230 VOLT APPLICATION. FOR 208 VOLT UNITS, CONNECT WIRE CODED (23) TO TRANSFORMER TERMINAL MARKED 208V.

5. FAN MULTI-SPEED MOTOR POWER LEADS ARE FIELD CONNECTED AS FOLLOWS:
   YELLOW LEAD AS SHOWN, BLACK LEAD (HIGH SPEED), BLUE LEAD, (MEDIUM SPEED), RED LEAD (LOW SPEED). SEE INSTALLATION INSTRUCTIONS FOR AIR FLOW AND STATIC PRESSURE DATA.

6. SEQUENCER TIMINGS INDICATED ARE MAXIMUM WITHIN THE SEQUENCE SHOWN.

7. MOTORS ARE INHERENTLY PROTECTED.

*SEE UNIT DATA PLATE FOR PROPER POWER SUPPLY.

| SEQUENCER TIMING (SEE NOTE 6) | | | | | | |
|---|---|---|---|---|---|---|
| SEQ. | TERM. | TIME ON | | | TIME OFF | |
| 1S | M1—M2 | OPEN | CLOSED | | CLOSED | OPEN |
| 1S | M3—M4 | OPEN | | CLOSED | CLOSED | OPEN |
| 2S | M1—M2 | OPEN | CLOSED | | CLOSED | OPEN |
| 2S | M3—M4 | OPEN | | CLOSED | CLOSED | OPEN |
| 2S | M5—M6 | OPEN | | CLOSED | CLOSED | OPEN |
| SECONDS | 0 | 24 | 110 | 0 | 40 | 110 |

372

## CONNECTION DIAGRAM

035-03044C REV. C

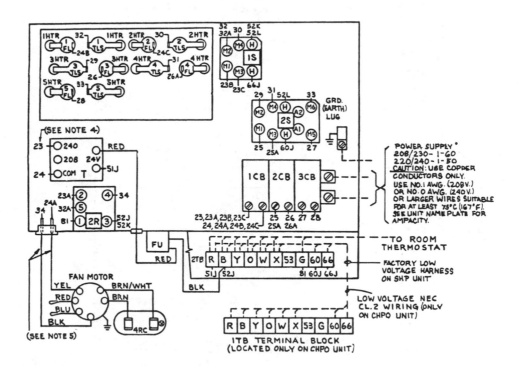

| | LEGEND |
|---|---|
| 1CB ETC. | CIRCUIT BREAKERS |
| 1FL ETC. | FUSIBLE LINK |
| 1HTR ETC. | ELEMENT, HEATER |
| 1S ETC. | SEQUENCER, HEATER |
| 1TLS ETC. | SWITCH, TEMP. LIMIT |
| T | TRANSFORMER |
| 2R | RELAY, FAN MOTOR |
| FU | FUSE 3.2 AMPS |
| 4RC | CAPACITOR, RUN, FAN MOTOR |
| ⊗ | IDENTIFIED TERMINAL ON RUN CAPACITOR |
| 2TB ☐ | TERMINAL BLOCK 24V |
| ———— | WIRING & DEVICES BY YORK |
| — — — | FIELD WIRING BY OTHERS—EXCEPT AS NOTED |

*SEE UNIT DATA PLATE FOR PROPER POWER SUPPLY.

# HEAT PUMP MODELS
## CHPI42 AND SHP42
## WITH HEAT ACCESSORY HPB20-6A

---

### ELEMENTARY DIAGRAM

035-03040C REV. C

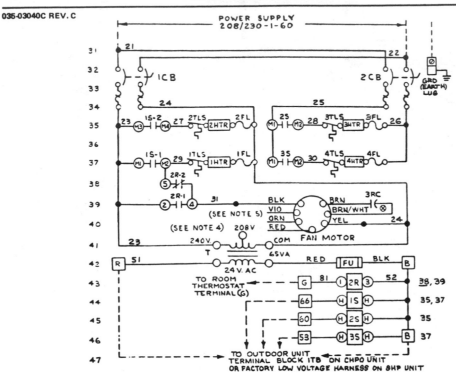

NOTES:

1. ALL FIELD WIRING TO BE IN ACCORDANCE WITH
   NATIONAL ELECTRIC CODE (N.E.C.), CANADIAN ELEC-
   TRIC CODE (C.E.C.), AND/OR LOCAL OR CITY CODES.
   PROVIDE FUSED DISCONNECTS OR CIRCUIT BREAKERS
   FOR ALL POWER SUPPLIES.

2. DRAWING PRACTICES AND SYMBOLS ARE IN ACCORD-
   ANCE WITH AIR CONDITIONING AND REFRIGERATION
   INSTITUTE (ARI) GRAPHIC ELECTRICAL STANDARDS.

3. IF ANY ORIGINAL WIRE MUST BE REPLACED, IT MUST
   BE REPLACED WITH TYPE AWM, 105°C WIRE OR EQUIV.

4. DIAGRAM SHOWN FOR 230 VOLT APPLICATIONS. FOR
   208 VOLT UNITS, CONNECT WIRE CODED (23) TO TRANS-
   FORMER TERMINAL MARKED 208V.

5. FAN MULTI-SPEED MOTOR POWER LEADS ARE FIELD
   CONNECTED AS FOLLOWS:
   YELLOW LEAD AS SHOWN, BLACK LEAD (HIGH SPEED),
   VIOLET LEAD (MED. HIGH SPEED), RED LEAD (LOW
   SPEED),
   ORANGE LEAD (MED. LOW SPEED).
   SEE INSTALLATION INSTRUCTIONS FOR AIR FLOW AND
   STATIC PRESSURE DATA.

6. SEQUENCER TIMINGS INDICATED ARE MAXIMUM WITH-
   IN THE SEQUENCE SHOWN.

7. SEQUENCER (3S) IS ONLY ENERGIZED WHEN COMPRES-
   SOR IS LOCKED-OUT OR SYSTEM IS ON EMERGENCY
   HEAT.

8. MOTORS ARE INHERENTLY PROTECTED.

| SEQUENCER TIMING (SEE NOTES 6 & 7) | | | | | | | |
|---|---|---|---|---|---|---|---|
| SEQ. | TERM. | TIME ON | | | TIME OFF | | |
| 1S | M1–M2 | OPEN | CLOSED | | CLOSED | | OPEN |
| 1S | M3–M4 | OPEN | | CLOSED | CLOSED | OPEN | |
| 2S | M1–M2 | OPEN | CLOSED | | CLOSED | OPEN | |
| 3S | M1–M2 | OPEN | CLOSED | | CLOSED | OPEN | |
| SECONDS | 0 | 24 | 110 | 0 | 40 | 110 | |

374

## CONNECTION DIAGRAM

035-03040C REV. C

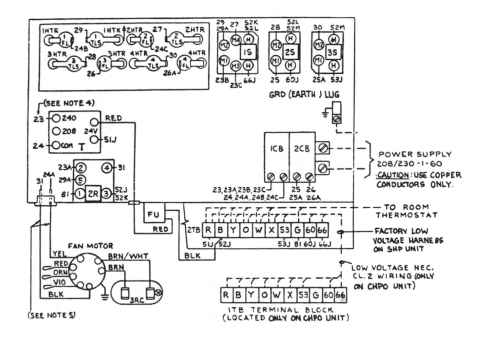

| LEGEND | |
|---|---|
| 1FL ETC. | FUSIBLE LINK |
| 1HTR. ETC. | ELEMENT, HEATER |
| 1S ETC. | SEQUENCER, HEATER |
| 1TLS ETC. | SWITCH, TEMP. LIMIT |
| T | TRANSFORMER |
| 2R | RELAY, FAN MOTOR |
| FU | FUSE, 3.2 AMPS |
| 3RC | CAPACITOR, RUN, FAN MOTOR |
| $\otimes$ | IDENTIFIED TERMINAL ON RUN CAPACITOR |
| 2TB ☐ | TERMINAL BLOCK 24V |
| 1CB ETC. | CIRCUIT BREAKER |
| ——————— | WIRING & DEVICES BY YORK |
| — — — — — | FIELD WIRING BY OTHERS—EXCEPT AS NOTED |

# HEAT PUMP MODELS
## CHPI42, 48, 60 AND SHP42, 48, 60
## WITH HEAT ACCESSORY AS SHOWN BELOW*
*CHPI42 OR SHP42 – WITH HPC15-6A
CHPI48, CHPI60, SHP48 OR SHP60 – WITH HPD15-6A

---

## ELEMENTARY DIAGRAM

035-03039C REV. D

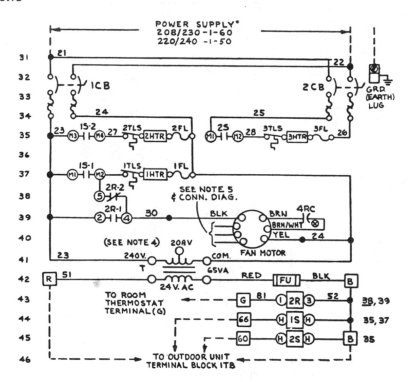

NOTES:

1. ALL FIELD WIRING TO BE IN ACCORDANCE WITH NATIONAL ELECTRIC CODE (N.E.C.), CANADIAN ELECTRIC CODE (C.E.C.), AND/OR LOCAL OR CITY CODES. PROVIDE FUSED DISCONNECTS OR CIRCUIT BREAKERS FOR ALL POWER SUPPLIES.

2. DRAWING PRACTICES AND SYMBOLS ARE IN ACCORDANCE WITH AIR CONDITIONING AND REFRIGERATION INSTITUTE (ARI) GRAPHIC ELECTRICAL STANDARDS.

3. IF ANY ORIGINAL WIRE MUST BE REPLACED, IT MUST BE REPLACED WITH TYPE AWM, 105°C WIRE OR EQUIV.

4. DIAGRAM SHOWN FOR 230 VOLT APPLICATION. FOR 208 VOLT UNITS, CONNECT WIRE CODED (23) TO TRANSFORMER TERMINAL MARKED 208V.

5. FAN MULTI-SPEED MOTOR POWER LEADS ARE FIELD CONNECTED.
   FOR MODELS CHIP42 & 48 CONNECT AS FOLLOWS:
   YELLOW LEAD AS SHOWN, BLACK LEAD (HIGH SPEED), VIOLET LEAD (MED. HIGH SPEED), RED LEAD (LOW SPEED),
   ORANGE LEAD (MED. LOW SPEED).

   FOR MODEL CHPI60 ONLY
   YELLOW LEAD AS SHOWN, BLACK LEAD (HIGH SPEED), BLUE LEAD (MEDIUM SPEED), RED LEAD (LOW SPEED).
   SEE INSTALLATION INSTRUCTIONS FOR AIR FLOW AND STATIC PRESSURE DATA.

6. SEQUENCER TIMINGS INDICATED ARE MAXIMUM WITHIN THE SEQUENCE SHOWN.

7. MOTORS ARE INHERENTLY PROTECTED.

*SEE UNIT DATA PLATE FOR PROPER POWER SUPPLY.

| SEQUENCER TIMING (SEE NOTE 6) | | | | | | |
|---|---|---|---|---|---|---|
| SEQ. | TERM. | TIME ON | | | TIME OFF | |
| 1S | M1–M2 | OPEN | CLOSED | | CLOSED | OPEN |
| 1S | M3–M4 | OPEN | | CLOSED | CLOSED | OPEN |
| 2S | M1–M2 | OPEN | | CLOSED | CLOSED | OPEN |
| SECONDS | 0 | 24 | 110 | 0 | 40 | 110 |

376

## CONNECTION DIAGRAM

035-03039C REV. D

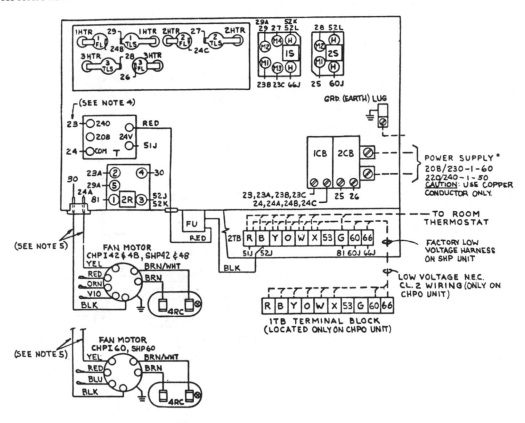

| | LEGEND | |
|---|---|---|
| 1FL ETC. | | FUSIBLE LINK |
| 1HTR ETC. | | ELEMENT, HEATER |
| 1S ETC. | | SEQUENCER, HEATER |
| 1TLS ETC. | | SWITCH, TEMP. LIMIT |
| T | | TRANSFORMER |
| 2R | | RELAY, FAN MOTOR |
| FU | | FUSE 3.2 AMPS |
| 4RC | | CAPACITOR, RUN, FAN MOTOR |
| ⊗ | | IDENTIFIED TERMINAL ON RUN CAPACITOR |
| 1CB ETC. | | CIRCUIT BREAKER |
| 2TB | ☐ | TERMINAL BLOCK 24V |
| ——————— | | WIRING & DEVICES BY YORK |
| – – – – – | | FIELD WIRING BY OTHERS—EXCEPT AS NOTED |

*SEE UNIT DATA PLATE FOR PROPER POWER SUPPLY.

# HEAT PUMP MODELS
## CHPI36 AND SHP36
## WITH HEAT ACCESSORY HPB15-6A

---

### ELEMENTARY DIAGRAM

---

035-03037C REV. D

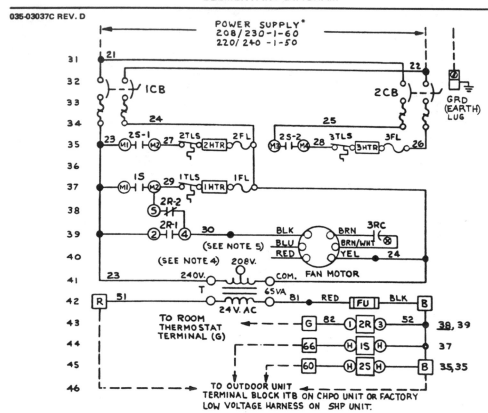

NOTES:

1. ALL FIELD WIRING TO BE IN ACCORDANCE WITH NATIONAL ELECTRIC CODE (N.E.C.), CANADIAN ELECTRIC CODE (C.E.C.), AND/OR LOCAL OR CITY CODES. PROVIDE FUSED DISCONNECTS OR CIRCUIT BREAKERS FOR ALL POWER SUPPLIES.

2. DRAWING PRACTICES AND SYMBOLS ARE IN ACCORDANCE WITH AIR CONDITIONING AND REFRIGERATION INSTITUTE (ARI) GRAPHIC ELECTRICAL STANDARDS.

3. IF ANY ORIGINAL WIRE MUST BE REPLACED, IT MUST BE REPLACED WITH TYPE AWM, 105°C WIRE OR EQUIV.

4. DIAGRAM SHOWN FOR 230 VOLT APPLICATION. FOR 208 VOLT UNITS, CONNECT WIRE CODED (23) TO TRANSFORMER TERMINAL MARKED 208V.

5. FAN MULTI-SPEED MOTOR POWER LEADS ARE FIELD CONNECTED AS FOLLOWS:
   YELLOW LEAD AS SHOWN, BLACK LEAD (HIGH SPEED), BLUE LEAD (MEDIUM SPEED), RED LEAD (LOW SPEED). SEE INSTALLATION INSTRUCTIONS FOR AIR FLOW AND STATIC PRESSURE DATA.

6. SEQUENCER TIMINGS INDICATED ARE MAXIMUM WITHIN THE SEQUENCE SHOWN.

7. MOTORS ARE INHERENTLY PROTECTED.

*SEE UNIT DATA PLATE FOR PROPER POWER SUPPLY.

| SEQUENCER TIMING (SEE NOTE 6) | | | | | | |
|---|---|---|---|---|---|---|
| SEQ. | TERM. | TIME ON | | TIME OFF | | |
| 1S | M1—M2 | OPEN | CLOSED | CLOSED | | OPEN |
| 2S | M1—M2 | OPEN | CLOSED | CLOSED | | OPEN |
| 2S | M3—M4 | OPEN | CLOSED | CLOSED | | OPEN |
| SECONDS | | 0    24 | 110 | 0 | 40 | 75 |

## CONNECTION DIAGRAM

035-03037C REV. D

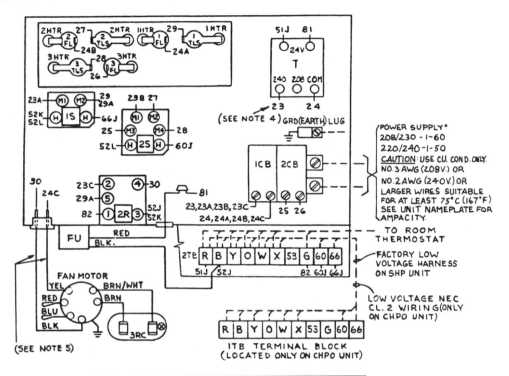

| LEGEND | |
|---|---|
| 1FL ETC. | FUSIBLE LINK |
| 1HTR ETC. | ELEMENT, HEATER |
| 1S ETC. | SEQUENCER, HEATER |
| 1TLS ETC. | SWITCH, TEMP. LIMIT |
| T | TRANSFORMER |
| 2R | RELAY, FAN MOTOR |
| FU | FUSE 3.2 AMPS |
| 3 RC | CAPACITOR, RUN, FAN MOTOR |
| ⊗ | IDENTIFIED TERMINAL ON RUN CAPACITOR |
| 1CB ETC. | CIRCUIT BREAKER |
| 2TB ☐ | TERMINAL BLOCK 24V |
| ──────── | WIRING & DEVICES BY YORK—EXCEPT AS NOTED |
| ─── ─── ─── | FIELD WIRING BY OTHERS |

*SEE UNIT DATA PLATE FOR PROPER POWER SUPPLY.

# HEAT PUMP MODELS
## CHPI30 AND SHP30
## WITH HEAT ACCESSORY HPA15-6A

---

### ELEMENTARY DIAGRAM

---

035-03036C REV. D

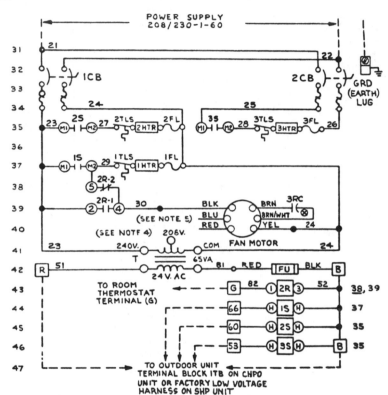

NOTES:

1. ALL FIELD WIRING TO BE IN ACCORDANCE WITH NATIONAL ELECTRIC CODE (N.E.C.), CANADIAN ELECTRIC CODE (C.E.C.), AND/OR LOCAL OR CITY CODES. PROVIDE FUSED DISCONNECTS OR CIRCUIT BREAKERS FOR ALL POWER SUPPLIES.

2. DRAWING PRACTICES AND SYMBOLS ARE IN ACCORDANCE WITH AIR CONDITIONING AND REFRIGERATION INSTITUTE (ARI) GRAPHIC ELECTRICAL STANDARDS.

3. IF ANY ORIGINAL WIRE MUST BE REPLACED, IT MUST BE REPLACED WITH TYPE AWM, 105°C WIRE OR EQUIV.

4. DIAGRAM SHOWN FOR 230 VOLT APPLICATION. FOR 208 VOLT UNITS, CONNECT WIRE CODED (23) TO TRANSFORMER TERMINAL MARKED 208V.

5. FAN MULTI-SPEED MOTOR POWER LEADS ARE FIELD CONNECTED AS FOLLOWS:
   YELLOW LEAD AS SHOWN, BLACK LEAD (HIGH SPEED), BLUE LEAD (MEDIUM SPEED), RED LEAD (LOW SPEED). SEE INSTALLATION INSTRUCTIONS FOR AIR FLOW AND STATIC PRESSURE DATA.

6. SEQUENCER TIMINGS INDICATED ARE MAXIMUM WITHIN THE SEQUENCE SHOWN.

7. SEQUENCER (3S) IS ONLY ENERGIZED WHEN COMPRESSOR IS LOCKED-OUT OR SYSTEM IS ON EMERGENCY HEAT.

8. MOTORS ARE INHERENTLY PROTECTED.

| SEQUENCER TIMING (SEE NOTE 6 & 7) | | | | | | |
|---|---|---|---|---|---|---|
| SEQ. | TERM. | TIME ON | | | TIME OFF | |
| 1S | M1–M2 | OPEN | CLOSED | | CLOSED | OPEN |
| 2S | M1–M2 | OPEN | | CLOSED | CLOSED | OPEN |
| 3S | M1–M2 | OPEN | | CLOSED | CLOSED | OPEN |
| SECONDS | 0 | 24 | 70 | 0 | 40 | 75 |

## CONNECTION DIAGRAM

035-03036C REV. D

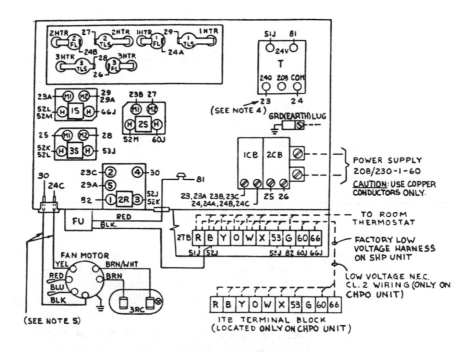

| | LEGEND | |
|---|---|---|
| 1FL ETC. | | FUSIBLE LINK |
| 1HTR ETC. | | ELEMENT, HEATER |
| 1S ETC. | | SEQUENCER, HEATER |
| 1TLS ETC. | | SWITCH, TEMP, LIMIT |
| T | | TRANSFORMER |
| 2R | | RELAY, FAN MOTOR |
| FU | | FUSE 3.2 AMPS |
| 3RC | | CAPACITOR, RUN, FAN MOTOR |
| ⊗ | | IDENTIFIED TERMINAL ON RUN CAPACITOR |
| 1CB ETC. | | CIRCUIT BREAKER |
| 2TB | ☐ | TERMINAL BLOCK 24V |
| | | WIRING & DEVICES BY YORK |
| | | FIELD WIRING BY OTHERS—EXCEPT AS NOTED. |

# Glossary

**Absolute humidity:** The weight in grains of water vapor actually contained in 1 cubic foot (0.0283 m³) of the air and moisture mixture.

**Absolute pressure:** The sum of gauge pressure plus atmospheric pressure.

**Absolute temperature:** The temperature at which all molecular motion of a substance stops. At this temperature the substance theoretically contains no heat.

**Absorbent:** A substance that has the ability to absorb another substance.

**Accessible hermetic:** A single unit containing the motor and compressor; this unit may be field serviceable.

**Accumulator:** A shell placed in the suction line to prevent liquid refrigerant from entering the compressor by vaporizing the liquid.

**Acid condition:** A condition in which the refrigerant and/or oil in the system is contaminated with other fluids that are acidic in nature.

**ACR tubing:** Tubing used in the refrigeration and air-conditioning industry. The ends are sealed and the tubing is dehydrated.

**Air:** An elastic gas. Air is a mechanical mixture of oxygen and nitrogen and slight traces of other gases; it also may contain moisture known as humidity. Dry air weighs 0.075 lb/ft³ (0.0012 g/m³). One Btu (252 calories) will raise the temperature of 55 ft³ (1.56 m³) of air 1 °F (0.56 °C).

**Air change:** The number of times in an hour the air in a room is changed either by mechanical means or by the infiltration of outside air leaking into the room through cracks around doors and windows and through other openings.

**Air cleaner:** A device designed to remove airborne impurities such as dust, fumes, and smoke. Air cleaners include air washers and air filters.

**Air coil:** The coils on heat pump systems through which air is blown.

**Air conditioning:** The simultaneous control of the temperature, humidity, air motion, and air distribution within an enclosure. Where human comfort and health are involved, a reasonable air purity with regard to dust, bacteria, and odors is also included. The primary requirement of a good air-conditioning system is a good heating system.

**Air-conditioning unit:** A device designed for the treatment of air; it consists of a means for ventila-

382

tion, air circulation, air cleaning, and heat transfer, with a control to maintain the temperature within the prescribed limits.

*Air diffuser:*   An air-distribution element or outlet designed to direct the flow of air in the desired patterns.

*Air handler:*   A unit consisting of the fan or blower, heat transfer element, filter, and housing components of an air-distribution system.

*Air infiltration:*   The leakage of air into a house through cracks and crevices and through doors, windows, and other openings by wind pressure and/or temperature difference.

*Air-sensing thermostat:*   A thermostat that operates as a result of air temperature. The sensing bulb is located in the air stream.

*Air, standard:*   Air with a temperature of 68 °F (20 °C), a relative humidity of 36%, and a pressure of 14.7 psi (1.03 kPa).

*Algae:*   A low form of plant life found in water. It is especially troublesome in water towers.

*Alternating current:*   Electrical current in which the direction of current reverses periodically, usually 60 times per second. Also called 60-cycle current.

*Ambient temperature:*   The temperature of the air that surrounds an object.

*Ammeter:*   An electric meter calibrated in amperes that is used to measure current flow in a circuit.

*Amperage:*   The unit of electrical current equivalent to a flow of 1 coulomb per second.

*Analyzer:*   A device used to check the condition of electrical components.

*Anemometer:*   An instrument used to measure the velocity of air motion.

*Apparatus dew point:*   The dew point of the air leaving the cooling coil.

*Atmospheric pressure:*   The pressure exerted by the atmosphere in all directions, as indicated by a barometer. Standard atmospheric pressure at sea level is considered to be 14.7 psi (1.03 kPa).

*Automatic control:*   The control of the various functions of a piece of equipment without manual adjustment.

*Automatic defrost:*   A method of automatically removing ice and frost from the evaporating coil.

*Back pressure:*   The pressure in the low-pressure side of the refrigerant circuit.

*Back seating:*   The wide-open position of a service valve, which usually closes off the gauge connection fitting.

*Baffle:*   A plate or wall for deflecting gases or fluids.

*Ball check valve:*   A valve assembly that allows the flow of fluids in only one direction.

*Barometer:*   An instrument that measures atmospheric pressure.

*Bearing:*   A low-friction device used for supporting and aligning moving parts.

*Bimetal:*   Two metals each having different rates of expansion for each change in temperature. They are fused together in such a manner as to form one piece. When they are heated or cooled, one will expand at a faster rate than the other, which causes a bending of the two metals. This action can be used to open or close electrical contacts or other suitable movement.

*Bleed:*   To release pressure slowly from a system or cylinder by opening a valve slightly.

*Bleed valve:*   A valve with a small opening through which pressure is permitted to escape at a slow rate when the valve is closed.

*Boiling point:*   The boiling temperature of a liquid under a pressure of 14.7 psi (1.03 kPa).

*Boyle's law:*   A law of physics that deals with the volume of gases as the pressure varies. If the temperature remains constant, the volume varies. If the pressure is increased, the volume is decreased; if the pressure is reduced, the volume is increased.

*Braze:*   To join metals with a nonferrous filler using heat between the temperature of 800 °F (427 °C) and the melting point of the base metal.

*British thermal unit (Btu):*   The quantity of heat required to raise the tempeature of 1 lb. (453.6 g) of water 1 °F (0.56 °C). (This measurement is approximate but is sufficiently accurate for any work discussed in this book.)

*Bypass:*   A pipe or duct usually controlled by a valve or a damper for short circuiting the flow of a fluid.

*Calibrate:*   To determine the position of indicators as necessary to obtain accurate measurements.

*Calorie:* The calorie, as used in engineering, is a large heat unit and is equal to the amount of heat required to raise 1 kg of water 1 °C.

*Capacitance:* The property of an electric current that permits the storage of electrical energy in an electrostatic field and the release of that energy at a later time.

*Capacitor:* An electrical device capable of storing electrical energy.

*Capacitor-start motor:* A motor that uses a capacitor in the starting circuit to increase the starting torque.

*Capacity:* The capacity of a refrigeration unit is the heat-absorbing capacity per unit of time. Usually measured in Btu per hour.

*Capillary tube:* A tube of small diameter used to regulate the flow of liquid refrigerant into the evaporator.

*Celsius scale:* A thermometer scale on which the freezing temperature is 0° and the boiling temperature is 100° when water is the medium tested. Formerly, centigrade scale.

*Centimeter:* A metric unit of linear measurement equal to 0.3937 in.

*Central fan system:* A mechanical indirect system of heating, ventilating, or air conditioning consisting of a central plant where the air is conditioned and then circulated by fans or blowers through a system of distributing ducts.

*Charge:* The amount of refrigerant in a system. Also, to put the refrigerant into a system.

*Charging board:* A specially designed panel fitted with gauges, valves, and refrigerant cylinders and used for evacuating and charging refrigerant and oil into a system.

*Check valve:* A device that allows the flow of a fluid in only one direction.

*Chill factor:* A calculated figure based on the dry bulb temperature and wind velocity.

*Circuit:* The tubing, piping, or electrical wiring that permits flow from the energy source through the circuit and back to the energy source.

*Circuit breaker:* A safety device used to protect an electrical circuit from overload conditions.

*Circuit, parallel:* The arrangement of electrical devices so that the current is equally divided between each circuit.

*Circuit, pilot:* A secondary circuit used to control or signal a device in the main circuit, usually of different voltage.

*Circuit, series:* The arrangement of electrical devices in such a way that the current passes through each of the devices one after the other.

*Closed circuit:* A completed electrical circuit through which the electrons are flowing.

*Closed cycle:* Any cycle in which the refrigerant is used over and over again.

*Coefficient of heat transmission (U):* The amount of heat transmitted from air to air in 1 h/ft$^2$ of the wall, floor, roof, or ceiling for a difference in temperature of 1 °F (0.56 °C) between the air inside and the air outside of the wall, floor, roof, or ceiling.

*Coefficient of performance (COP):* The ratio of work performed to energy used.

*Cold:* The absence of heat.

*Comfort zone:* The range of effective temperatures over which the majority of adults feel comfortable.

*Compound gauge:* An instrument used to indicate the pressure both above and below atmospheric pressure. Commonly used to measure low-side pressures.

*Compression ratio:* The ratio of the volume of the clearance space to the total volume of the cylinder. It is also used as the ratio of the absolute suction pressure to the absolute discharge pressure.

*Compression system:* A refrigeration system in which the pressure-imposing element is mechanically operated. Distinguished from an absorption system, which uses no compressor.

*Compressor:* The device used for increasing the pressure on the refrigerant.

*Compressor displacement:* The volume in inches (millimeters) represented by the area of the piston head multiplied by the length of the piston stroke.

*Condensate:* The moisture resulting from the removal of heat from a vapor to bring it below the dew-point temperature.

*Condensing medium:* A substance, such as air or water, that is used to remove heat from the condensing coil to change the refrigerant vapor to a liquid.

**Condensing pressure:** The pressure inside the condensing coil at which the refrigerant vapor gives up latent heat of vaporization and changes to a liquid. This pressure varies with the ambient temperature.

**Condensing temperature:** The temperature inside the condensing coil at which the vaporous refrigerant gives up latent heat of vaporization and becomes a liquid. This temperature varies with the pressure.

**Conduction:** The transmission of heat through and by means of matter.

**Conductivity:** The amount of heat in Btus (calories) transmitted in 1 hour through 1 ft² (0.0929 m²) of a homogeneous material 1 in. (25.4 mm) thick for a difference in temperature of 1 °F (0.56 °C) between the two surfaces of the material.

**Conductor:** A substance or body capable of conducting electrical or heat energy.

**Constrictor:** A device used to restrict the flow of a vapor or liquid.

**Contactor:** An electromagnetic relay that is used for repeatedly starting and stopping an electric device such as a compressor motor, fan motor, or heat strip.

**Contaminant:** A substance, such as dirt or moisture, that is foreign to the refrigerant or the oil in the system.

**Control:** Any manual or automatic device used for the regulation of a machine in normal operation. It is usually responsive to temperature or pressure, but not to both at the same time.

**Control system:** All the components used for the automatic control of a given process.

**Control valve:** A valve that regulates the flow of a medium that affects the controlled process. The valve is controlled by a remote signal from other devices, which use pneumatic, electric, or electrohydraulic devices for power.

**Convection:** The transmission of heat by the circulation of a liquid or gas such as air. When it occurs naturally, it is caused by the difference in weight between hotter or colder fluids.

**Cooling anticipator:** A nonadjustable heating resistor that is placed in parallel with the cooling circuit inside the thermostat. It produces a false heat in the thermostat during the equipment off cycle, to cause a shorter off cycle, to maintain a more even temperature inside the space.

**Cooling tower:** A device that cools to the wet bulb temperature by means of evaporation of water.

**Corrosion:** The deterioration of a metal by chemical action.

**Counterflow:** The opposing direction of fluids, the coldest portion of one meeting with the warmest portion of the other.

**Critical pressure:** The compressed condition of a refrigerant at which the vapor and liquid have the same properties.

**Critical temperature:** The temperature above which a vapor cannot be liquified.

**Cross charged:** The combining of two fluids to create a desired pressure-temperature curve.

**Cubic feet per minute (cfm):** The volume of air passing through a piece of equipment. The movement is expressed as cubic feet per minute.

**Current:** The flow of electrical energy in a conductor, which occurs when the electrons change positions.

**Cut in:** The temperature or pressure at which a set of contacts completes an electrical circuit.

**Cut out:** The temperature or pressure at which a set of electrical contacts opens an electrical circuit.

**Cycle:** The complete course of operations of a refrigeration system. This includes the four major functions of compression, condensation, expansion, and evaporation.

**Defrost:** To remove accumulated ice from an evaporating coil.

**Defrost cycle:** The refrigeration cycle in which the ice accumulation is melted from the coil.

**Defrost timer:** A device connected into the electrical circuit to start the defrost cycle and keep it on until the ice has melted from the coil.

**Degree:** The unit of measure on a temperature scale.

**Degree day:** A unit that represents one degree of difference between the inside temperature and the average outdoor temperature for one day. Normally used for estimating the fuel requirements of a building.

**Degree of superheat:** The difference between the boiling point of a refrigerant and the actual temperature above the boiling point.

**Dehumidifier:** A device used to lower the moisture content of the air passing through it.

**Dehumidify:** To remove water or moisture from the atmosphere; to remove water vapor or moisture from any material.

**Dehydrated oil:** A lubricant from which the moisture has been removed to an acceptable level.

**Desiccant:** A substance used to collect and hold moisture in a refrigeration system. The most commonly used desiccants are activated alumina and silica gel.

**Design pressure:** The highest pressure expected to be reached during normal operation. It is usually the operating pressure plus a safety factor.

**Dew point:** The temperature at which a vapor begins to condense. Usually 100% humidity.

**Differential:** The difference between the cut in and cut out of a pressure or temperature control.

**Discharge line:** A tube through which the refrigerant is discharged from the compressor to the condensing coil inlet.

**Discharge pressure:** The pressure produced by the compressing action of the compressor and is measured at the outlet of the compressor. Sometimes also called the head pressure.

**Double pole:** A contact arrangement that includes two separate contact forms, that is, two single-pole contact assemblies.

**Double throw:** A contact arrangement in which each contact form included is a break-make; that is, one contact opens its connection to another contact and then closes its connection to a third contact.

**Drier:** A device used to remove moisture from a refrigerant.

**Drip pan:** A pan or trough used to collect condensate from an evaporating coil.

**Dry bulb:** A thermometer used to measure the ambient air temperature.

**Dry bulb temperature:** The actual temperature of the air, as opposed to wet bulb temperature. The temperature of the air indicated by any thermometer not affected by the moisture content of the air.

**Duct:** A tube or channel through which air is forced in a forced-air system.

**Effective area:** The gross area of the grill minus the area of the vanes or bars. The actual flow area of an air inlet or outlet.

**Effective temperature:** The overall effect on a human being of the air temperature, humidity, and air movement.

**Electric defrosting:** A method of defrosting an evaporating coil by using electric current to heat the surface.

**Electric heating:** A heating system in which the source of heat is electricity.

**Electrolysis:** The chemical reaction of two substances due to the flow of electricity through them.

**Electrolytic capacitor:** A plate or surface capable of storing small electrical charges.

**Electronic leak detector:** An electronic instrument that senses refrigerant vapor in the atmosphere. An electronic flow change indicates a leak.

**Enthalpy:** The actual or total heat contained in a substance. It is usually calculated from a base. In refrigeration work the base temperature is accepted as $-40\,°F$ ($-40\,°C$).

**Environment:** The conditions of the surroundings.

**Evacuation:** The removal of air and moisture from a refrigeration system.

**Evaporation:** The change of state of a liquid to a vapor. The greatest amount of heat is absorbed in this process.

**Exhaust valve:** The outlet port that allows the compressed vapor to escape from the cylinder. Also called the discharge valve.

**Expansion valve:** The device in a refrigeration system that reduces a high-pressure refrigerant to a low-pressure refrigerant.

**External equalizer:** The tube connected to the low-pressure side of the diaphragm in the thermostatic expansion valve. It lets the evaporating coil outlet pressure help control the operation of the valve.

**Fahrenheit scale:** The scale on a standard U.S. thermometer with the boiling point of water at $212\,°F$ and the freezing point at $32\,°F$.

**Fail-safe control:** A device that opens an electric current when the sensing element senses an abnormal condition.

**Fan:** An enclosed propeller that produces motion in air. Commonly designates anything that causes air motion.

**Farad:** The unit of electrical capacity of a capacitor.

**Feet per minute (fpm):** A velocity measurement of a moving airstream. Usually expressed as fpm.

**Filter:** A device used to remove the solid particles from a fluid by straining.

**Fin:** The sheet-metal extension on evaporating and condensing coil tubes.

**Flash gas:** The gas that is the result of the instantaneous evaporation of refrigerant in a pressure-reducing device. It cools the remaining refrigerant to the evaporation temperature that exists at the reduced pressure.

**Flooded system:** A system in which the refrigerant enters the evaporating coil from a pressure-reducing valve and in which the evaporating coil is partly filled with liquid refrigerant.

**Flush:** To remove foreign materials or fluids from a refrigeration system through the use of refrigerant or other fluids.

**Foaming:** The formation of foam in the crankcase as a result of liquid refrigerant being absorbed in the oil and rapidly evaporating. This is most likely to occur on compressor startup.

**Foam leak detector:** Soap bubbles or a special foaming liquid brushed over suspected areas to locate leaks.

**Forced-feed oiling:** A lubrication system that uses a pump to force the oil to the surfaces of the moving parts.

**Freeze-up:** The formation of ice in the flow-control device as a result of the presence of moisture in the system. Also, a frost formation on an evaporating coil that reduces the flow of air through the fins.

**Freezing:** The change of state from a liquid to a solid.

**Freezing point:** The temperature at which a liquid will change to a solid when heat is removed. The temperature at which a given substance freezes.

**Fuse:** An electrical safety device consisting of a strip of fusible metal used to prevent overloading of an electrical circuit.

**Fusible:** Capable of being melted.

**Fusible plug:** A safety plug used in refrigerant containers that melts at a high temperature to prevent excessive pressure from bursting the container.

**Gas:** The vapor state of a substance.

**Gasket:** A resilient or flexible material used between mating surfaces to provide a leakproof seal.

**Gauge:** An instrument used for measuring pressures both above and below atmospheric pressure.

**Gauge manifold:** A manifold that holds both the pressure and compound gauges, the valves that control the flow of fluids through the manifold ports, and charging hose connections.

**Gauge port:** The opening or connection provided for the installation of gauges.

**Gauge pressure:** A measure of pressure taken with a gauge. Pressure measured from atmospheric pressure as opposed to absolute pressure.

**Grain:** A unit of weight equal to 1/7000 lb (0.06480 g), which is used to indicate the amount of moisture in the air.

**Grill:** A perforated covering for an air inlet or outlet usually made of wire screen, processed steel, or cast iron.

**Ground coil:** A heat exchanger that is buried in the ground; usually the outdoor coil on a heat pump system.

**Ground wire:** An electrical wire that will safely conduct electricity from a structure or piece of equipment to the ground in case of an electrical short.

**Halide refrigerants:** The family of synthetic refrigerants that contain halogen chemicals.

**Halide torch:** A device used to detect refrigerant leaks in a system. A burner equipped with a source of fuel, a mixing chamber, a reactor plate, and an exploring tube. The reactor plate surrounds the flame. When the open end of the exploring tube is held near a refrigerant leak, some of the refrigerant is drawn into the mixing chamber where its presence changes the color of the flame.

**Halogens:** Substances that contain fluorine, chlorine, bromine, and iodine.

**Hanger:** A device used to support lines by attachment to a wall.

*Head pressure:* The pressure against which the compressor must pump the vapor.

*Head pressure control:* A pressure-operated control that opens the electrical circuit when the head pressure exceeds preset limits.

*Heat:* A form of energy produced through the expenditure of another form of energy.

*Heat content:* The amount of heat, usually stated in Btus per pound (calories per gram), absorbed by a refrigerant in raising its temperature from a predetermined level to a final condition and temperature. Where change of state is encountered, the latent heat necessary for the change is included.

*Heat exchanger:* Any device that removes heat from one substance and adds it to another.

*Heat intensity:* The heat concentration in a substance; indicated by the temperature of the substance through use of a thermometer.

*Heat load:* The amount of heat, measured in Btus (caloreis) or watts, that must be added or removed by a piece of equipment during a 24-hour period.

*Heat of compression:* The heat developed within a compressor when a vapor is compressed as in a refrigeration system.

*Heat of fusion:* The amount of heat required to change a solid to a liquid or a liquid to a solid with no change in temperature. Also called latent heat of fusion.

*Heat of the liquid:* The heat content of a liquid or the heat necessary to raise the temperature of a liquid from a predetermined level to a final temperature.

*Heat of vapor:* The heat content of a vapor or the heat necessary to raise the temperature of a liquid from a predetermined level to the boiling point, plus the latent heat of vaporization necessary to convert a liquid to a vapor.

*Heat pump:* A compression cycle system used to supply heat to a space by reversing the flow of refrigerant from the cooling cycle.

*Heat transfer:* The movement of heat from one body to another. The heat may be transferred by radiation, conduction, or convection.

*Heat unit:* Usually refers to a Btu (calorie).

*Heating anticipator:* An adjustable resistance heater placed in the heating circuit inside the thermostat to provide a false heat during the on cycle of the equipment. It provides a shorter on cycle to provide a more even temperature inside the space.

*Hermetic compressor:* A unit in which the compressor and motor are sealed inside a housing. The motor operates in an atmosphere of refrigerant.

*Hertz (Hz):* The correct terminology for designating electrical cycles per second.

*Hg (mercury):* A silver-white, heavy, liquid metal. It is the only metal that is a liquid at room temperature.

*High-pressure cut out:* An electrical control switch operated by the pressure in the high-pressure side of the system; it automatically opens an electrical circuit when a predetermined pressure is reached.

*High side:* The part of the refrigeration system that contains the high-pressure refrigerant. Also refers to the outdoor unit, which consists of the motor, compressor, outdoor coil, and receiver mounted on one base.

*High-side charging:* The process of introducing liquid refrigerant into the high side of a refrigeration system. The acceptable manner for placing the refrigerant into the system.

*High-vacuum pump:* A vacuum pump capable of creating a vacuum in the range of 1000 to 1 micrometers.

*Holding charge:* A partial charge of refrigerant placed in a piece of refrigeration equipment after dehydration and evacuation either for shipping or testing purposes.

*Holding coil:* That part of a magnetic starter or relay that causes the device to operate when energized.

*Hot gas:* The refrigerant vapor leaving the compressor.

*Hot-gas defrost:* A method of evaporating coil defrosting that uses the hot discharge vapor to remove frost from the coil.

*Hot-gas line:* The line that carries the hot compressed vapor from the compressor to the condensing coil.

*Humidifier:* A device used to add vapor to the air or to any material.

*Humidistat:* An automatic control that is sensitive

to humidity and is used for the automatic control of relative humidity.

*Humidity:* Moisture in the air.

*Humidity, absolute:* The weight of water vapor per unit volume of space occupied; expressed in grains of moisture per cubic foot of dry air.

*Humidity, relative:* The amount of moisture in the air expressed in terms of percentage of total saturation of the existing dry bulb temperature.

*Hunting:* The fluctuation caused by the controls attempting to establish an equilibrium against difficult conditions.

*Inches of water gauge:* A means of measuring small pressures or vacuums by use of a manometer filled with water.

*Infiltration:* The leakage of air into a building or space.

*Inhibitor:* A substance that prevents a chemical reaction such as corrosion or oxidation to metals.

*Instrument:* A term broadly used to designate a device that is used for measuring, recording, indicating, and controlling equipment.

*Insulation, electrical:* A substance that has almost no free electrons.

*Insulation, thermal:* A material that has a high resistance to heat flow.

*Interlock:* A device that prevents certain parts of an air-conditioning or refrigeration system from operating when other parts of that system are not operating.

*Junction box:* A box or container that houses a group of electrical terminals or connections.

*Kelvin scale:* A thermometer scale that is equal to the Celsius scale, and according to which absolute zero is 0 degrees, the equivalent of $-273.16\,^\circ\text{C}$. Water freezes at 173.16 K and boils at 373.16 K on this scale.

*Kilowatt (kW):* A unit of electrical power indicating 1000 watts.

*Kilowatt-hour (kWh):* A unit of electrical energy equal to 1000 watt-hours.

*King valve:* A service valve on the liquid receiver outlet.

*Lag:* A delay in the response to some demand.

*Latent heat:* The heat added to or removed from a substance to change its state but which cannot be measured by a change in temperature.

*Latent heat of condensation:* The heat removed from a vapor to change it to a liquid with no change in temperature.

*Latent heat of vaporization:* The quantity of heat required to change a liquid to a vapor with no change in temperature.

*Leak detector:* A device used to detect leaks; usually a halide torch, soap bubbles, or an electronic leak detector.

*Liquid:* A substance in which the molecules move freely among themselves, but do not tend to separate as in a vapor.

*Liquid absorbent:* A liquid chemical that has the ability to take on or absorb other fluids.

*Liquid charge:* Usually refers to the power element of temperature controls and thermostatic expansion valves. The power element and remote bulb are sometimes charged with a liquid rather than a vapor.

*Liquid filter:* A very fine strainer used to remove foreign matter from the refrigerant.

*Liquid indicator:* A device located in the liquid line with a glass window through which the flow of liquid may be observed.

*Liquid line:* The line carrying the liquid refrigerant from the receiver or condensing coil to the evaporator.

*Liquid receiver:* A cylinder connected to the condensing coil outlet and used to store liquid refrigerant.

*Liquid receiver service valve:* A two- or three-way manually operated valve located at the receiver outlet; it is used for installation and service operations. Also called a king valve.

*Liquid sight glass:* A glass bull's-eye installed in the liquid line to permit visual inspection of the liquid refrigerant. Used primarily to detect bubbles in the liquid, indicating a shortage of refrigerant in the system. Also, a liquid indicator.

*Liquid strainer:* *See* Liquid filter.

*Liquid-vapor valve:* A dual hand valve used on refrigerant cylinders to release either vapor or liquid from the cylinder.

*Liter:* A metric unit of volume measurement equal to 61.03 in³.

*Load:* The required rate of heat removal. Heat per unit of time that is imposed on the system by a particular job.

*Locked rotor amps (LRA):* The amperage flowing in the circuit to a motor-driven apparatus when the rotor of the motor is locked to prevent its movement. This amperage is typically as much as six times the full-load current of the motor and four to five times the full-load current in hermetic compressors.

*Low-pressure control:* A pressure-operated switch in the suction side of a refrigeration system that opens its contacts to stop the compressor at a given cut-out setting.

*Low side:* The parts of a refrigeration system in which the refrigerant pressure corresponds to the evaporating coil pressure.

*Low-side charging:* The process of introducing refrigerants into the low side of the system. Usually reserved for the addition of a small amount of refrigerant after repairs.

*Low-side pressure:* The pressure in the low side of the system.

*Magnetic across-the-line starter:* A motor starter or switch that allows full line voltage to the motor windings when engaged.

*Manifold, service:* A chamber equipped with gauges, manual valves, and charging hoses; it is used in servicing refrigeration units.

*Manometer:* An instrument for measuring small pressures. Also, a U-shaped tube partly filled with liquid.

*Manual shutoff valve:* A hand-operated device that stops the flow of fluids in a piping system.

*Mechanical efficiency:* The ratio of work done by a machine to the energy used to do it.

*Mechanical refrigeration:* A term usually used to distinguish a compression system from an absorption system.

*Megohm:* A measure of electrical resistance. One megohm equals 1 million ohms.

*Megohmmeter:* An instrument used to measure extremely high resistances.

*Melt:* To change state from a solid to a liquid.

*Melting point:* The temperature, measured at atmospheric pressure, at which a substance will melt.

*Meter:* An instrument used for measuring. Also, a unit of length in the metric system.

*Metric system:* The decimal system of measurement.

*Micro:* One-millionth of a specified unit.

*Microfarad:* A unit of electrical capacitance equal to 1/1,000,000 of a farad.

*Micrometer:* A unit of length in the metric system equal to 1/1000 mm.

*Micrometer gauge:* An instrument used for measuring vacuum that is very close to a perfect vacuum.

*Micron:* A metric unit of measurement measuring 1/25,400 of an inch used to measure high vacuums.

*Motor:* A device that converts electrical energy to mechanical energy.

*Motor burnout:* A shorted condition in which the insulation of an electric motor has deteriorated by overheating.

*Motor, capacitor:* A single-phase induction-type motor that utilizes an auxiliary starting winding (a phase winding) connected in series with a capacitor to provide better starting or running characteristics.

*Motor, capacitor start:* An induction motor having separate starting winding. Similar to the split-phase motor except that the capacitor start motor has an electrical capacitor connected to the starting winding for added starting torque.

*Motor, capacitor start and run:* A motor similar to the capacitor start motor except that the capacitor and start winding are designed to remain in the circuit at all times, thus eliminating the switch used to disconnect the start winding.

*Motor control:* A device used to start and/or stop an electric motor or a hermetic motor compressor at certain temperature and/or pressure conditions.

*Motor, shaded pole:* A small induction motor with a shading pole used for starting; has a very low starting torque.

*Motor, split-phase:* An induction motor with a separate winding for starting.

*Motor starter:* A series of electrical switches that are normally operated by electromagnetism.

*Movable contact:* The member of a contact pair that is moved directly by the actuating system.

*Muffler:* A device used in the hot vapor line to silence the compressor discharge surges.

*Natural convection:* The movement of a fluid that is caused by temperature differences.

*Neutralizer:* A substance used to counteract the action of acids.

*Nominal-size tubing:* Tubing with an inside diameter the same as iron pipe of the same inside diameter.

*Noncondensable gas:* Any gas, usually in a refrigeration system, that cannot be condensed at the temperature and pressure at which the refrigerant will condense, and therefore requires a higher head pressure.

*Normal charge:* A charge that is part liquid and part vapor under all operating conditions.

*Normally closed contacts:* A contact pair that is closed when the device is in the de-energized condition.

*Normally open contacts:* A contact pair that is open when the device is in the de-energized condition.

*Odor:* The contaminants in the air that affect the sense of smell.

*Off cycle:* The period when equipment, specifically a refrigeration system, is not in operation.

*Ohm (R):* A unit of measurement of electrical resistance. One ohm exists when 1 volt causes 1 ampere to flow through a circuit.

*Ohmmeter:* An instrument used to measure electrical resistance in ohms.

*Ohm's law:* A mathematical relationship between voltage, current, and resistance in an electric circuit. It is stated as follows: voltage *(E)* = amperes *(I)* × ohms *(R)*.

*Oil binding:* A condition in which a layer of oil on top of liquid refrigerant may prevent it from evaporating at its normal pressure and temperature.

*Oil, compressor lubricating:* A highly refined lubricant made especially for refrigeration compressors.

*Oil, entrained:* Oil droplets carried by high-velocity refrigerant vapor.

*Oil filter:* A device in the compressor used to remove foreign material from the crankcase oil before it reaches the bearing surfaces.

*Oil level:* The level in a compressor crankcase at which oil must be carried for proper lubrication.

*Oil pressure failure control:* A device that acts to shut off a compressor whenever the oil pressure falls below a predetermined point.

*Oil pressure gauge:* A device used to show the pressure of oil developed by the pump within a refrigeration compressor.

*Oil pump:* A device that provides the source of power for forced-feed lubrication systems in refrigeration compressors.

*Oil separator:* A device that separates oil from the refrigerant and returns it to the compressor crankcase.

*Oil sight glass:* A glass bull's-eye in the compressor crankcase that permits visual inspection of the compressor oil level.

*Oil sludge:* Usually a thick, slushy substance formed by contaminated oils.

*Oil trap:* A low spot, sag in the refrigerant lines, or space where oil will collect. Also, a mechanical device for removing entrained oil.

*On cycle:* The period when the equipment, specifically refrigeration equipment, is in operation.

*Open circuit:* An electrical circuit that has been interrupted to stop the flow of electricity.

*Open compressor:* A compressor that uses an external drive.

*Operating cycle:* A sequence of operations under automatic control intended to maintain the desired conditions at all times.

*Operating pressure:* The actual pressure at which the system normally operates.

*Output:* The amount of energy that a machine is able to produce in a given period of time.

*Outside air, fresh air:* Air from outside the conditioned space.

*Overload:* A load greater than that for which the system or machine was designed.

*Overload protector:* A device designed to stop the motor should a dangerous overload condition occur.

*Overload relay:* A thermal device that opens its

contacts when the current through a heater coil exceeds the specified valve for a specified time.

*Oxidize:* To burn, corrode, or rust.

*Oxygen:* An element in the air that is essential to animal life.

*Package units:* A complete set of refrigeration or air-conditioning components located in the refrigerated space.

*Packless valve:* A valve that does not use a packing to seal around the valve stem.

*Performance:* A term frequently used to mean output or capacity, as performance data.

*Performance factor:* The ratio of heat removed by the refrigeration system to the heat equivalent of the energy required to do the job.

*Perimeter duct system:* An air-conditioning sytem in which the associated ductwork is placed around the outside of the building and the air is distributed to the outer walls.

*Pilot control:* A valve arrangement used in an evaporator pressure regulator to sense the pressure in the vapor line and to regulate the action of the main valve.

*Pitch:* The slope of a pipe line used to enhance drainage.

*Pitot tube:* A device that measures air velocity.

*Plenum chamber:* An air compartment to which one or more pressurized distributing ducts are connected.

*Potential relay:* A starting relay that opens its contacts on high voltage on its coil and closes them on low voltage.

*Potentiometer:* A wire-wound coil used for measuring or controlling by sensing small changes in electrical resistances.

*Power element:* The sensitive element of a temperature-operated control.

*Power factor:* The correction factor for the changing of current and voltage of alternating-current electricity.

*Pressure:* The force exerted per unit of area.

*Pressure, absolute:* The pressure measured above an absolute vacuum.

*Pressure, atmospheric:* The pressure exerted by the earth's atmosphere. Under standard conditions at sea level, atmospheric pressure is 14.7 psia or 0 psig (0 kPa).

*Pressure, condensing:* See Pressure, discharge.

*Pressure, crankcase:* The pressure in the crankcase of a reciprocating compressor.

*Pressure, discharge:* The pressure against which the compressor must deliver the refrigerant vapor.

*Pressure drop:* The loss of pressure due to friction or lift.

*Pressure gauge:* An instrument that measures the pressure exerted by the contents of a container.

*Pressure, gauge:* The pressure existing above atmospheric pressure. Gauge pressure is, therefore, 14.7 psi (0 kPa) less than the corresponding absolute pressure.

*Pressure, heat diagram:* A graph representing a refrigerant's pressure, heat, and temperature properties. Also, a Mollier diagram.

*Pressure, limiter:* A device that remains closed until a predetermined pressure is reached and then opens to release fluid to another part of the system or opens an electrical circuit.

*Pressure motor control:* A control that opens and closes an electrical circuit as the system pressures change.

*Pressure, saturation:* The pressure at which vapor at any specific temperature is saturated.

*Pressure, suction:* The pressure forcing the vapor to enter the suction inlet of a compressor.

*Pressure switch:* A switch operated by a rise or fall in pressure.

*Pressure–temperature relationship:* The relationship of temperature and corresponding pressure of a given type of refrigerant. As either the temperature or the pressure changes, the other will make a corresponding change at each and every change.

*Pressure tube:* A small line carrying pressure to the sensitive element of a pressure controller.

*Process tube:* A length of tubing fastened to the dome of a hermetic compressor; used when servicing the unit.

*Psychrometer:* A device that uses both a wet bulb and a dry bulb thermometer. It is used to measure the relative humidity of a given quantity of air. Most types include a scale for converting to the percent of relative humidity present.

*Psychrometric chart:* A chart used to determine all the properties of a given volume of air under a given set of conditions.

*Pump:* Any one of various machines used to force fluids through pipes from one place to another.

*Pump, centrifugal:* A pump that produces fluid velocity by centrifugal force.

*Pumpdown:* The reduction of pressure within a refrigeration system.

*Purge:* The discharge of impurities and noncondensable gases to the atmosphere.

*Quick-connect coupling:* A device that permits fast and easy connection of two refrigerant lines by use of compression fittings.

*Range:* The limits of the settings for pressure or temperature control.

*Rankine scale:* The name given to the absolute Fahrenheit scale. The zero point on this scale is − 460°F.

*Rating:* The assignment of capacity.

*Reciprocating:* Back-and-forth motion in a straight line.

*Recirculated air:* Return air that is passed through the conditioning unit before being returned to the conditioned space.

*Recording ammeter:* An instrument that uses a pen to record on a piece of moving paper the amount of current flowing to a unit.

*Recording thermometer:* A temperature-sensing instrument that uses a pen to record on a piece of moving paper the temperature of a conditioned space.

*Reed valves:* A piece of thin, flat, tempered steel plate fastened to the valve plate.

*Refrigerant:* A substance that absorbs heat as it expands or evaporates. In general, any substance used as a medium to extract heat from another body.

*Refrigerant control:* A control that meters the flow of refrigerant from the high side to the low side of the refrigeration system. It also maintains a pressure difference between the high- and low-pressures sides of the system while the unit is running.

*Refrigerant velocity:* The movement of vaporous refrigerant required to entrain oil mist and carry it back to the compressor.

*Refrigerating capacity:* The rate at which a system can remove heat. Usually stated in tons or Btus per hour.

*Refrigerating effect:* The amount of heat a given quantity of refrigerant will absorb in changing from a liquid to a vapor at a given evaporating pressure.

*Refrigeration:* In general, the process of removing heat from an enclosed space and maintaining that space at a temperature lower than its surroundings.

*Refrigeration cycle:* The complete operation involved in providing refrigeration.

*Refrigeration tables:* Tables that show the properties of saturated refrigerants at various temperatures.

*Register:* A combination grill and damper assembly placed over the opening at the end of an air duct to direct the air.

*Reheat:* To heat air after dehumidification if the temperature is too low.

*Relative humidity:* The amount of water vapor actually present in a given quantity of air compared to the amount of moisture it could hold at those same conditions. Usually expressed as a percentage.

*Relay:* A device that is made operative by a variation in the condition of one electric circuit to effect the operation of other devices in the same or another circuit.

*Relay, control:* An electromagnetic device that opens or closes contacts when its coil is energized.

*Relay, thermal overload:* A thermal device that opens its contacts when the current through a heater coil exceeds the specified value for a given time.

*Remote bulb:* A part of the expansion valve. The remote bulb assumes the temperature of the suction vapor at the point where the bulb is secured to the suction line. Any change in the suction vapor superheat at the point of bulb application tends to operate the valve in a compensating direction to restore the superheat to a predetermined valve setting.

*Remote bulb thermostat:* A control whose sensing element is located separate from the mechanism it controls.

*Remote system:* A heat pump system whose outdoor unit is located away from the conditioned space.

*Restrictor:* A reduced cross-sectional area of pipe that produces resistance or a pressure drop in a refrigeration system.

*Return air:* Air taken from the conditioned space and brought back to the conditioning equipment.

*Reverse cycle defrost:* A method of reversing the flow of refrigerant through an evaporating coil for defrosting purposes.

*Reversing valve:* A valve that reverses the direction of refrigerant flow depending on whether heating or cooling is needed.

*Rotary blade compressor:* A rotary compressor that uses moving blades as cylinders.

*Rotary compressor:* A mechanism that pumps fluids by means of a rotating motion.

*Running time:* The amount of time that an outdoor unit operates per hour.

*Running winding:* The electrical winding in a motor through which current flows during normal operation of the motor.

*Saddle valve:* A valve equipped with a body that can be connected to a refrigerant line. Also, a tap-a-line valve.

*Safety control:* Any device that allows the refrigeration or air-conditioning unit to stop when unsafe pressures or temperatures exist.

*Safety factor:* The ratio of extra strength or capacity to the calculated requirements to ensure freedom from breakdown and ample capacity.

*Safety plug:* A device that releases the contents of a container to prevent rupturing when unsafe pressures or temperatures exist.

*Safety valve:* A quick-opening safety valve used for the fast relief of excessive pressure in a container.

*Saturation:* A condition existing when a substance contains the maximum amount of another substance that it can hold at that particular pressure and temperature.

*Schrader valve:* A spring-loaded valve that permits fluid to flow in only one direction when the center pin is depressed.

*Seat:* The portion of a valve against which the valve button presses to effect a shutoff.

*Seat, front:* The part of a refrigeration valve that forms a seal with the valve button when the valve is in the closed position.

*Self-contained air-conditioning unit:* An air conditioner containing a condensing unit, evaporating coil, fan assembly, and complete set of operating controls within its casing.

*Semihermetic compressor:* A hermetic compressor on which minor field service operations can be performed.

*Sensible heat:* Heat that causes a change in temperature of a substance but not a change of state.

*Sensible heat ratio:* The percentage of total heat removed that is sensible heat. It is usually expressed as a decimal and is the quotient of sensible heat removed divided by total heat removed.

*Sensor:* An electronic device that undergoes a physical change or a characteristic change as surrounding conditions change.

*Sequence controls:* A group of controls that act in a series or in a timed order.

*Service valve:* A shutoff valve intended for use only during shipment, installation, or service procedures.

*Shell and coil:* A designation for heat exchangers, condensers, and chillers consisting of a tube coil within a shell or housing.

*Shell and tube:* A designation for heat exchangers, condensers, and chillers consisting of a tube bundle within a shell or casing.

*Short circuit:* An electrical condition occurring when part of the circuit is in contact with another part and causes all or part of the current to take a wrong path.

*Short cycle:* Starting and stopping that occurs too frequently. A short on cycle and a short off cycle.

*Shroud:* A housing over a coil or fan to increase air flow.

*Silica gel:* An absorbent chemical compound used as a drying agent. When heat is applied, the moisture is released and the compound may be reused.

*Silver brazing:* A brazing process in which the brazing alloy contains some silver.

*Single-phase motor:* An electric motor that operates on a single-phase alternating current.

*Single-pole, double-throw switch (SPDT):* An

electric switch with one blade and two sets of contact points.

*Single-pole, single-throw switch (SPST):* An electric switch with one blade and one set of contact points.

*Sling psychrometer:* A device with a dry bulb and a wet bulb thermometer that is moved rapidly through the air to measure humidity.

*Slugging:* A condition in which a quantity of liquid enters the compressor cylinder, causing a hammering noise.

*Solder:* To join two metals by the adhesion process using a melting temperature less than 800 °F (427 °C).

*Solenoid valve:* An electromagnetic coil with a moving core that operates a valve.

*Specific heat:* The heat required to raise the temperature of 1 pound of a substance 1 °F (0.56 °C).

*Specific volume:* The volume per unit of mass. Usually expressed as cubic feet per pound (cubic meters per gram).

*Splash system, lubrication:* The method of lubricating moving parts by agitating or splashing the oil around in the crankcase.

*Split-phase motor:* A motor with two windings. Both windings are used in the starting of the motor. One is disconnected by a centrifugal switch after a given speed is reached by the motor. The motor then operates on only one winding.

*Split system:* A refrigeration or air-conditioning system in which the outdoor unit is located apart from the indoor coil.

*Squirrel cage:* A centrifugal fan.

*Standard atmosphere:* A condition existing when the air is at 14.7 psia (0 kPa). of pressure, 68 °F (20 °C) temperature, and 36% relative humidity.

*Standard conditions:* Conditions used as a basis for air-conditioning calculations. They are temperature of 68 °F (20 °C), pressure of 29.92 in. of mercury (Hg) (1.03 kPa), and a relative humidity of 30%.

*Starting relay:* An electrically operated switch used to connect or disconnect the starting winding of an electric motor.

*Start winding:* A winding in an electric motor used only briefly during the starting period to provide the extra torque required during this period.

*Static head:* The pressure due to the weight of a fluid in a vertical column or, more generally, the resistance due to lift.

*Static pressure:* The force per unit area as measured inside the air ducts of an air distribution system.

*Strainer:* A device, such as a screen or filter, used to remove foreign particles and dirt from the refrigerant.

*Stratification of air:* A condition existing when there is little or no air movement in the room.

*Subcooling:* The cooling of a liquid refrigerant below its condensing temperature.

*Subcooling coil:* A supplementary coil in an evaporative condenser, usually a coil or loop immersed in the spray water tank, that reduces the temperature of the liquid leaving the condenser.

*Sublimation:* The change of state from a solid to a vapor without the intermediate liquid state.

*Suction (vapor) line:* The pipe that carries the refrigerant vapor from the evaporating coil to the compressor coil.

*Suction pressure:* Low-side pressure. Same as evaporating pressure.

*Suction service valve:* A two-way manually operated valve installed on the compressor inlet and used during service operations.

*Suction side:* The low-pressure side of the system, extending from the flow-control device through the evaporating coil and to the compressor suction valve.

*Suction temperature:* The boiling temperature of the refrigerant corresponding to the suction pressure of the refrigerant.

*Superheat:* A temperature increase above the saturation temperature or above the boiling point.

*Superheated vapor:* A vapor whose temperature is higher than the evaporation temperature at the existing pressure.

*Switch, auxiliary:* An accessory switch available for most damper motors and control operators that can be arranged to open or close an electric circuit whenever the control motor reaches a certain position.

*Switch, disconnect:* A switch usually provided for a motor that completely disconnects the motor from the source of electric power.

*Temperature:* The measure of heat intensity.

*Temperature, ambient:* The temperature of the air around the object under consideration.

*Temperature, condensing:* The temperature of the fluid in the condensing coil at the time of condensation.

*Temperature, discharge:* The temperature of the vapor leaving the compressor.

*Temperature, dry bulb:* The temperature of the air measured with an ordinary thermometer. It indicates only sensible heat changes.

*Temperature, entering:* The temperature of a substance as it enters an apparatus.

*Temperature, evaporating:* The temperature at which a fluid boils under the existing pressure.

*Temperature, final:* The temperature of a substance as it leaves an apparatus.

*Temperature, saturation:* The boiling point of a refrigerant at a given pressure. It is considered to be the evaporating coil temperature in refrigeration.

*Temperature, suction:* The temperature of the vapor as it enters the compressor.

*Temperature, wet bulb:* The temperature of the air measured with a thermometer having a bulb covered with a moistened wick.

*Test charge:* An amount of refrigerant vapor forced into a refrigeration system to test for leaks.

*Test light:* A light provided with test leads that is used to test electrical circuits to determine if electricity is present.

*Therm:* A symbol used in the industry representing 100,000 Btu (25,200 kcal).

*Thermal overload element:* The alloy piece used for holding an overload relay closed; it melts when the current draw is too great.

*Thermal relay:* A heat-operated relay that opens or closes the starting circuit to an electric motor. Also called a hot wire relay.

*Thermistor:* Basically, a semiconductor whose electrical resistance varies with the temperature.

*Thermodisc defrost control:* An electrical bimetal disk switch that is controlled by changes in temperature.

*Thermometer:* An instrument used for measuring sensible temperature.

*Thermometer well:* A small pocket or recess in a pipe or tube designed to provide good thermal contact with a test thermometer.

*Thermostat:* A device used to control equipment in response to temperature change. A temperature-sensitive controller.

*Thermostatic expansion valve:* A flow-control device operated by both the temperature and pressure in an evaporating coil.

*Time-delay relay:* A relay actuated after a predetermined time from the point of impulse.

*Timer:* A clock-operated mechanism used to open and close an electric circuit on a predetermined schedule.

*Ton of refrigeration:* A unit of refrigeration capacity measuring 200 Btu (50.4 kcal) per minute, 12,000 Btu (3024 kcal) per hour, or 288,000 Btu (72,576 kcal) per day. It is so named because it is equivalent in cooling effect to melting 1 ton of ice in 24 hours.

*Total heat:* The sum of sensible and latent heat contained in a substance.

*Total pressure:* The sum of both the static pressure and the velocity pressure of a moving stream of air at the point where the measurement is taken.

*Transformer:* An electrical device that transforms electrical energy from one circuit to another by electrical induction.

*Tube within a tube:* Heat-exchange surfaces or condensers constructed of two concentric tubes.

*Tubing:* A pipe with a thin wall.

*Unitary system:* A combination heating and cooling system that is factory assembled in one package and is usually designed for conditioning one room or space.

*Urethane foam:* A type of insulation that is foamed between the shell and the liner of cabinets.

*Useful oil pressure:* The difference in pressure between the discharge and suction sides of the compressor oil pump.

*Vacuum:* A reduction in pressure below atmospheric pressure. Usually stated in inches of mercury or microns.

*Vacuum pump:* A pump used to exhaust a system. Also, a pump designed to produce a vacuum in a closed system or vessel.

*Valve:*   A device that controls the flow of fluid.

*Valve, cap seal:*   A manual valve whose stem is protected by a tightly fitting cap.

*Valve, charging:*   A valve located on the liquid line, usually near the receiver, through which refrigerant may be charged into the system.

*Valve, check:*   A valve that permits flow in one direction only; it is designed to close against backward flow.

*Valve, cylinder discharge:*   The valve in a compressor through which the vapor leaves the cylinder.

*Valve, cylinder suction:*   The valve in a compressor through which the vapor enters the cylinder.

*Valve, expansion:*   A type of refrigerant flow-control device that maintains a constant pressure or temperature in the low side of the system.

*Valve plate:*   The part of the compressor that contains the compressor valves and ports and is located between the compressor body and cylinder head.

*Valve port:*   The passage in a valve that opens and closes to control the flow of fluid in accordance with the relative position of the valve button to the valve seat.

*Valve purge:*   A valve through which noncondensable gases may be purged from the condenser or receiver.

*Vapor:*   A fluid in the gaseous state formed by evaporation of the liquid.

*Vapor barrier:*   A sheet of thin plastic or metal foil used in an air-conditioned structure to prevent vapor from penetrating the insulation.

*Vaporization:*   The changing of a liquid to a vapor.

*Vapor pressure:*   The pressure imposed by a vapor.

*Vapor, saturated:*   A vapor in a condition that will result in condensation into droplets of liquid if the vapor temperature is reduced.

*Velocimeter:*   An instrument used to measure air speeds on a direct-reading scale.

*Velocity:*   The speed or rapidity of motion.

*Velocity pressure:*   The pressure of a moving fluid which can cause an equivalent velocity as required to force the same fluid through an orifice such that all of the pressure energy is converted to kinetic energy.

*Vent:*   A port or opening through which pressure is relieved.

*Voltmeter:*   An instrument used to measure electrical voltage.

*Water-cooled condenser:*   A condenser cooled by water rather than air.

*Water treatment:*   The treatment of water with chemicals to reduce its scale-forming properties or to change other undesirable characteristics.

*Watt:*   A unit of electrical power.

*Wet bulb temperature:*   The temperature shown by a thermometer whose bulb is covered with a wet wick, or sock.

# Index

# HAVE YOU ORDERED YOUR COPIES OF THESE BILLY C. LANGLEY TITLES?

**TO ORDER:** Please complete the coupon below and mail to Prentice Hall, Book Distribution Center, Route 59 at Brook Hill Drive, West Nyack, NY 10995.
**Or if you prefer,** call (201) 767–5937 to place your order.

☐ (R0204–9) AIR CONDITIONING AND REFRIGERATION TROUBLESHOOTING HANDBOOK, 1980, cloth, $42.67

☐ (R0417–7) BASIC REFRIGERATION, 1982, cloth, $34.00

☐ (R0887–1) COMFORT HEATING, Third Edition, 1985, cloth, $32.00

☐ (17167–8) CONTROL SYSTEMS FOR AIR CONDITIONING AND REFRIGERATION, 1985, cloth, $28.67

☐ (R1036–4) COOLING SYSTEMS TROUBLESHOOTING HANDBOOK, 1986, cloth, $37.33

☐ (24751–0) ELECTRIC CONTROLS FOR REFRIGERATION AND AIR CONDITIONING, Second Edition, 1988, cloth, $30.00

☐ (R1600–7) ELECTRICITY FOR REFRIGERATION AND AIR CONDITIONING, 1984, paper, $27.67

☐ (R1790–6) ESTIMATING AIR CONDITIONING SYSTEMS, 1983, cloth, $32.00

☐ (38576–5) HEAT PUMP TECHNOLOGY: SYSTEMS DESIGN, INSTALLATION AND TROUBLESHOOTING, Second Edition, 1989, cloth, $25.50

☐ (R2805–1) HEATING SYSTEMS TROUBLESHOOTING HANDBOOK, 1988, cloth, $37.33

☐ (R5578–1) PLANT MAINTENANCE, 1986, cloth, $38.33

☐ (R5638–3) PRINCIPLES AND SERVICE OF AUTOMOTIVE AIR CONDITIONING, 1984, cloth, $32.00

☐ (R6629–1) REFRIGERATION AND AIR CONDITIONING, Third Edition, 1986, cloth, $36.33

☐ (82336–9) SOLID STATE ELECTRONIC CONTROLS FOR AIR CONDITIONING AND REFRIGERATION, 1989, cloth, $21.00

Please send me the book(s) checked above. After I've had the chance to examine the book(s) for 15 days, I'll either send the indicated price, plus a small charge for shipping and handling, or return the book(s) and owe nothing. I understand that the publisher will refund the purchase price and permit me to retain, as a complimentary copy, any book that is later ordered in quantities of 10 or more copies by any organization with which I am affiliated.

☐ BILL ME under the terms outlined above.

☐ PAYMENT ENCLOSED—publisher will pay all shipping and handling charges, same full refund, same return privilege as guaranteed above.

NAME/TITLE _____

DEPT/ORGANIZATION _____

ADDRESS _____

CITY _____ STATE _____ ZIP _____

DEPT. 1                                                                      D–DSBM–RP(8)

Prices subject to change without notice.